# Civil engineering construction contracts

# Civil engineering construction contracts

Michael O'Reilly

SECOND EDITION

Thomas Telford

Published by Thomas Telford Publishing, Thomas Telford Ltd, 1 Heron Quay,
London E14 4JD

URL: http://www.t-telford.co.uk

Distributors for Thomas Telford books are
USA: ASCE Press, 1801 Alexander Bell Drive, Reston, VA 20191-4400
Japan: Maruzen Co. Ltd, Book Department, 3–10 Nihonbashi 2-chome, Chuo-ku,
Tokyo 103
Australia: DA Books and Journals, 648 Whitehorse Road, Mitcham 3132, Victoria

First published 1996
This edition 1999

Appendix 2 is Crown copyright and reproduced by kind permission of Her Majesty's
Stationery Office

A catalogue record for this book is available from the British Library

ISBN: 0 7277 2785 0

Typeset by Pier Publishing, Brighton
Printed and bound in Great Britain by Bookcraft (Bath) Ltd

# Contents

# Preface

This book is designed principally for construction professionals, although I hope that non-specialist lawyers may also find it a useful introduction to the subject. The first edition, published in 1996, was, I think, fairly well received, and I am grateful to Thomas Telford for pushing me to write a second edition.

In the three years since the first edition, construction law has undergone a revolution. There have been many changes, some major, some minor: amongst them, the introduction of the Arbitration Act 1996, the new CIMAR Rules, the Woolf Reforms, the change of name and practice of the courts which deal with construction, FIDIC's re-invention of its standard form contracts and the House of Lords' decision to overturn 14 years of *Crouch*. These are more than sufficient to warrant a new edition.

But it is Part II of the Housing Grants, Construction and Regeneration Act 1996 which is really at the heart of the revolution. This legislation came into force on 1 May 1998, and applies to all British construction and allied contracts entered into after that date. The statute imposes a variety of procedures associated with payment, but most important of all, it provides a speedy means of enforcing payment rights. Statutory adjudication has been criticised by many, often unfairly, but the courts have, in recent months, shown a robust determination to make the legislation work.

As a consequence of all these changes, most standard contract drafting bodies have had to revise their forms. It is not easy to ensure that one always has the latest edition to hand, but I hope that I have had some success in keeping up. The book contains the full text of several new documents, including the ICE Conditions of Contract, 7th Edition, and the new ICE Adjudication and Conciliation procedures.

Many other changes are described in the text. One, in particular, deserves mention, as it has caused some consternation as the book goes to press in late 1999. I had anticipated that the Contract (Rights of Third Parties) Bill would have been enacted by now; in the event, several new amendments of significance for construction have been laid down for further debate. In the text of the book, I have, nevertheless, called the Bill 'The Contract (Rights of Third Parties) Act 1999' in anticipation of its imminent enactment and I express some confidence as to its eventual content. Readers will understand that I do not have a crystal ball, and I hope that I may be forgiven if my presumption should backfire on me.

# Table of cases

# Table of statutes and statutory instruments

---

1    At the date of going to press this had still not received Royal Assent.

# 1

# Introduction to civil engineering construction contracts

A civil engineering construction contract is one in which the contractor's principal obligation is to carry out works of civil engineering construction. The contract may also oblige the contractor to carry out ancillary obligations, such as the maintenance or operation of the works.

Civil engineering contracts are, by and large, made, interpreted and enforced in the same way as any other commercial contract. Contracts for work in Britain made after April 1998, however, must comply with the requirements of Part II of the Housing Grants, Construction and Regeneration Act 1996. These create a statutory right to adjudication clause, a right to payments by instalment and a prohibition on 'pay–when–paid' clauses. Contracts which fail to meet any of these requirements become subject to the relevant provisions of the Scheme for Construction Contracts (1998).

## 1. The use and importance of contracts in civil engineering

The vast majority of civil engineering work is performed under contract. A contract is simply an agreement which obliges the parties to do specified things. Most importantly, in the case of a construction contract, it requires the contractor to build the works and requires the employer to pay for them.

Contracts have a number of different functions. In the case of a construction contract, they include:

(1)  specifying the work to be done by the contractor (or sub-contractor etc.), including the required quality and time for completion of various parts of the work

(2)  defining what amount is to be paid, how any additional or reduced payments are to be computed and when payments are to be made

(3)  defining which party is responsible for events occurring outside the parties' direct control which affect the work; such events may include bad weather, access difficulties, local authority restrictions, changes in the law, unexpectedly poor ground, etc.

(4)  defining who has responsibility for undertaking the various administrative or dispute resolution functions which may be required, including obtaining consents, giving instructions, making decisions about claims, appointing adjudicators, arbitrators, etc.

### *The contractual setting*

A construction contract rarely exists in isolation. Sub-contracts, supply contracts, insurance contracts, performance bonds and so on create a network of contractual obligations. In some major projects the situation can be very complex indeed; as well as sub-contracts etc. there may be a variety of finance agreements, contracts for the supply of a wide variety of professional services, direct warranties issued by professionals in favour of financiers, developers and prospective users.

A contract provides a self-contained statement of obligations as between its own parties. The basic position in English law is that only the parties can sue upon it and, as a result, we can extract the contract from the web of relationships and analyse it as an independent entity.[1] Nevertheless, it should be recognised that commercial risks frequently arise from associated contracts. For example, the amount of damages for delay which a sub-contractor may have to pay to a contractor is likely to be closely related to the amount that the employer is entitled to recover from the contractor for the same delay.

## 2.  The variety of construction contracts

Although there have frequently been calls by policy advisors for the construction industry to adopt a single standard form of contract (e.g. Sir

---

1    See, however, the Contracts (Rights of Third Parties) Act 1999, discussed in Chapter 6.

Michael Latham, *Constructing the Team*, HMSO, 1994), the parties—and employers in particular—have always been keen to reserve the right to adopt forms of contract which suit their own commercial outlook and approach to risk.

Since the parties are free, within very wide parameters, to contract on any basis which to them seems convenient, and because the construction scenarios that can be conceived are infinitely various, a wide range of contractual arrangements and contract types have been developed. They may be classified in a number of ways.

A review of a variety of particular contract types, and their legal consequences, is undertaken in Chapter 12.

## 3. Housing Grants, Construction and Regeneration Act 1996

The Housing Grants, Construction and Regeneration Act 1996 applies to contracts in writing, entered into after 1st May 1998, which relate to 'construction operations' in England, Wales or Scotland irrespective of whether or not the law of England and Wales or Scotland otherwise applies. 'Construction operations' are given an extended meaning which includes not only contracts for the carrying out of construction operations, but also contracts for the arrangement and design of construction contracts. Accordingly, most contracts for the design, management and carrying out of construction in Britain are included. There are a number of specific exceptions. These include drilling or tunnelling for the purpose of extracting minerals, access and machinery for works relating to the basic utilities and to the chemical, food and pharmaceutical industries. Where only part of the contract relates to construction operations, that part only is subject to the Act.

For other contracts, such as those relating to work abroad, the position is as it was prior to the coming into force of the Act.

An analysis of the detailed provisions of the Act will be undertaken in chapters on payment and dispute resolution.[2] The professional bodies which issue the major standard form contracts have all drawn up amendments which purport to comply with the Act. In most cases, the amendments clearly comply, but in one or two instances it is not entirely clear that they do. For example, the amendments to the ICE Conditions of

---

2    Chapters 4 and 11 respectively.

Contract (which are embodied in the ICE Conditions of Contract, 7th Edition) set up a fiction that no dispute arises for a period of one month following the referral of a 'matter of dissatisfaction' to the Engineer. This is a sensible provision in that it avoids precipitate adjudication. But it seems to contravene the stipulation in Section 108 that the parties can refer a matter to adjudication 'at any time'.


## 4. Further issues relating to the procurement of works of civil engineering

It is important to recognise that civil engineering contracts exist within a procurement and legal environment which affects the way in which they are let, interpreted and operated. The following examples suffice to demonstrate this point.

*European procurement law*  European directives have been published aimed at enhancing competition across the European Union.[3] A central theme is that public sector contracts should be open to all Community members on a common basis and that such contracts should not stipulate specifications which inherently favour goods or supplies from one locality or state.[4] Public sector contracts valued in excess of prescribed limits must comply with advertising and tendering procedures in accordance with the directives.[5] Contractors who have been disadvantaged may seek damages.[6]

*Planning and compulsory purchase*  Construction is 'development' within the meaning of the Town and Country Planning Act 1990 and requires planning consent. Where construction requires the purchase of land, this

---

3   See The Public Supply Contracts Regulations 1991, SI 1991 No. 2679; The Public Works Contracts Regulations 1991, SI 1991 No. 2680; The Public Services Contracts Regulations 1993, SI 1993 No. 3228; The Utilities Supply and Works Contracts Regulations 1992, SI 1991 No. 3279.

4   See, for example, *The Commission of the European Communities v. Ireland* (1988) 44 BLR 1; *Ballast Nedam Groep NV v. Belgian State* (1997) 88 BLR 32 (Court of Justice of the European Communities).

5   Also see the Local Government Act 1999, which provides that local authorities must have regard to principles of 'best value'.

6   For a general outline see Morgan H, Working with the Works Directive, *Construction Law Review* 1998, 29 and the case of *R. v. Portsmouth City Council* ex parte *Peter Coles, Colwick Builders and George Austin Ltd* (1996) 81 BLR 1 (CA); Pigott A, Recoverable damages for breaches of EU Procurement legislation, *Construction Law Review* 1998, 33; Medhurst D, *EU procurement law*, Blackwell Sciences, 1997; Craig RW, *Procurement Law for Construction and Engineering Works and Services*, Blackwell Science, 1999.

must be authorised either by specific legislation (e.g. the Channel Tunnel Rail Link Act 1996) or more general legislation and is normally governed by the Compulsory Purchase Act 1965.

*Safety law*  Obligations in regard to occupation of any site are imposed by the Occupier's Liability Acts 1957 and 1984. The 1957 Act relates to invitees and the 1984 Act makes provision for the safety of trespassers. Safety of the construction itself is to be managed in accordance with the Health and Safety at Work etc. Act 1974 and the Construction (Design and Management) Regulations 1994.

*Welfare*  The various employment legislation applies no less to civil engineering than to other sectors. The Working Time Regulations[7] provide that workers cannot work in excess of the maximum prescribed hours; this affects traditional shift (and particularly night shift) work in the civil engineering industry. The Data Protection Acts 1984 and 1998 apply to information held.

*International contracts*  Where a contract contains an international connection, English law does not necessarily apply.[8] Where there are international elements it is prudent for the parties explicitly to agree which substantive law is to apply. This is normally done in a single clause, using wording such as: 'The law applicable to this contract is the law of England' (or Spain, Ukraine, etc.). Note that the law chosen must be that of a 'legal country' rather than political state. If the parties choose the 'law of the UK' or the 'law of the USA' this causes ambiguity because the UK contains three legal countries—England and Wales, Scotland and Northern Ireland and each state in the USA has its own law. Where the parties choose a law, this choice is respected by the courts of most jurisdictions. Where there is no express choice, the applicable law is that of the country with which the contract is most closely connected.[9] It is quite possible for parties to agree that a contract which concerns the nationals of one country only will be subject to the law of a different country. Most countries, however, have rules forbidding the use of this 'flag of convenience' device for avoiding

---

7   SI 1998 No. 1833.
8   See generally the Contracts (Applicable Law) Act 1990. This does not apply to arbitration.
9   See, for example, *JMJ Contractors* v. *Marples Ridgeway* (1985) 31 BLR 104. Here the law applicable to a sub-contract made in the FCEC form was held to be Iraqi as this was the law applicable to the main contract. This case was decided prior to the enactment of the 1990 Act but remains indicative of the reasoning process which will be employed.

safety, welfare and similar obligations. In this regard, it is worth noting that the Housing Grants, Construction and Regeneration Act 1996 applies to construction within Britain, irrespective of the applicable law of the contract.

*Environmental impact and pollution*   New projects and developments to existing construction falling within particular classes require a prior environmental impact assessment (EIA).[10] Pollution, noise and waste disposal (including landfill tax) are controlled by the Control of Pollution Act 1974, the Environmental Protection Act 1990 and the Pollution Prevention and Control Act 1999.

*Specialist structures*   Highway engineers, for example, must consider the Highways Act 1980 and the New Roads and Street Works Act 1991.[11] Reservoirs capable of holding more than 25 000 m$^3$ of water are subject to periodic inspection under the Reservoirs Act 1975.

---

10   See the Town and Country Planning (Environmental Impact Assessment) (England and Wales) Regulations 1999, SI 1999, No. 293.
11   See Sauvain SJ, *Highway Law*, 2nd edn, Sweet & Maxwell, 1997. The Streetworks Bill 1999 is currently (August 1999) before Parliament. When enacted it will require an undertaker of streetworks to supply specific information to the public.

# 2

## General principles of contract law

### 1. Formation of contract

In many cases, it is important to determine (*a*) whether or not there is a contract, (*b*) if so, when it was formed and (*c*) what terms it contains. The fact that parties are proceeding with the works does not necessarily mean that a contract has come into existence.

In principle, a contract comes into existence when the following requirements are met:

(1) the parties have reached an agreement
(2) the parties intend their agreement to be legally enforceable.

#### Agreement

*General* Determining whether or not there has been agreement involves a question of fact. It involves an objective assessment[1]. The test is whether or not an independent observer, appraised of the background facts known to the parties, would consider there to have been agreement, and if so, what agreement. The subjective state of the negotiators' minds is irrelevant. Where, however, there is a mutual mistake concerning an important element of the proposed agreement, this may prevent agreement.[2]

*Offer and acceptance* Where the agreement is reduced to writing there is usually little doubt about the existence and content of the agreement. On

---

1    *Tamplin v. James* (1880) 15 ChD 215 (CA).
2    *Bell v. Lever Brothers* [1932] AC 161 (HL).

other occasions, there may be doubts about whether, when and, if so, what agreement was reached. In order to investigate such a situation, the law uses a two-stage analysis: (*a*) offer; and (*b*) acceptance.

Once an offer has been made it may be accepted by the party to whom it is addressed. If it is unconditionally accepted there is agreement and a contract comes into existence.

An offer may only be accepted by the person to whom it is addressed (the offeree). No acceptance is possible after an offer ceases to exist. The offer may be destroyed by (*a*) being withdrawn by the offeror; (*b*) lapsing at a time specified by the offeror or at a reasonable time after being made; (*c*) being superseded by a subsequent offer (by the offeror or by the offeree);[3] or (*d*) being rejected by the offeree.

*Typical stages in the contract formation process*   The formation of a straightforward civil engineering contract may proceed in the following stages:

(1)   *Invitation to tender.* This is generally a pre-offer invitation,[4] in which the employer invites one or more contractors to tender quotations for a specified piece of work. There is generally no obligation for the person inviting tenders to accept the lowest or any offer, although there may be an obligation to consider all tenders properly submitted.[5] An invitation to tender may amount to an offer if the invitation clearly states that a contract will be awarded to the person submitting the tender which most fully meets a stated criterion.[6]

(2)   *Tender or quotation.* This is generally an offer by the contractor to undertake the work for the price or rates quoted and upon the terms contained in his offer. Where the document is styled an 'estimate' it may, nevertheless, amount to an offer;[7] it is a matter of interpreting the document.

(3)   *Letter of intent.* Upon receipt of tenders, the employer will often write to a tenderer in terms such as 'we intend to place a firm order with you

---

3   A request for clarification or further information may not amount to a counter-offer: *Stevenson* v. *McLean* (1880) 5 QBD 346.

4   *Spencer* v. *Harding* (1870) LR 5 CP 561. This is frequently termed an 'invitation to treat'.

5   *Blackpool and Fylde Aero Club* v. *Blackpool Borough Council* [1990] 1 WLR 1195 (CA).

6   Such as the lowest tender. See, for example, *Harvela Investments Ltd* v. *Royal Trust Company of Canada Trust* [1986] AC 207 (HL).

7   *Croshaw* v. *Pritchard* (1899) 16 TLR 45; see also *Cana Construction* v. *The Queen* (1973) 21 BLR 12 (Canadian Supreme Court) where an employer was responsible for inaccuracies in a document styled an estimate.

shortly' and it may continue 'please commence the works as soon as possible'. Such a communication does not generally amount to an acceptance as it is not unconditional.[8] Where the contractor commences work pursuant to such a letter of intent, he will be entitled to be paid a reasonable sum for the work he has actually performed in the event that no contract eventuates.[9]

(4)  *Acceptance.* This is the unconditional acceptance of the tender.[10]

*Apparently inconclusive negotiations*   Civil engineering contracts tend to involve complex technical specifications and logistical arrangements. Frequently, the parties embark upon lengthy and detailed negotiations. It is not uncommon for the negotiations to appear inconclusive. The question arises whether or not a contract then comes into existence at all. In practice, the law does not require absolute agreement on every detail. Provided that the parties have indicated a willingness to contract[11] and are in substantial agreement on the main issues,[12] a contract will often be created.[13] Where the parties each put forward their own preferred terms, the question whether or not a contract has in fact come into existence and, if so, upon what terms, is a question to be decided in the light of all the circumstances.[14] Likewise where the parties commence work in accordance with

---

8   *British Steel Corporation* v. *Cleveland Bridge and Engineering Co. Ltd* [1984] 1 All ER 504; see also *Wilson Smithett* v. *Bangladesh Sugar* [1986] 1 Lloyd's Rep. 378 for a case where a contract came into existence.

9   See Chapter 8, Section 3. British Steel Corporation v. Cleveland Bridge and Engineering Co. Ltd [1984] 1 All ER 504; see also *Wilson Smithett* v. *Bangladesh Sugar* [1986] 1 Lloyd's Rep. 378 for a case where a contract came into existence.

10   Note that the term 'acceptance' is often used loosely to mean 'acceptance pending final contact': see M Harrison & Co (Leeds) Ltd v. Leeds City Council (1980) 14 BLR 123.

11   This will normally be so where they have entered into negotiations about important matters such as price, quality and/or time. It will, however, be negatived where they set up some barrier, such as using the phrase 'subject to contract' which normally has the effect of putting the contract on hold: *Regalian Properties plc* v. *London Docklands Development Corporation* [1995] 1 WLR 212.

12   In many cases, the price will be an essential term; but this will not always be so: see *British Steel Corporation* v. *Cleveland Bridge and Engineering Co. Ltd* [1984] 1 All ER 504.

13   See the following where the conditions necessary for the formation of a contract during and after negotiations are discussed at length: G. *Percy Trentham* v. *Archital Luxfer* [1993] 1 Lloyd's Rep. 25 per Steyn L.J. at 25; *Pagnan SpA* v. *Feed Products Ltd* [1987] 2 Lloyd's Rep. 601 per Bingham J. at 612; J *Murphy & Sons Ltd* v. *ABD Daimler-Benz Transportation (Signal) Ltd*, TCC, 2 December 1998.

14   *Butler Machine Tool Co. Ltd* v. *Ex-Cell O Corporation (England) Ltd* [1979] 1 WLR 401. This case is frequently said to be authority for the proposition that the 'last shot wins'. This was a potential test, but no ruling of principle was laid down. See also *Chichester Joinery* v. *John Mowlem* (1987) 42 BLR 100 for an example of the possible stages of negotiation through which a construction contract may progress.

draft terms without subsequent communication on the terms, the question whether or not those terms govern the relationship between the parties is to be decided by reference to all the circumstances; but in appropriate circumstances, acceptance may be inferred from conduct alone.[15] Acceptance, cannot, however, be inferred from mere silence;[16] where one party indicates that he will assume his terms to be accepted unless he hears to the contrary, and he hears nothing, this will not suffice.

*Failure for lack of certainty*   Even where parties believe they have agreed, that agreement must be reasonably clear and certain; otherwise there is, in fact, no agreement. Where a term which is at the very heart of the agreement contended for has no ascertainable meaning, there can no agreement.[17] Where, for example, the specification in an informal contract is extremely vague and yet a price has been agreed, so that it is not possible to say even in broad terms what is included in the price, no contract will be formed.[18] The law attempts to save an agreement, so if the uncertain term can be severed from the agreement without destroying its basis, then this will be done.[19]

## The parties' intention to be bound by the agreement

*Expressed or implied intention not to be bound at law*   The parties are entitled to agree that their agreement is not subject to legal enforcement. They may do this expressly, as where they state that their agreement is 'subject to contract' or where there is a partnership agreement in which the parties expressly disclaim any rights at law. Likewise, where the agreement is made in a social context, the law will ordinarily infer that the parties did not intend their agreement to be enforceable at law.[20]

*Consideration*   Where one party appears, by the terms of an agreement, to be receiving 'something for nothing' the law presumes that the other does not intend the agreement to be binding at law. The law applies an objective test of intention: has the party who relies on the agreement provided a commercial input into the agreement. This commercial input is known as

---

15   *Brogden v. Metropolitan Railway Co.* (1877) 2 App Cas 666 (HL).
16   *Felthouse v. Bindley* (1862) 11 CB (NS) 869.
17   *Scammel v. Ouston* [1941] AC 251 (HL).
18   *Southway Group Ltd v. Wolff and Wolff* (1991) 57 BLR 33.
19   *Nicolene v. Simmonds* [1953] 1 QB 543 (CA).
20   *Rose and Frank Co. v. Crompton Brothers* [1925] AC 445 (HL).

'consideration'. The consideration may consist of the promise to pay money, do work, forego an existing legal right or any other matter that might be said to have a commercial value.[21] Since the law is not concerned with ensuring that commercial contracts are fair, but that they were intended to be binding, the question whether or not the consideration is commercially adequate does not arise. A token consideration suffices.[22] In a civil engineering contract, consideration is generally easy to find: the contractor agrees to the works and the employer agrees to pay for it—each is providing consideration.

## 2. Variation of contract

*Distinction between variation of contract and variation to scope of the works to be performed under the contract*   Parties may agree to vary their contract, that is to vary the very basis of their obligation one to the other. This variation of contract must be distinguished from a variation of the work to be performed under the contract. The latter occurs where the parties agree in their contract that the employer's agent may instruct a change in the scope or specification of the works. This involves no change in the underlying contract.

*General principle*   At any stage after making a contract, the parties may agree to modify or vary the terms or, indeed, to discharge the contract altogether. The same legal elements are required as are required to form a contract in the first instance, namely agreement, intention to create legal relations and consideration.

*Pre-existing contractual obligation as consideration*   Where a contractor falls behind in his performance and the employer agrees to increase the price payable in return for the contractor improving his performance so as to comply with the contract, the contractor appears to be providing no new consideration for the extra money. His only commercial input is his existing

---

21   A well known definition is

> A valuable consideration, in the sense of the law, may consist either in some right, interest, profit, or benefit accruing to the one party, or some forbearance, detriment loss or responsibility, given suffered or undertaken by the other

(*Currie* v. *Misa* [1875] LR 10 Ex. 153 at 163).

22   For example, a peppercorn rent is one where a very small rent (originally, it may literally have been a peppercorn) provides the owner with no substantial benefit, but provides the tenant with the substantial benefit of his contract for possession of the property.

obligation to comply with the contract. Where, however, the new agreement is reached without duress and where the employer experiences a
benefit, this will be sufficient consideration to support the contractor's
claim for the agreed extra money.[23]

*Duress*   Any agreement must be genuine and not obtained by duress.
Where a contract has been part-performed and one party uses a leverage it
has obtained by that part-performance to extract a variation to the contract
terms which are to his benefit, this will render the revised agreement voidable. In such cases the 'innocent party' may either affirm or avoid the new
arrangement created by the variation.[24]

## 3. Identifying the terms of the contract

### Identifying the express terms

*General*   Where the contract is in writing, the terms of the agreement are
usually readily identified. In other cases, there may be a dispute about
which terms are included. Even where there is a written agreement, one
party may claim that the contract has omitted an agreed term, that the
written document was supplemented by additional oral terms or that other
terms must be 'implied' to give the agreement business efficacy.

*The terms of an oral or partly oral agreement*   Most civil engineering contracts are made in or evidenced in writing. It is not uncommon, however,
for a party to claim that there are oral terms. Evidential difficulties may
arise from conflicting recollections of what was said, but there is no objection in principle to an oral term. When proved, it carries the same weight as
an equivalent written term. Where there has been a series of negotiations,
involving a number of offers and counter-offers, an objective test is used to
determine which terms of previously-made offers are to be carried over into
any subsequent offer. Clearly, terms in a later offer destroy earlier inconsistent terms. But terms advanced during early negotiations may never have
been contradicted and the question may arise as to whether or not they
have survived the negotiations to become contract terms.[25]

---

23   *Williams and Roffey Bros v. Nicholls (Contractors) Ltd* [1991] 1 QB 1; Re. Selectmove Ltd [1995] 1
     WLR 474.
24   *North Ocean Shipping Co. Ltd v. Hyundai Construction Co. Ltd* [1979] QB 705.
25   *Schawel v. Reade* [1913] IR 81, for a case where an early statement survived into the eventual
     contract.

*Terms incorporated by reference*   Terms may be imported by reference to a document; second- and third-remove incorporation is common. If a document A is incorporated into the contract and A specifically incorporates document B, which in turn specifically incorporates document C, then the terms of C will form part of the contract, as well as those of A and B.

> Where parties by an agreement import the terms of some other document as part of their agreement those terms must be imported in their entirety ... but subject to this: that if any of the imported terms in any way conflict with the expressly agreed terms, the latter must prevail over what would otherwise be imported.[26]

*Terms omitted by mistake from a formal document*   Rectification is a discretionary remedy whereby a written contract document is amended to reflect the agreement reached by the parties.

> The purpose of rectification is to amend an instrument to conform to the intention of the parties and to enable it to take effect as if originally so expressed.[27]

It may be granted in two distinct situations: (*a*) Where a party can show convincingly that the document as drawn up does not reflect the objective intention of the parties at the time of its execution;[28] or (*b*) where one party is labouring under a mistake and the other knows of,[29] conceals and takes advantage of that mistake, rectification may be granted where it would be inequitable to enforce the agreement as contained in the executed document.[30] An arbitrator may, in general, grant rectification providing the arbitration agreement is sufficiently wide;[31] it is thought that he will not be able to rectify an agreement where this has the effect either of clothing him with jurisdiction or depriving him of it.

---

26   *Modern Buildings Wales Ltd* v. *Limmer and Trinidad Ltd* [1975] 1 WLR 1281 per Buckley L.J. at 1289.

27   *Bank of Scotland* v. *Brunswick Development (1987) Ltd*, 29 April 1999, per Lord Hoffman (HL).

28   *Joscelyne* v. *Nissen* [1970] 2 QB 86 (CA).

29   There must normally be actual knowledge. Thus mistakes in tendered prices which slip through without being noticed will remain binding. Where, however, the defendant must very strongly have suspected the mistake of the other, rectification may be ordered: *Commission for the New Towns* v. *Cooper (Great Britain) Ltd* [1995] 2 WLR 677 (CA).

30   *Roberts & Co.* v. *Leicestershire County Council* [1961] Ch 555. Here the employer amended the completion time without advising the contractor; the evidence was that the employer knew that the contractor had not noticed and rectification was granted.

31   *Ashville Investments* v. *Elmer Construction* [1989] QB 488 (CA).

## Implied terms

In addition to the expressly agreed terms, terms may be implied. Implied terms cannot contradict express terms; but where they are implied they carry the same weight as express terms and may found a claim for breach of contract. They may be implied by statute or by ordinary operation of law.

*Terms implied by statute*   A number of statutes affect the content of commercial contracts. The Supply of Goods and Services Act 1982, for instance, implies terms into all contracts for the supply of services such as civil engineering contracts, except where these terms are inconsistent with the agreed express terms. This statute implies obligations to perform the work with reasonable care and skill, to do the work in a reasonable time and to pay a reasonable remuneration for the work.[32] Where it applies, the Late Payment of Commercial Debts (Interest) Act 1998 implies a term as to interest on debts arising under the contract unless there is a contractual agreement for a 'substantial remedy' as to interest.[33]

*Terms implied by operation of law*

> Where there is, on the face of it, a complete bilateral contract, the courts are sometimes willing to add terms to it, as implied terms. This is very common in mercantile contracts where there is an established usage. In that case, the courts are spelling out what both parties know and would, if asked, unhesitatingly agree to be part of the bargain. In other cases, where there is an apparently complete bargain, the courts are willing to add a term on the ground that without it the contract will not work.[34]

Terms will thus be implied if they either (*a*) 'go without saying'[35] or (*b*) are

---

32   Supply of Goods and Services Act 1982 Sections 13, 14 and 15, respectively.
33   Section 9 of the Late Payment of Commercial Debts (Interest) Act 1998 defines 'substantial remedy'. See Chapter 4, Section 7, for discussion of this statute.
34   *Liverpool City Council* v. *Irwin* [1976] 1 All ER 39 (HL) per Lord Wilberforce at 43.
35   In *Shirlaw* v. *Southern Foundries Ltd* [1939] 2 KB 206 (CA), MacKinnon L.J. at 227 set down a frequently cited expression of this:

> Prima facie that which in any contract is left to be implied and need not be expressed is something so obvious that it goes without saying; so that, if, while the parties were making their bargain, an officious bystander were to suggest some express provision for it in their agreement, they would testily suppress him with a common 'Oh, of course!'.

See also *Reigate* v. *Union Manufacturing Co.* [1918] 1 KB 592 (CA) where Scrutton L.J. at 605 made a not dissimilar observation.

necessary to give the contract proper business efficacy.[36] They will not otherwise be implied because they are reasonable or will promote efficiency.[37] For instance, it will be implied that the employer may not hinder or prevent the contractor's performance; it will not, however, be implied that third parties will not prevent performance. It will be implied that in so far as is necessary the employer will cooperate with the contractor; but it will not be implied that the employer will organise his work so as best to accommodate the contractor. Terms will not, of course, be implied if they conflict with the express terms.

## 4. The interpretation of contracts

The interpretation of a contract means the determination of its true objective meaning at law. The subjective intentions or views of the parties either before or after making the contract are irrelevant. It becomes necessary to interpret the contract when, for instance, there is a dispute about the meaning of a term or where mutually contradictory terms exist.

*The main principle*   Modern cases make it absolutely plain that the interpretation of a commercial contract involves no recourse to arcane legal principles but involves the application of a business common sense.

> Interpretation is the ascertainment of the meaning which the document would convey to a reasonable person having all the background knowledge which would reasonably have been available to the parties in the situation in which they were at the time of the contract.[38]

*The meaning of a contract and the meaning of its words*   The meaning of words in a contract clause may be determined from dictionaries and

---

36   *The Moorcock* (1889) 14 PD 64 (CA) per Bowen L.J. at 68:

> In business transactions such as this, what the law desires to effect by the implication is to give such business efficacy to the transaction as must have been in tended at all events by both parties who are business men; not to impose on one side all the perils of the transaction, or to emancipate one side from all chances of failure, but to make each party promise in law as much, at all events, as it must have been in the contemplation of both parties that he should be responsible for in respect of those perils and chances.

37   *Trollope & Colls Ltd v. North West Metropolitan Regional Hospital Board* [1973] 1 WLR 601 per Lord Pearson at 609 (HL).

38   *Investors Compensation Scheme Ltd v. West Bromwich Building Society* [1998] 1 WLR 898 (HL) per Lord Hoffman at 913. Lord Hoffman went on to give further principles which inform the remainder of this section.

grammars. Their meaning may, however, be altered when the contract is taken as a whole and set against the relevant background facts which both parties knew or should have known.

> It is, of course, well known that the context in which particular words are used may be of great importance with the result that language, taken out of d chances.context and construed on its own, may appear to have one meaning, but assumes a different meaning when it is read in the context of a complete contractual document.[39]

The true interpretation of a contract involves ascertaining the meaning of the contract read as a whole. This may properly involve examining the purpose of a contract term[40] and its underlying function in allocating risk.[41]

*Business common sense to prevail unless the parties' different intention is clear*    The law is slow to hold that the parties have used the wrong words to express their intention; but the law does not require judges to attribute to the parties an intention which they plainly could not have had.

> If detailed semantic analysis of words in a commercial contract is going to lead to a conclusion that flouts business common sense, it must be made to yield to business common sense.[42]

The possibility should not be ignored, however, that the parties may wish some unusual result, which would strike others as unbusiness-like or even unreasonable; provided the result is clearly expressed and does not lead to illegality, the intention of the parties will be enforced.

> The fact that a particular construction leads to a very unreasonable result must be a relevant consideration. The more unreasonable the result the more unlikely it is that the parties can have intended it, and if they do intend it the more necessary it is that they shall make that intention abundantly clear.[43]

*Evidence of negotiations*    Although the facts at the time of making the agreement are relevant to and admissible on the interpretation of the contract, evidence of negotiations prior to the formation of the agreement is

---

39    *Giffen (Electrical Contractors) Ltd v. Drake & Scull Engineering Ltd* (1993) 37 ConLR 84 per Sir Thomas Bingham MR at 90 (CA).
40    *Prenn v. Simmonds* [1971] 1 WLR 1381 (HL)
41    See, for example, *Balfour Beatty Building Ltd v. Chestermount Properties Ltd* (1993) 9 Constr. L.J. 117, per Coleman J at 127.
42    *Antaios Compania Naviera SA v. Salen Rederierna* [1985] 1 AC 191 (HL); *Davy Offshore Ltd v. Emerald Field Contracting Ltd* (1991) 55 BLR 1.
43    *Schuler (L) AG v. Wickman Machine Tool Sales Ltd* [1974] AC 235 per Lord Reid at 251.

inadmissible to prove the meaning of the contract.[44] The explanation for this rule is as follows. Where proposed terms are offered and the offeree rejects them, the rejected terms add nothing to an understanding of the terms which are eventually accepted. By definition, the rejected proposals are irrelevant; they represent the direction in which one party wished, unsuccessfully, to steer the negotiations. Where, however, a fact is established during negotiations, this assists in understanding the agreement which was eventually made.

*Subsequent conduct*   Once the contract is made, the way in which the parties operate it is not evidence of its true objective meaning. Such conduct may, however, create an estoppel or waiver which will prevent a party relying on the true objective meaning if this would result in injustice.

*Special meanings*   Words in a civil engineering contract which bear a special meaning to those working in the civil engineering industry will be interpreted with that specialist meaning unless it is objectively clear that the ordinary meaning was intended.

*Deletions in contract documents*   There is some high authority that it is permissible to look at deletions to assist in ascertaining the meaning of a contract.[45] It is not clear that this is correct. Deletions necessarily represent prior negotiations and the same reasons for not considering them apply.

*Priority*   The general rule of interpretation is that all terms are to be construed in the context of the contract as a whole; this gives rise to a presumption that no term is to carry greater weight than any other. Where, however, there is a plain contradiction, priority must be established. Many contracts give internal rules as to the order of priority which settle the matter.[46] Where this is not the case, determining priority between competing terms is assisted by a number of rules. For example, the terms of incorporating documents prevail over terms in those which they incorporate and especially created documents prevail over standard form documents.

---

44   *Prenn v. Simmonds* [1971] 1 WLR 1381 (HL); *The Bank of New Zealand v. Simpson* [1900] AC 183 (PC).

45   *Motram Consultants Ltd v. Bernard Sunley & Sons* [1975] 2 Lloyd's Rep. 197 per Lord Cross at 209 (HL).

46   The ICE Conditions of Contract do not; in fact Clause 5 states that all the Contract documents are mutually explanatory of one another.

*Previously considered expressions*   Where the court has previously considered a formula, it may well be that the meaning becomes virtually fixed. However, contracts must always be interpreted as a whole and so the same words can have different meanings when set in different contexts.[47]

*Historical rules*   A number of rules—such as the *contra proferentem* and *ejusdem generis* rules—have been applied historically. It is thought that these play a limited role in modern interpretation. Where they assist in reaching a business common sense interpretation, they may be useful; to the extent that they undermine the intention of the parties, as gleaned from the contract as a whole, then they probably have no application.

(1)   The *contra proferentem* rule. Where one party proffers his own form of contract, ambiguities upon which he relies are to be construed against him;[48] where the contract is in a standard form accepted in the industry there is some doubt concerning the applicability of the *contra proferentem* rule.

(2)   The *ejusdem generis* rule. This rule provides that where words are used including or excluding specific matters, followed by more general words (e.g. 'or any other cause', 'or any other delay') then only matters which are of a similar description to the specific examples are so included or excluded.

## 5. Exclusion, limitation and indemnity clauses

A clause in a contract purporting to exclude a party's liability for any event is an exclusion clause. A clause which purports only to limit liability is termed a limitation clause. Generally, exclusion clauses are construed strictly against the person relying upon them; limitation clauses receive less strict interpretation.[49]

*The Unfair Contract Terms Act 1977*   Exclusion and limitation clauses are subject to the provisions of The Unfair Contract Terms Act 1977. The Act's significance is limited in civil engineering practice,[50] apart from two

---

47   In *Mitsui v. AG for Hong Kong* (1986) 33 BLR 1 (PC), Lord Bridge said that any comparison between the wording of contracts could be 'positively misleading'.

48   See, for example, *Rosehaugh Stanhope v. Redpath Dorman Long* (1990) 50 BLR 75 (CA).

49   Generally, see *Photo Production v. Securicor Transport* [1980] AC 827 (HL).

50   In *Photo Production v. Securicor Transport* [1980] AC 827 (HL), Lord Wilberforce was of the clear opinion that in commercial contracts the Act would have limited impact: see especially 843.

points. Firstly, it restricts a party's ability to exclude his liability for his own negligence; liability for death or personal injury cannot be restricted at all and other liability cannot be restricted 'except in so far as the term or notice satisfies the requirement of reasonableness'.[51] Secondly, it restricts a party's right to exclude his liability for breach of contract where he is contracting on his own standard terms; any such clause must satisfy the test of reasonableness. This may become an issue where an employer maintains a standard contract.[52]

## 6. Contracts and formality

### The need for writing

Some contracts, notably those of guarantee and for interests in land need to be evidenced in writing; otherwise they are unenforceable. Excepting these, contracts may be in any form. They are equally binding whether they are made orally or by a highly formal sealed document.

### Contracts under seal

The principal effect of executing a contract under seal is that the limitation period (i.e. the time following a breach of contract in which the innocent party is entitled to commence proceedings to enforce the agreement) is extended from six to twelve years.[53] Before being sealed the contract is known as a 'simple' contract; when sealed, the contract is said to be 'a specialty', 'a deed' or 'under seal'. Standard form civil engineering contracts frequently entitle one or both parties to require the other to execute the contract under seal.[54]

---

51   Subsection 2 (2); *Rees Hough Ltd* v. *Redland Reinforced Plastics Ltd* (1984) 27 BLR 136; *Barnard Pipeline Technology Limited* v. *Marston Construction Co. Ltd* (1992) CILL 743.

52   See *Chester Grosvenor* v. *Alfred McAlpine* (1991) 56 BLR 115. In *British Fermentation Products Ltd* v. *Compair Reavell Ltd*, TCC, 8 June 1999, Judge Bowsher said

> I shall not attempt to lay down any general principle as to when or whether the Unfair Contract Terms Act applies in the generality of cases where use is made of Model Forms drafted by an outside body. However, if the Act ever does apply to such Model Forms, it does seem to me that one essential for the application of the Act to such forms would be proof that the Model Form is invariably or at least usually used by the party in question. ... I leave open the question what would be the position where there is such proof, and whether such proof either alone or with other features would make section 3 of the Act applicable.

53   Limitation Act 1980.

54   For example, see Clause 9 of the ICE Conditions of Contract, 7th Edition.

## 7. Ending contracts

A contract may be ended in two distinct senses. In its hard sense the contract may be eradicated so that the parties may not rely on its terms in future. In its soft sense, neither party is entitled to perform further obligations under the contract, but the terms of the contract survive and govern the rights of the parties to damages or other entitlements. Termination in its hard sense occurs where the contract is avoided or frustrated, or if the parties vary the agreement so that all claims, present and future, under the prior agreement are compromised. Termination in its soft sense is achieved by purported performance, termination in accordance with the terms of the contract or an accepted fundamental breach.

### *Agreement to discharge or vary the contract*

An agreement to discharge or vary the contract operates to substitute a new agreement for the one formerly in existence. Any agreement to discharge or vary the contract must be made by the parties (or their authorised agents) and be supported by consideration. Where both parties have unperformed obligations, sufficient consideration is supplied where each foregoes the right to insist on the other's further performance. If one party has completed his obligations, the other's agreement to tender a lesser performance provides no consideration.[55] Where a party accepts less than complete performance from the other in return for some undertaking from that other, this is often referred to as an 'accord and satisfaction'. Such agreements are typically made to avoid disputes being litigated or arbitrated and the consideration provided often consists of one party foregoing the right to pursue a bona fide claim.

### *Accepted fundamental breach or accepted repudiation*

A party to a contract may indicate that he no longer intends to comply with or be bound by the contract in one of two ways: (*a*) his serious breach which strikes at the purpose of the contract, or (*b*) by renouncing the contract. In either case, the 'innocent party' may elect whether or not to continue with his own performance of the contract. The contract is not automatically

---

55    *D & C Builders v. Rees* [1966] 2 QB 617 (CA).

terminated, and will continue to exist if the innocent party affirms it or fails clearly to indicate that he accepts the repudiation.[56] The innocent party may be entitled to continue to perform the agreement, even if the other party has no further interest in the subject matter of the agreement,[57] though this rule does not apply when the work is to be performed on the land of the party in breach[58] and may be unsound generally.

A repudiation, often referred to as an 'anticipatory breach', consists not of the failure to perform the obligation but, rather, of the renunciation of the agreement itself. The innocent party may elect to terminate the agreement immediately upon the repudiation and does not have to wait until the time for the performance of the obligation in question.[59] The test is whether the allegedly repudiatory conduct amounts to a renunciation of the contract or an absolute refusal to perform it.[60] Thus where a party relies in good faith on his own interpretation of the contract to support his refusal to perform obligations in particular circumstances, this will not amount to a repudiation, even if his interpretation turns out to be mistaken.[61] Where a party refuses to perform part only of the contract, this will be a repudiation of the whole contract, where the part which is refused is sufficiently central to the contract that it would be unjust to require the innocent party to accept his remedy in damages.[62]

A number of reported cases consider which breaches are sufficiently grave to entitle the innocent party to elect to terminate the agreement. The grant of only partial possession of the site,[63] failure to pay for deliveries of construction materials[64] and sub-letting in breach of the contract[65] have each been considered in this context. However, each contract and each breach is unique; accordingly, reliance on historical instances provides at best a poor guide and may be positively misleading. Basic principles should be used in preference to specific examples. The important question in every

---

56  *Vitol SA v. Norelf Ltd* [1995] 3 WLR 549 (CA).
57  *White & Carter (Councils) Ltd v. McGregor* [1962] AC 413 (HL).
58  *London Borough of Hounslow v. Twickenham Garden Developments Ltd* [1971] Ch. 233; *Mayfield Holdings Ltd v. Moana Reef Ltd* [1973] 1 NZLR 309 (New Zealand Supreme Court); *Tara Civil Engineering Ltd v. Moorfield Developments Ltd* (1989) 46 BLR 72.
59  *Hochster v. De La Tour* (1853) 2 E&B 678.
60  *Mersey Steel and Iron Co. v. Naylor Benzon & Co.* (1884) 9 App Cas 434 (HL).
61  *Woodar Investment Development Ltd v. Wimpey Construction UK Ltd* [1980] 1 WLR 571 (HL).
62  *Decro-Wall International SA v. Practitioners in Marketing Ltd* [1971] 2 All ER 216 (CA) per Buckley L.J.
63  *Carr v. JA Berriman Pty Ltd* (1953) 27 ALJ 273.
64  *Mersey Steel and Iron Co. v. Naylor, Benzon & Co.* [1884] 9 App Cas 434 (HL).
65  *Thomas Feather & Co. (Bradford) Ltd v. Keighley Corporation* [1953] 53 LGR 30.

case is whether the breach is so serious that it strikes at the purpose of the contract and/or constitutes a renunciation of it.

## Determination under the terms of the contract[66]

Determination clauses enable one or both parties to treat the contract at an end upon the happening of some specified event or circumstance. This event or circumstance is normally a serious breach, a serious delay in performance or the contractor's insolvency; however, it need not be directly related to the performance of the project. The determination clause may also provide for compensation or transfer of property upon termination; such provisions will be unenforceable if they amount to a penalty.

## Frustration

A contract is frustrated when the circumstances of its performance change so radically that the contract becomes an agreement to do something wholly different from that envisaged at the time the agreement was originally made. The frustrating event must not be within the control of either party. Instances considered by the courts have included where the subject matter of the project is destroyed[67] and where equipment to be used for performing the project is requisitioned by the state for military purposes.[68] If the work merely becomes more difficult to perform as a result of shortages of labour and materials it will not be considered as frustrated.[69] If the allegedly frustrating events are provided for in the agreement there can be no frustration.[70] If the parties contract in the knowledge that a particular and real risk is ever-present, the occurrence of that risk after the date of the contract will not frustrate it.[71] Unless otherwise stated in the contract,[72] the Law Reform (Frustrated Contracts) Act 1943 provides for the effect of frustration on property transferred during the subsistence of the contract.

---

66  See Chapter 7, Section 3, for a more detailed discussion of this topic.
67  *Taylor* v. *Caldwell* (1863) 32 LJQB.
68  *Metropolitan Water Board* v. *Dick, Kerr and Co. Ltd* [1918] AC 119 (HL).
69  *Davis Contractors Ltd* v. *Fareham Urban District Council* [1956] AC 696 (HL).
70  *McAlpine Humberoak* v. *McDermott International* 58 BLR 1 (CA).
71  In *Amalgamated Investment and Property Co.* v. *John Walker & Sons Ltd* [1976] 3 All ER 509. The risk here was the building on a site acquired under the contract for development might be listed as a building of historical interest by the Secretary of State.
72  Clause 63 of the ICE Conditions of Contract provides its own rules for property passed during the subsistence of the contract.

# 8. Legal personality and agency

## Legal personality

The law recognises two categories of legal person: (*a*) natural persons (i.e. human beings) and (*b*) corporations.

Natural persons of the age of majority may enter freely into contracts. Corporations may be created by charter or by statute (e.g. local authorities) or as companies by registration under the Companies Acts. Corporations are legally distinct from their members. Thus, while a company is owned by its shareholders, it is, in law, an independent person. The liabilities of the companies do not, therefore, devolve upon the shareholders; creditors must sue the company itself rather than the shareholders if they wish to seek redress against the company. The distinction here is particularly important when the corporation is a limited liability company in liquidation, for here the personal liability of the shareholders is limited to the value of the initial stake in the company allotted to them. The directors of the company also escape liability for the company debts provided they have not been reckless or fraudulent in their dealings.[73]

Non-incorporated associations, such as clubs, partnerships, unincorporated joint ventures, etc., have no legal personality and so cannot formally contract in their own right but as a collection of persons (natural or legal or both).[74] Special rules exist in many cases enabling, for example, a partnership to sue and be sued in the name of the partnership. And many joint ventures are set up using a limited liability company as a vehicle with participants holding shares in proportion to their stake in the venture.

## Agency

Most commercial organisations operate through agents. The statement 'Company X contracts with Company Y' means that authorised representatives of X and Y have agreed, on behalf of X and Y respectively, that X and Y would be bound by the contract. In the law of agency, the authorised representatives are known as agents and the companies they represent are the principals. The key question relates to the authority of agents.

---

73 See also *Williams* v. *Natural Life Health Foods Ltd*, 30 April 1998 (HL) where it was decided—overturning the Court of Appeal—that a director could not normally be made liable for the representations of his company.

74 *Bradley Egg Farm* v. *Clifford* [1943] 2 All ER 378 (CA).

Where a person A (for agent) purports to act on behalf of a principal (P) in contract negotiations, A will bind P where he has been expressly authorised to do so. In addition, where A holds a position of apparent authority in relation to P (e.g. where A is P's managing director) he will have authority to bind P whenever it would be usual for someone of his rank to have such authority.[75] Where A acts without actual authority he may become personally liable.[76] Where A fraudulently represents the scope of his authority, P's liability (if any) will depend on all the circumstances.[77]

Generally speaking, Engineers and other contract administrators will not be authorised to vary a contract on behalf of their clients.[78] Some civil engineering contracts deal specifically with the extent and limitations on the Engineer's authority.[79]

---

75    This is frequently termed ostensible or apparent authority.
76    See, for example, *Yonge* v. *Toynbee* [1910] 1 KB 215 (CA).
77    *Armagas* v. *Mundogas* [1986] 2 WLR 1063 (HL).
78    *Toepfer* v. *Warinco* [1978] 2 Lloyd's Rep. 569.
79    For example, Clause 2(i)(c) of the ICE Conditions of Contract, 7th Edition.

# 3

# The scope and quality of the agreed work

The contractor's principal obligation under the contract is to perform the work described in the original contract, as properly varied during the course of the project. In this chapter, the definition of the extent and quality of the contractor's obligations will be considered.

## 1. Documents used to describe the scope of civil engineering works

A civil engineering contract generally contains a series of documents which describe what has to be done. The detail varies from a simple description given in an oral contract to a very detailed set of documents in a major civil engineering project. In each case, the scope of the contractor's obligations is determined by interpreting these terms.

For a standard construct-only contract, the scope and quality of the works may be mentioned in any or all of the following:

(1) conditions of contract
(2) specification and drawings
(3) bills of quantities
(4) programme and/or method statement.

Where the contract is to be let on a design and build basis, the specification is normally replaced by documents known as 'employer's requirements' and 'contractor's proposals' which set out what is to be achieved; also the bill of quantities will be replaced by a 'pricing document' or 'pricing analysis' which sets out how the overall price has been put together.

For contracts that increase the contractor's obligations yet further to include maintenance, operation, training, etc., there will be further documents establishing the scope of these additional obligations.

*Conditions of contract*   The conditions of contract normally consist of a standard form set of conditions (amended or unamended).[1] Where exotic contract types (e.g. concession contracts[2]) are used, the contract may be specially drafted for the project. The conditions of contract will state the extent of the contractor's responsibility for design, coordination with third parties, protection of the works, etc.

*Drawings*   Drawings are invariably used to show the geographical scope and positional interrelationships between the items of work.

*Specifications*   Specifications set out the quality required. Where, or to the extent that no specification is provided, it will be implied into the contract that work is to be done with proper skill and care, using good quality materials which are reasonably suitable for their purpose.[3] The specification documents tend also to contain a variety of requirements and stipulations as to the manner of working.[4] A number of standard forms of specifications are published which relate to specific sectors of the civil engineering industry, including specifications for the water industry, tunnelling works, etc.[5] It is common for large public and quasi-public sector bodies to have in-house variants of standard specifications which they use on their own projects.

*Bills of quantities*   Bills of quantities are lists of items with associated quantities. The effect of the bill of quantities within the contract is a matter of interpreting the contract as a whole in each case. Thus, the effect of the bill

---

1   A survey of available standard form contracts is given in Appendix 1.

2   See Chapter 16.

3   *Young & Marten Ltd* v. *McManus Childs Ltd* [1969] 1 AC 454 (HL); *Gloucestershire County Council* v. *Richardson* [1969] 1 AC 480 (HL).

4   Parties frequently use the 'specification' to deal with all manner of sundry matters. In some cases, one finds some of the most important clauses in the specification. In contracts, such as the ICE Conditions of Contract 7th Edition, which do not specify an order of priority for documents, specification clauses can have important effects; where they are inconsistent with terms found in the main conditions of contract, the specification clauses may take precedence.

5   For example, the *Civil Engineering Specification for the Water Industry*, 4th edn, 1993, Water Research Centre.

of quantities in one ICE 7th Edition contract may differ from its effect in another ICE 7th Edition contract because of amendments to the Conditions of Contract or even clauses introduced into the specification. As a result, caution is required and the following comments should be taken as indicative only:

(1) *Contracts for a lump sum price.* Items required to complete the works must generally be provided despite their being omitted from the bill;[6] if there is no mechanism in the contract for recovering payment for these extra items, the contractor will have to pay for them. There may also be a presumption[7] that the quantities do not form a term in lump sum contracts unless the contract states otherwise; accordingly, the contractor will not receive an additional payment if the quantities required are greater than stated in the bill.

(2) *Measure and value contracts.* Estimated quantities are set out for each class of work. When tendering, the contractor quotes a rate for each class of work. The bill total is the sum of all the products of rates and estimated quantities; but the sum payable is the product of the actual quantities and rates. The process by which the quantity of each item is determined is called 'measurement', which may be physical measurement on site or the computation of area or volumetric quantities from survey data. If an item of work is to be done for which there is no agreed rate, nor agreed mechanism for calculating its value, the contractor is entitled to be paid a reasonable rate/sum.[8] The quantities in the bill are expressed to be estimates; any errors or omissions are to be corrected by the Engineer and any items required to be added in will be paid for in accordance with the contract.[9] Accordingly, where items have been accidentally omitted from the bill, the contractor is compensated.

---

6    *Williams* v. *Fitzmaurice* (1858) 3 H & N 844 per Channel B. 'It was a contract for the erection of a house and though the flooring was not mentioned in express terms, it was necessarily implied'.

7    This cannot be stated with any certainty since the principal authorities are from the 19th and early 20th centuries and the decisions seem somewhat lacking in harmony. Generally see: *Hudson's Building Contracts*, 4th edn, Vol. 2: *Patman and Fotheringham* v. *Pilditch* (1904) p. 368; *Priestly* v. *Stone* (1888) p. 134; *Re Ford and Bemrose* (1902) p. 324 (CA). See also *Sharpe* v. *San Paulo Railway Co.* (1873) LR 8 Ch App 597.

8    *Re. Walton-on-the-Naze* UDC v. *Moran* (1905) *Hudson's Building Contracts*, 4th edn, Vol. 2, p. 376.

9    ICE Conditions of Contract, 7th Edition, Clause 55.

*Methods of measurement*   The item descriptions found in bills of quantities cannot fully describe what is included. For example, an item 'excavation' without further elaboration can reasonably mean any of the following: (*a*) excavation and nothing more, (*b*) excavation and carry away spoil to tip, (*c*) excavation, support sides and carry spoil away to tip, pay the tipping charges, etc. Likewise, it is often difficult to decide how to measure an item. When excavating rock, for instance, the material bulks up to almost one and a half times its *in situ* volume. When measuring the volume excavated, which volume is to be used? In order to ensure that the short item description in the bill is fully defined and the parties agree on how to measure the items, a 'method of measurement' may be used. For example, in the ICE Conditions of Contract, the bills are deemed to be prepared in accordance with the Civil Engineering Standard Method of Measurement.[10]

*Schedules of rates etc.*   In addition to documents described as bills of quantities, similar documents described as schedules of rates,[11] schedules of prices,[12] etc., are frequently used. None of these terms are terms of art and their effect is determined by construing the agreement in each case.

*Design and construct documents*   In a design and construct contract, the specification is usually replaced by documents known as 'employer's requirements' and 'contractor's proposals'. The former set out what the employer requires from the completed project. The latter sets out how the contractor proposes to achieve it. The total specification is both together.

## 2.  Questions and ambiguities regarding the scope of the contract works

*General*   The scope of the contractor's obligation is determined by interpreting the terms set out in the drawings, specifications, etc. The contractor is only obliged to execute that work which, on a true interpretation of the contract, forms part of the contract work. The contractor is entitled to be paid additionally for any work he is asked to perform outside

---

10   Methods of Measurement are frequently used with bills of quantities. These contain lists of item descriptions and define what is included in each item. Many define items using codes so that there can be no confusion as to which item is being referred to.

11   This term is frequently used for term contracts since the work to be done and, hence, the quantities are a matter of great uncertainty; all that is known are the rates.

12   This term is frequently used for lump sum contracts where the items are listed without quantities.

the true scope of the contract as priced. Likewise, he is not liable for risks which falls outside his scope of responsibility.

*Ambiguities and contradictions: determining the true objective intention of the parties*   The documents describing the works may contain contradictions and ambiguities. The main source of difficulty is where the contract is for a lump sum price and the documents are ambiguous. Consider, for example, a bridge contract where the drawings show electrical lighting and guard rails but these items are not decribed in a list of items included within the contract documents; it is not immediately obvious whether or not the lighting and rails are to be provided within the price. The ambiguity will have to be resolved by determining the true objective intention of the parties. This will be achieved by examining the contract as a whole and deciding whether the parties agreed that the contractor should provide a completed job, including the lighting and rails or only the heavy civils work, leaving others to complete these items as part of another contract.

*Measure and value contracts*   In a measure and value contract, which has a comprehensive method of measurement, the work will be paid for in any event. Accordingly, disputes over the scope of the project works are usually unnecessary.

*Aids to interpretation*   Some contracts contain mechanisms which assist in resolving these problems. For example, an order of priority between classes of documents may be defined. The contract may also describe in broad terms what has to be completed, such as

> The contractor will, notwithstanding any omission or ambiguity in the documentation, deliver to the employer, for the agreed price, a fully operational bridge, complete with all ancillary mechanical, electrical, safety and other ordinary features.

*All things reasonably necessary*   In general the contractor must supply all items which are reasonably to be defined as necessary to complete his obligations. Thus in *Williams* v. *Fitzmaurice* [13] a builder agreed to make a house and in particular

> to do and perform all the works of every kind mentioned and contained in the foregoing specification, according in every respect to the drawings

---

13    (1858) 3 H & N 844.

furnished ... the house to be completed and dry and fit for occupation by August 1st 1858.

The specification described the floor-joists but not the flooring itself. The builder refused to supply and install the flooring without additional payment. Pollock CB said

> I had some doubt whether the specification was not to be regarded as the contract between the parties; but upon the whole the whole facts being disclosed it appears to me that no person can entertain any reasonable doubt that it was intended that the [builder] should provide the flooring as well as the other materials requisite for the building and that it was merely by inadvertance that no mention of flooring was made in the specification.

*Where the work cannot be accomplished as originally envisaged*    Where site conditions are worse than expected, the contractor is not entitled to stop work or to seek additional payment unless the contract expressly provides for it.[14] Where, on a true interpretation, the contractor's obligation is to achieve a completed job, he will not be entitled to be paid extra if it cannot be accomplished as originally envisaged.[15]

*Wholly unexpected work*    In a bridge building contract, where the work could not be accomplished using the method of construction proposed by the eminent engineer who drew up the contract drawings, the court held that on a true interpretation of the contract, the contractor's obligation was to complete the bridge at the same price using revised methods. The House of Lords did, however, accept that there may come point at which was had to be done was so radically different that it would amount to a different contract.

> If ... it was additional or varied work, so peculiar, so unexpected, and so different from what any person reckoned or calculated upon, that it is not within the contract at all, then it appears to me, one of two courses might have been open to him; he might have said: I entirely refuse to go on with the contract. ... I never intended to construct this work upon this new and unexpected footing. Or he might have said, I will go on with this, but this is not the kind of extra work contemplated by the contract, and if I do it, I must be paid a *quantum meruit* for it.[16]

---

14    *Bottoms* v. *Mayor of York* (1892) *Hudson's Building Contracts*, 4th edn, Vol. 2, p. 208.
15    *Sharpe* v. *San Paulo Railway Co.* (1873) LR 8 Ch App 597.
16    *Thorn* v. *Mayor and Commonalty of London* [1876] 1 AC 120 (HL), per Lord Cairns.

In many modern standard form contracts an express term is included allowing the contractor to be paid if the works cannot be constructed in accordance with the drawings and specifications.

*Extent of obligations to design and to select materials*  Questions as to the contractor's responsibility for design and other services are to be determined by a true construction of the contract documents. Where it is clear that the contractor takes on a design and construct obligation, he will be obliged to provide works which are fit for their purpose, unless the contract provides otherwise.[17] This will be so, even where the employer specifies key aspects of the design.[18] Where the employer specifies materials, it will not be implied that the contractor warrant by their suitability; but he will still warrant that they are of good quality.[19]

# 3. Varying the scope or description of the works

The contractor's obligations in terms of the scope of the work to be done may change with time. This might happen in two ways. The first is where the contract itself is varied. The second is where the contract provides a mechanism for the scope or specification of the work to be varied and that mechanism is operated.

## Variations of the contract

If the employer, or his authorised representative, allows a relaxation to the specification so that a set of mutual benefits (thus providing sufficient

---

17   In *IBA v. EMI Electronics Ltd and BICC Construction Ltd* (1980) 14 BLR 1 (HL), the litigation resulting from the collapse of the Emley Moor television mast, for example, Lord Scarman said:

In the absence of a clear contractual indication to the contrary, I can see no reason why one who in the course of his business contracts to design, supply and erect a television aerial mast is not under an obligation to ensure that it is reasonably fit for the purpose for which he knows it was intended to be used.

18   *Cammell Laird Co. Ltd* v. *Manganese Bronze and Brass Co. Ltd* [1934] AC 402 (HL). The shipyard drew up outline designs for a propellor; the thickness of the medial lines was given and the note on the drawing read 'edges to be brought up to fine lines.' The manufacturer cast the propellor. When put into operation, it did not perform satisfactorily and the ship was refused Lloyd's classification. Held, the manufacturer was obliged to provide a propellor which was fit for its purpose. *Lynch* v. *Thorne* [1956] 1 WLR 303 (CA) appears to be at odds with this; here, a builder drew up a design for a house with 9 inch solid walls and began to construct it. He sold the house during construction, having advised the purchaser of the design. The solid walls were unsuitable and allowed damp penetration. Held, the builder complied with design and was not liable.

19   *Young & Marten Ltd* v. *McManus Childs Ltd* [1969] 1 AC 454 (HL).

consideration) is derived from the relaxation, in circumstances where it is clear that both the contractor and employer intend the relaxation to be binding, then a variation of the contract will ensue. A compliance with the relaxed specification will be a compliance with the new contract. Sufficient consideration is ordinarily demonstrated where each party receives a bene-fit from the new arrangement.[20] Variations to the contract are rare in practice, not least because most contract adminstrators do not have the requisite authority to vary the contract.[21]

## Variation to the scope or specification of the work

*The need for express power*   The employer will ordinarily have no right under the contract to specify how the work is to be done[22] or to order that the work be varied. In practice, of course, most contractors will allow the employer to order variations. In this case, the original work will be paid for in accordance with the contract rates and the work outside the contract will be paid at any agreed rate, or, if none is agreed, at a reasonable rate.[23] There is also the related situation, where the employer wishes to omit work. In the absence of a contractual power to omit agreed work, the employer will be in breach of contract if he instructs the contractor to omit any work. An employer does not, of course, have to suffer a contractor fixing any work to his land, even work to which he has already agreed.[24] But if he refuses to allow the contractor to undertake the originally agreed work, this will enti-tle the contractor to damages.

*The effect of express powers to vary and a proper variation order*   Most civil engineering contracts provide an express power for the employer or his

---

20   *Williams and Roffey Bros v. Nicholls (Contractors) Ltd* [1991] 2 WLR 1 QB 1.
21   *Toepfer v. Warinco* [1978] 2 Lloyds Rep. 569 at 577 per Brandon J.:

> It is well-established that an architect or engineer has no implied authority from the building owner by whom he is employed to vary or waive the terms of a building contract.

22   *Greater London Council v. Cleveland Bridge Engineering Co. Ltd* (1984) 34 BLR 50:

> In the absence of anything to the contrary, a contractor is entitled to plan and perform the work as he pleases

(per Staughton J.). The employer can, however, insist that the performance be reasonable, e.g. that it is done at a reasonable hour, without causing a nuisance, etc.
23   *Sir Lindsay Parkinson & Co. Ltd v. Commissioners of Works* [1949] 2 KB 632 (CA).
24   *London Borough of Hounslow v. Twickenham Garden Developments Ltd* [1971] Ch 233; but see also *White & Carter (Councils) Ltd v. McGregor* [1962] AC 413 (HL).

agent to vary the scope of the work or the specification of the work, with appropriate money and time adjustments. Where properly operated, the contractor's obligation becomes the performance of the newly defined work. Some contracts contain a term such that if the works cannot legally or physically be accomplished in accordance with the specification then the contractor is entitled to a variation and here the specification may be varied automatically.

*Agent's authority where the contract does not provide for variations*   Where the contractor acts on an engineer's instruction to provide additional work, the employer will not be responsible for paying for that additional work, unless the engineer was properly authorised or the employer knew that the work was in progress and took no steps to alert the contractor.[25] In such cases, the engineer may become personally liable for payment.

*The degree of variation permitted*   Where the employer or his agent is entitled to order variations, the degree of permitted variations will, in the first instance, be a matter of interpreting the variations clause. There will, however, be a presumption that the clause is to be interpreted so as not to deprive the contractor of the reasonable benefit of the contract as a whole. Thus, where the variations clause apparently gives the employer absolute power to omit work, he will not be able to do so in order to re-let to another contractor unless there are clear words to this effect.[26] Where there is no stated permitted value or extent, the general rule is that reasonable variations connected with the works may be ordered; entirely new works cannot be imposed on the contractor.[27] In a contract for dredging works, the agreed method of dumping of dredged material became impossible; it was subsequently decided that an artificial island was to be formed. The question for

---

25  *Stockport MBC v. O'Reilly* [1978] 1 Lloyd's Rep. 595.

26  *Carr v. JA Berriman Pty Ltd* (1953) 27 ALJ 273. Here the variations clause provided

> The architect may in his absolute discretion ... issue ... written instructions ... in regard to the ... omission ... of any work.

Fullagar J. said

> [A] power in the architect to hand over at will any part of the contract to another contractor would be a most unreasonable power, which very clear words would be required to confer.

27  *Thorn v. Mayor and Commonalty of London* (1876) 1 App Cas 120 (HL); *Blue Circle Industries v. Holland Dredging Co. (UK) Ltd* (1987) 37 BLR 40.

the court was whether this change was such that it could fall within the scope of the variation clause.

> The original dredging contract provided that the spoil from excavating should be deposited in 'areas within Lough Larne to be allocated ... until approval by the local authorities'. In the event, as a result of local pressures and the attitude of the licensing authority this term became impossible to fulfill legally. The only alternatives were dumping at sea or the creation of an artificial bund with the formation of an artificial island. Either of these two solutions were wholly outside the scope of the original dredging contract and therefore had [the contractors] not been willing they could not, in my judgment, have been obliged to carry out the work as a variation. ... In my judgment [counsel's] submission that the island contract is separate from the dredging contract is correct.[28]

*Variations to be in a prescribed format*    Where the contract requires that instructions or variation orders be given in a prescribed format or manner, any purported instruction or variation order which is not so given may be ineffective, and the contractor may not be entitled to be paid for it,[29] unless the employer has waived his right to insist on compliance with the strict terms of the agreement.[30] However, even where the contract provides that variation orders must be in a particular format, the absence of that format may not be fatal where the primary question relates to whether or not the work instructed or ordered falls within the scope of the original contract. This is particularly so if the contract contains an arbitration clause entitling the arbitrator to review the decisions of the person vested with authority to order variations.[31]

*Variations requested by the contractor*    Where the variation mechanism within the contract is properly operated, it matters not, in principle, for whose benefit the variation was required. Accordingly, a contractor may, in appropriate circumstances, recover additional payment for a variation, where that variation is made at his request and for his benefit.[32] Where the

---

28    *Blue Circle Industries v. Holland Dredging Co. (UK) Ltd* (1987) 37 BLR 40, per Purchas L.J. The contract was on the ICE Standard Form 5th Edition which included a variation clause, Clause 51.

29    *Forman & Co. Proprietary Ltd v. The Ship 'Liddlesdale'* [1900] AC 190 (PC).

30    *Meyer v. Gilmer* (1899) 19 NZLR 129 (New Zealand) where the requirement for written variation orders was held to be waived on account of the attendance of the employer at the meetings where the oral variation orders were given.

31    *Brodie v. Cardiff Corporation* [1919] AC 337 (HL).

32    *Simplex Piling v. St Pancras Borough Council* (1958) 14 BLR 80.

contractor is put on notice that the option is his and he will not be paid additional sums, he will not be entitled to be paid.[33]

*Pricing of variations*    Civil engineering contracts typically provide rules for the valuation of variations. Where no rules are agreed, the valuation will depend on the interpretation of the agreement. Broadly speaking, if the varied work is of a type which is identical or similar to that already priced in the contract, then the contract prices will apply to the varied work. Where the work is of a different type or is undertaken in different circumstances, and a price may readily be determined by analogy with work for which prices are given, then this analogous price will be payable. Where there is little relation between the varied work and that originally to be supplied, a reasonable price will be payable. Contractors frequently submit claims based on daywork rates published by various bodies; these are only applicable where the contract expressly permits them or they are shown to represent reasonable rates and the construction method and time taken are each reasonable.

## 4. The required quality and performance of the works

### In the absence of express agreement

In the absence of express provisions, a number of terms may be implied.

*Quality of materials*    A contractor will warrant the quality of materials, even where the employer specifies the material to be used and the defects are hidden at the time of the purchase.[34] The matter is largely one of policy and convenience. In the ordinary case, the contractor has an action against the supplier while the employer does not; the presumed intention is that the risk should fall on the contractor. Where the parties know that the

---

33   *Howard de Walden Estates Ltd* v. *Costain Management Design Ltd* (1991) 55 BLR 124. Some standard form contracts provide that where the variation is required as a result of the contractor's default, no account may be taken of it in the valuation (e.g. ICE Conditions of Contract, Clause 51(3)); even this will not, it is thought, prevent the Contractor recovering for a variation which is simply made to accommodate the Contractor and which is not necessitated as a result of his prior default.

34   *Young & Marten Ltd* v. *McManus Childs Ltd* (1969) 9 BLR 77 (HL). The employer specified a particular brand of roof tiles, available only from one supplier. There were latent defects which did not come to light until the tiles had been on the roofs for a year. Held, the contractor was in breach of his warranty as to quality.

contractor's recourse against the supplier will be limited, the parties' presumed intention may be that the contractor is not liable for hidden defects,[35] although he will continue to be responsible for defects which should have been identified through good workmanship.

*Suitability of the materials used for their purpose*   A contractor will ordinarily warrant that materials are suitable for their purpose.[36] This presumption may be displaced where the employer specifies the materials to be used, for here the employer does not rely on the contractor's skill.[37] Nevertheless, where the 'specification' is, in fact, no more than a 'suggestion for the contractor's consideration' and the employer relies on the contractor's skill, the contractor will be responsible for the suitability of those materials.[38]

*Quality of the workmanship*   A contractor must do the work in a 'good and workmanlike manner'.[39]

*Suitability of the overall construction for its known purpose*   Where the work is taken on a design and construct basis, the contractor will ordinarily warrant that the final product is suitable for its known purpose.[40] Where, the

---

35   In *Young & Marten Ltd* v. *McManus Childs Ltd* (1969) 9 BLR 77 (HL), Lord Reid said at 88:

>   It would make a difference if that manufacturer was only willing to terms which excluded his ordinary liability under the Sale of Goods Act, and that fact was known to the employer and contractor when they made their contract. For it would be unreasonable to put on the contractor a liability for latent defects when the employer had chosen the supplier with knowledge that the contractor could have no recourse against him.

>   In *Gloucestershire County Council* v. *Richardson* [1969] AC 480 (HL), the employer required the contractor to enter into a nominated sub-contract with a supplier who excluded his liability for quality. The contractor had no discretion to refuse to enter the contract. Held, in these special circumstances, there was no warranty of quality.

36   *GH Myers & Co.* v. *Brent Cross Service Co.* [1934] 1 KB 46 per du Parcq at 55:

>   [A] person contracting to do work and supply materials warrants that the materials which he uses will be of good quality and reasonably fit for the purpose for which he is using them unless the circumstances of the contract are such as to exclude any such warranty.

37   In *Young & Marten Ltd* v. *McManus Childs Ltd* (1969) 9 BLR 77 (HL), Counsel conceded and the House of Lords accepted that where an employer specifies the material, a warranty as to suitability is excluded.

38   *Steel Company of Canada* v. *Willand Management* [1966] 58 DLR (2d) 595.

39   This is the expression commonly used by lawyers to express this term. It is found, for example, in Lord Denning's judgment in *Hancock* v. *Brazier (Anerley) Ltd* [1966] 1 WLR 1317 (CA).

40   *IBA* v. *EMI Electronics Ltd and BICC Construction Ltd* (1980) 11 BLR 14 BLR 1.

works only have only one proper purpose,[41] then defects which render the works unsuitable for that purpose are clearly breaches. Where the works may have more than one purpose (e.g a warehouse which can be used for storing a variety of items) it is thought that the law will not burden the contractor with having to ensure suitability for unusual purposes.

## Express provisions agreed in the contract

*General*   Construction contracts normally contain long, detailed specifications which set out the required quality for all materials to be used. Standard specifications are published for specific areas of work. These frequently form the core specification supplemented by bespoke clauses or amendments. Specification clauses can be 'performance' or 'recipe'. In the former, a level of performance of the product in the construction is required; in the latter a specific means of achieving the construction is specified. Those drawing up contracts should ensure that the specifications used are appropriate and that all materials to be used are covered.

*Performance testing*   In many projects and particularly in projects for process, water supply or power generation works, the performance of the works is generally tested. Any failure to meet performance targets is coupled with liquidated damages provisions.

*Where the works do not comply with the quality defined in the specification*   Where the works are not performed in strict accordance with the original specification and there is no properly authorised variation to cover them, this will be a breach of contract by the contractor. Any non-compliance is normally inadvertent. It may, however, be deliberate as where the contractor believes that what is supplied is 'as good as' what is specified or he considers that the employer's agent has approved the work. Claims that the work is 'as good as' are rarely meritorious[42] unless the contract, on a true construction, specifies a minimum standard and what is provided is clearly better. Wherever it is reasonable to do so, the employer will be entitled to have the work re-done; but where this is unreasonable he will be able to claim damages for the reduced performance or value of the construction.[43]

---

41   For example, *Viking Grain Storage* v. *TH White* (1985) 33 BLR 103 (grain storage depot); *IBA* v. *EMI Electronics Ltd and BICC Construction Ltd* (1978) 11 BLR 29 (HL) (television aerial mast).

42   *Forman & Co. Proprietary Ltd* v. *The Ship 'Liddlesdale'* [1900] AC 190 (PC); but see also *Ata Ul Haq* v. *City Council of Nairobi* (1959) 28 BLR 76 (PC).

43   *Ruxley Electronics* v. *Forsyth* [1995] 3 WLR 118 (HL).

To succeed in a claim that the employer's agent approved the non-compliant work, the employer's agent must have been authorised to vary the contract or to waive[44] the employer's contractual rights and he must have done so expressly. In most civil engineering situations, the employer's agent is rarely authorised to vary the contract or to waive the employer's rights under it.[45] Clearly, the fact that the engineer merely witnesses the non-compliant work cannot be sufficient to exculpate the contractor; generally, the engineer is employed to protect the employer's interests, not to protect the contractor.[46] However, an engineer is ordinarily entitled to vary the specification; where the engineer's variation falls within the compass of the variation provisions, disconformities may become compliant.[47]

44   *WJ Alan & Co.* v. *El Nasr Export and Import Co.* [1972] 2 All ER 127 (CA).
45   *Toepfer* v. *Warinco* [1978] 2 Lloyd's Rep. 569:

    It is well established that an architect or engineer has no implied authority from the building
    owner by whom he is employed to vary or waive the terms of a building contract

(per Brandon J. at 577).
This is so even where the contract stipulates that the work is to be done to a specification and also
to the satisfaction of the engineer, and the engineer indicates his approval: *National Coal Board* v.
*William Neill & Sons (St Helens) Ltd* (1983) 26 BLR 81.
46   *East Ham Corporation* v. *Bernard Sunley Ltd* [1966] AC 406 (HL). See also *AMF International Ltd*
v. *Magnet Bowling* [1968] 1 WLR. 1028:

    In general an architect owes no duty to a builder to tell him promptly during the course of
    construction, even as regards permanent work, when he is going wrong; he may, if he wishes,
    leave that to the final stages notwithstanding that the correction of a fault then may be much
    more costly to the builder then had his error been pointed out earlier

(per Mocatta J. at 1053).
47   *Shanks & McEwan (Contractrors) Ltd* v. *Strathclyde Regional Council* (1994) CILL 916 (Scottish
Court of Session).

# 4

## Payment in accordance with the contract

### 1. The price

*General*   The price is the amount payable under the contract. Where there is a fixed scope of work, a fixed price may be agreed. Where the scope is uncertain—either because the scope is not well-defined or because the employer has reserved the right to vary its scope—a method for computing the price may be agreed. Although there are recognised categories of method for computing the price (e.g. based on a lump sum, measure and value, cost reimbursement, target price, etc.), the calculation in each case depends upon a true interpretation of the particular contract. Consequently, the descriptions 'lump sum contract' or 'measure and value contract' give useful indications as to the general type of payment mechanism; but the details of the computation can only be found in the detailed terms of the contract.

*Where there is no express agreement as to price*   In many cases, agreement on price (or its method of computation) is fundamental to the formation of a contract, so that in the absence of such agreement there is no contract.

> No doubt in the vast majority of business transactions, particularly those of substantial size, the price will indeed be an essential term, but in the final analysis it must be a question of construction whether it is so.[1]

---

1   Goff J. in *British Steel Corporation* v. *Cleveland Bridge and Engineering Co. Ltd* [1984] 1 All ER 504.

Where the parties achieve a binding contract without agreement as to price, the law will imply an obligation for the employer to pay a reasonable price.[2] Where there is no contract, but the contractor, nevertheless, carries out work at the request of the employer, the contractor will be able to recover a reasonable amount under the principles of 'restitution'.[3]

*Computing the price*   The agreed terms generally deal at some length with the price, its computation and how it may be varied to account for claims, variations, etc. The main types of price formula used are:

(1) *Lump sum.*[4] Here the contractor undertakes to perform a defined scope of work for a stated amount—the lump sum. The lump sum and the defined scope of the work are opposite sides of the same coin. The lump sum is for the totality of work falling within the agreed scope, but no more. The agreed scope is the description of work to be done for the lump sum. If the contractor performs work less than the agreed scope, he will not be entitled to the full lump sum. Where there are no variations to the scope of the work, the price is equal to the lump sum. Where the contract entitles the employer to order variations to the work and such variations are ordered, the price will be higher or lower than the lump sum depending on whether work has been added or deducted. A bill of quantities or a schedule of rates may be included in the contract for the purpose of valuing variations. Under lump sum contracts, many disputes over price are, in essence, disputes about whether or not a particular item of work falls within the agreed contract scope and, hence, about whether there has been a variation to the scope which attracts additional payment.

   Lump sum arrangements come closer to a fixed price than other methods of computing the price and are favoured by employers seeking price certainty. There are, however, wide differences between contracts. Where the employer retains the risk for unforeseen ground conditions,[5] for example, the price of a lump sum groundworks contract can be subject to significant increases.

---

2   Supply of Goods and Services Act 1982.
3   See Chapter 8, Section 3.
4   The GC/Works/1 1998 Without Quantities, the IChemE Red Book, the JCT 98 (versions 'without quantities' and 'with quantities') and the FIDIC Turnkey contracts are examples.
5   As in the GC/Works/1 Single Stage Design and Build (1998) Contract and the ICE Design and Construct Conditions of Contract.

(2) *Measure and value.*[6] Here the price is computed by multiplying units of work required by agreed unit rates. The rates are usually set out in bills of quantities. The components of work to be included in the rates may be closely defined by reference to a 'method of measurement'.[7] Such contracts often provide that where there are significant changes in quantities or working conditions, the rates are to be modified to account for this.

Traditional civil engineering contracts tend to use a measure and value system. The amount of earthworks, the different types of sub-soil materials and the volumes of fill materials or mass concrete are all highly uncertain. Furthermore, much of the work is done under the close direction of the Engineer who may, for example, order extra height on temporary flood protection banks, additional grouting to seal a dam or deeper foundations, etc., as suits the conditions on site. Measure and value arrangements provide fairness for the contractor coupled with flexibility for the Engineer.

(3) *Cost reimbursement.*[8] Here the contractor is reimbursed for all his reasonable direct expenses and outgoings in connection with the works, together with an additional sum representing the contractor's fee. The term 'cost' is usually defined in the contract. Where it is not, it is thought that costs arising from unreasonable extravagance will not be costs of the contract work, but other costs at the contractor's risk, and hence irrecoverable.

Cost reimbursement contracts are frequently used where the scope of the work is ill-defined. They enable construction to commence before the design is finalised.

(4) *Target cost.*[9] This is a variant on the cost reimbursable contract involving the use of a target out-turn cost. The parties agree a target cost at the outset. If the contractor exceeds the target he pays a 'penalty' on his fee. If he performs the work at a keener cost than the target he receives a bonus on his fee. Target cost arrangements are a type of 'incentive contract' where there is an explicit attempt to align the interests of the contractor and employer.

---

6    This is also called 'measurement' or 'admeasurement'. Examples are the ICE Conditions of Contract, 7th Edition, and the FIDIC Red Book Conditions of Contract.

7    A number of standard methods of measurement are published: for instance see *Civil Engineering Standard Method of Measurement*, 3rd edn (CESSM 3), Thomas Telford.

8    The IChemE Green Book is an example. They are also used frequently for management contracts; see for example the Engineering and Construction Contract Option E.

9    See, for example, the Engineering and Construction Contract Option C.

*Adjustment of price for claims other than variations*   Civil engineering standard forms make specific provision for adjusting the price when various
situations arise. Each individual contract has its own regime and each contract must be interpreted individually. In civil engineering contracts there
may typically be a clause entitling the contractor to his additional costs for:
(*a*) dealing with unforeseeable ground conditions and (*b*) for increases in
the market price of particular materials. Where the contract makes such
provision, the price will be adjusted accordingly.

*Retention monies*   Civil engineering contracts almost invariably provide
for payments to be subject to an agreed amount—often 5%— of 'retention'.
The purpose of this is to ensure that the employer has a fund available in
the event that the contractor defaults and is either unable to complete the
works or leaves the works with defects. Traditionally, half of the retained
monies are paid upon substantial completion and the remainder is paid
after a set time (often associated with an agreed period during which the
contractor is both obliged and entitled to rectify defects following substantial completion). Some employers place retained monies into a separate
fund, held on trust for the contractor so that it is secure in the event of the
employer's bankruptcy. Where the contract obliges the employer to do this,
the contractor may obtain an injunction requiring it to be done.

*Bonuses*   Civil engineering contracts occasionally provide for bonuses
(e.g. for early completion), but this is not commonplace. A bonus is a means
of gaining some alignment in the interests of the employer and contractor.
As contract risk theory develops and more employers appreciate that such
an alignment is beneficial to them, the use of bonuses may increase.

*Tax and VAT*   Civil engineering construction may be zero rated or standard rated for VAT purposes, depending on the nature of the works and the
nature of the employer. The contract will normally state expressly that
prices are exclusive of VAT, which will be charged at the prevailing rate.
Payments made to sub-contractors will involve a deduction for tax, unless
the contractor can provide a valid exemption certificate.

## 2. Payment process—the provisions of the HGCRA 1996

*General*   Civil engineering contracts have, for the most part, almost
invariably contained provisions for payment by instalment. Where it

applies, the Housing Grants, Construction and Regeneration Act 1996 (the HGCRA 1996)[10] stipulates statutory minimum requirements as to payments by instalment, notices as to payment and a prohibition, for most purposes on conditional payment clauses. Failure to meet these activates the relevant provisions of the Scheme for Construction Contracts.[11]

*Terminology*   The HGCRA 1996 describes both parties as 'a party'. Thus, by Section 109, for instance: 'A party to a construction contract is entitled to payment by instalments …'. This cannot, of course, ordinarily apply to an employer, but the sense is clear. In this chapter, the terms 'employer' and 'contractor' are used to designate the paying and receiving parties respectively.

*Application of the HGCRA 1996*   The Act applies to contracts in writing,[12] entered into on or after 1st May 1998, which relate to 'construction operations' in England, Wales or Scotland irrespective of whether or not the law of England and Wales or Scotland otherwise applies.[13] 'Construction operations' are given an extended meaning which includes not only contracts for the carrying out of construction operations, but also contracts for the arrangement and design of construction contracts.[14] Accordingly, the Act applies to most civil engineering management, design and construction contracts for work in Britain. There are a number of specific exceptions which are important for civil engineering. These include drilling or tunnelling for the purpose of extracting minerals and access and machinery for works relating to the basic utilities and to the chemical, food and pharmaceutical industries.[15] In addition, Private Finance Initiative projects are excluded. Where only part of the contract relates to construction operations, that part only is subject to the Act.

---

10   The full text of Part II of the Act is set out in Appendix 2, together with the text of the Scheme for Construction Contracts. A fuller analysis of the operation of the adjudication provisions of the Act is given in Chapter 15.

11   Scheme for Construction Contracts (England and Wales) Regulations 1998, SI 1998 No. 649. Separate Regulations are made for Scotland.

12   Section 107 of the Housing Grants, Construction and Regeneration Act 1996. Note that the expression 'in writing' is given an extended meaning and includes, for example, contracts evidenced in writing: Section 107(2)(c).

13   Section 104(6), (7) of the Housing Grants, Construction and Regeneration Act 1996.

14   Section 104(1), (2) and Section 105 of the Housing Grants, Construction and Regeneration Act 1996.

15   Section 105(2) and Section 105 of the Housing Grants, Construction and Regeneration Act 1996.

## Payment by instalments

*The entitlement to payment by instalment*   Section 109 of the HGCRA 1996
provides:

(1) A party to a construction contract is entitled to payments by instal-
    ments, stage payments or other periodic payments for any work under
    the contract unless—
    (a) it is specified in the contract that the duration of the work is to be
        less than 45 days, or
    (b) it is agreed between the parties that the duration of the work is esti-
        mated to be less than 45 days.
(2) The parties are free to agree the amounts of the payments and the inter-
    vals at which, or circumstances in which, they become due.
(3) In the absence of such agreement, the relevant provisions of the
    Scheme for Construction Contracts apply.

*The parties' freedom to agree the intervals or circumstances*   The Act does
not limit the maximum instalment period; it may extend to the entire duration of the project. This is unobjectionable because the Act does attempt to
prescribe what commercial arrangements parties may adopt; it simply
requires that the payment process be stated explicitly, so that both
employer and contractor agree the expected cash flow. In practice, parties
may agree time periods or milestones etc. as triggers for payment.

*Where there is no agreement as to the instalment period*   Where there is no
agreement, the Scheme provides that the period is one of 28 days[16] (called
in the Scheme 'a relevant period').

## Computing the time for payment and notices

Section 110(1) provides, *inter alia*, that

> Every construction contract shall—(*a*) provide an adequate mechanism for
> determining what payments become due under the contract, and when, and
> (*b*) provide for a final date for payment in relation to any sum which becomes
> due.

Where the contract fails to provide an adequate mechanism, the Scheme is
activated.

---

16   Scheme for Construction Contracts (England and Wales) Regulations 1998, Part II, Para-
     graph 12.

*Agreeing an adequate mechanism*   Section 110 requires that there be an adequate mechanism for computing two dates in relation to each instalment:

(1)   the date when payments become due
(2)   a final date for payment in relation to any sum which becomes due.

Where these dates are not defined, the Scheme is activated. The agreed mechanism need not be sophisticated.[17] All generally used standard forms are likely to comply. One potential problem relates to payments triggered by the completion of defined elements of work; where an employer is responsible for a delay to the trigger date and there is no provision for this in the contract, the mechanism may not be adequate since otherwise, the employer may take advantage of his own culpable default.

*The Scheme's provisions for the various periods and dates associated with payment*   The Scheme provides the following:

(1)   *Instalment period.* Where there is no agreement as to the instalment period (called by the Scheme a 'relevant period'), it is 28 days.[18] The Scheme does specify when the time of reckoning should begin, but presumably it is computed in 28 day periods from the date defined by the contract as the 'commencement date' or where there is none defined, the date when work actually commenced.
(2)   *Due date for payment.* The accrued value of work done during each 28 day period becomes due seven days after the period ends or the making of a claim by the contractor.[19] The claim must be in writing, setting out the amount and basis of computation.[20] It is thought that the submission of an ordinary application for payment under most standard form contracts will suffice.
(3)   *Final date for payment.* The final date for payment of each instalment is 17 days from the date upon which the amount becomes due.[21] In the ordinary course of events, then, this will be 24 days after the end of an installation interval.

---

17   The default option in the Scheme is itself extremely simple.
18   Scheme for Construction Contracts (England and Wales) Regulations 1998, Part II, Paragraph 12.
19   Scheme for Construction Contracts (England and Wales) Regulations 1998, Part II, Paragraph 4.
20   Scheme for Construction Contracts (England and Wales) Regulations 1998, Part II, Paragraph 12.
21   Scheme for Construction Contracts (England and Wales) Regulations 1998, Part II, Paragraph 8(2).

(4) *The date of final payment.* The duration of the works is unlikely to be an exact multiple of 28 days. There will also be matters of negotiation and clarification to deal with at the end of the work. In order to make allowance for this, the Scheme provides that the final payment is 30 days after the date the works are completed.[22]

## Computing the amount due in each instalment

Section 110 requires that the contract shall contain an adequate mechanism for determining what payment is due in each instalment. The Act does not define 'adequate'. Current standard forms drawn up to meet this requirement probably comply.

Where the mechanism is not agreed or is inadequate, the Scheme applies a simple calculation for computing the amount which is due in respect of a period of payment,[23] viz: the total value of work performed from the start of the contract until the end of the stage/period minus the amounts already paid. The 'value of work' is defined as either:[24] (*a*) the amount determined in accordance with the contract; or where there is no such provision (*b*) the cost of any work to which the Act applies plus the same rate[25] of profit/overhead as included in the contract price. Clearly, where it is not specified, the level of profit/overhead, can be a source of dispute. The Scheme also states that the valuation shall exclude materials brought onto site or other special elements unless the contract expressly specifies that these elements will be included.[26]

## Notices as to payment or withholding payment

The HGCRA 1996 and the Scheme provide for the issue of notices. These are:

(1)   notice specifying the amount of payment made or proposed to be made
(2)   notice of intention to withhold payment
(3)   notice of intention to suspend performance.

---

22    Scheme for Construction Contracts (England and Wales) Regulations 1998, Part II, Paragraph 5.
23    Scheme for Construction Contracts (England and Wales) Regulations 1998, Part II, Paragraph 2.
24    Scheme for Construction Contracts (England and Wales) Regulations 1998, Part II, Paragraph 12.
25    In fact the Scheme says that what should be added is an 'amount equal to any overhead or profit included in the contract price'. Literally taken this seems to mean the full amount, but it is submitted that the only commonsense interpretation is that the same rate of profit/overhead should be applied.
26    Scheme for Construction Contracts (England and Wales) Regulations 1998, Part II, Paragraph 2(2), (3).

These notices are of some importance, especially where the contractor is considering suspending his performance of the work.

*Notice of payment made or proposed to be made*   Section 110(2) provides:

(2)   Every construction contract shall provide for the giving of a notice by a party not later than five days after the date on which a payment becomes due from him under the contract, or would have become due if—
  (a)   the other party had carried out his obligations under the contract, and
  (b)   no set-off or abatement was permitted by reference to any sum claimed to be due under one or more other contracts,
  specifying the amount (if any) of the payment made or proposed to be made, and the basis on which that amount was calculated.

The notice clearly needs to specify two matters only:

(1)   the amount made or proposed to be made; and
(2)   the basis on which it was calculated.

It is thought that the hypothetical assumptions in paragraphs (a) and (b) of the sub-section do not affect the content of the notice. They affect only its timing. Assumption (a) provides that the notice cannot be delayed because the employer claims that the contractor has been dilatory or is in default. Where payments are to be made by reference to the completed stages of work, its seems that 'the date on which a payment … would have become due' is to be computed as the date when the stage would have been completed if the contractor had not been in default. Assumption (b) preserves the need for the notice even where the employer claims a set-off which exceeds the amount due under the contract; otherwise it would be arguable that no net sum would be due and no notice need be served.

*Notice of intention to withhold payment*   Section 111 provides:

(1)   A party to a construction contract may not withhold payment after the final date for payment of a sum due under the contract unless he has given an effective notice of intention to withhold payment …
(2)   To be effective such a notice must specify—
  (a)   the amount proposed to be withheld and the ground for withholding payment, or
  (b)   if there is more than one ground, each ground and the amount attributable to it,
  and must be given not later than the prescribed period before the final date for payment.

The section also provides that the 'prescribed period' can be agreed by the parties and that

> The notice mentioned in section 110(2) may suffice as a notice of intention to withhold payment if it complies with the requirements of this section.

Whether or not a Section 110(2) notice is sufficient for the purpose of Section 111 may itself be disputed. Since the latter notice is a condition to the contractor suspending work, it is suggested that a prudent employer who proposes that the Section 110(2) be also a Section 111 notice should make this clear.

## 3. Conditional payment clauses

Section 113 of the HGCRA 1996 provides:

> 113 – (1) A provision making payment under a construction contract conditional on the payer receiving payment from a third person is ineffective, unless that third person, or any other person payment by whom is under the contract (directly or indirectly) a condition of payment by that third person, is insolvent.

Conditional payment clauses (e.g. pay–when–paid clauses) are generally ineffective. The one exception is where the person due to pay the payer is insolvent. The section provides a lengthy definition of all the circumstances which amount, for the purposes of this section, to insolvency. A main contractor cannot rely upon the employer's insolvency to refuse payment to a sub-contractor unless (*a*) he has not, in fact, been paid and (*b*) there is a pay–when–paid clause.

## 4. Interest

Most civil engineering contracts contain provisions for the payment for interest for late payment.[27] Where a contract does not contain provisions relating to interest, or where the provisions do not deliver a 'substantial remedy', the Late Payment of Commercial Debts (Interest) Act 1998 may apply. This provides that there is an implied term giving a right to simple interest. The current rate provides a serious incentive to pay.[28] This statute

---

27    For example, ICE Conditions of Contract, 7th Edition, Clause 60(6).
28    The rate is given in the Late Payment of Commercial Debts (Interest) Act 1998 (Rate of interest) (No. 2) Order 1998, SI 1998 No. 2765, as 8% over the official dealing rate as set from time to time by the Bank of England's Monetary Policy Committee.

is being introduced piecemeal and its precise status at any time should be checked.[29] Where a dispute is referred to an adjudicator, he will have power to decide issues of interest within the terms of the contract, including, where appropriate, those implied by the Late Payment of Commercial Debts (Interest) Act 1998.

Where a dispute is referred to an arbitrator or proceedings are brought before the court, these tribunals will have power to award interest under their governing legislation.[30] Where there is a contractual provision, agreed or implied, this will prevail over the arbitrator's discretion.

## 5. Referring payment matters to adjudication

The parties to a contract to which the HGCRA 1996 applies have the right to refer disputes or differences to adjudication, including those as to whether or not payment is due and when it is due.[31]

---

29   The Act came into force on 1 November 1998 for contracts between small business suppliers and
     (a) UK public authorities or (b) large business purchasers: SI 1998 No. 2765, The Late Payment of
     Commercial Debts (Interest) Act 1998 (Commencement No. 1) Order 1998. The terms 'small
     business supplier' etc. are defined in the order.
30   Arbitration Act 1996, Section 49; Supreme Court Act 1981 Section 35A.
31   See Chapter 11.

# 5

## Time for performance

### 1. The importance of time obligations

The time for performance of a civil engineering contract is important in a number of respects. The later the completion date, the later the employer will be able to benefit from his investment. The works may need to be operational on a particular date. And, more routinely, most civil engineering projects involve complex coordination of contractors, sub-contractors and supply deliveries which demands that time be managed to ensure proper co-ordination of these activities.

### 2. General obligations as to time

*Where there is no express agreement*   In the absence of express provisions in the contract as to time, the contractor's obligation is to complete the work within a reasonable time. If he fails to do so, he will be liable for damages.

*What is a reasonable time?*   A reasonable time is to be determined as a combination of (*a*) those factors  of which the parties were objectively aware of at the time of making the contract and (*b*) those events which occurred afterwards over which the contractor had no control.[1] The time required is, of course, not independent of the level of resources applied; where the terms of the contract, and/or the facts known to both parties at the time of the contract, evidence the presumed intention of the parties as

---

1    *Hick v. Raymond & Reid* [1893] AC 22 at 32 (HL).

to the appropriate level of resources, this may assist in computing the reasonable time.

*Where there is an agreement only to complete by a certain date*   Where there is an agreement to complete by a certain date, later completion will be a breach of contract, entitling the employer to damages. Where, however, the project has been delayed by the employer's default, the completion date is no longer applicable and time is said to be 'at large', that is, the time obligation becomes an obligation to complete the works in a reasonable time.[2] This reasonable time will not, however, be unrelated to the originally agreed time since the level of resources will remain the same.

*Extension of time clauses*   Standard form contracts invariably provide for 'extensions of time clauses'. These enable the completion date to be adjusted in the event of defaults by the employer or any variation to the works. An extension of time clause avoids time being put at large. This is particularly important where the completion obligation is tied to a liquidated damages provision; if time is put at large, and so there is no certain date from which liquidated damages can run, then the provision becomes inoperable. As a consequence, no liquidated damages can be claimed and the employer has to prove his loss.

*Time is ordinarily not of the essence*   An obligation whose breach is so fundamental that it entitles the other party to elect to refuse further performance is said to be essential or 'of the essence'. In cases involving sales of goods, time has often been held to be of the essence. The courts have, however, been unwilling to treat time as being of the essence in civil engineering contracts. Unlike a sale of goods, construction works become fixed to the employer's land and cannot readily be returned. A time obligation can, however, be made 'of the essence' by notice to the contractor in good time requiring him to complete within a reasonable time.[3]

### The contractor's time obligation is also his licence

> In the contract one finds the time limited within which the builder is to do this work. That means not only that he is to do it within that time, but it means that he is to have that time within which to do it. To my mind that limitation of time is clearly intended, not only as an obligation, but as a

---

2   Unless there is provision for extensions of time.
3   *Carr v. J A Berriman Pty Ltd* (1953) 27 ALJ 273.

benefit to the builder ... in my judgment, where you have a time clause and a penalty clause, it is always implied in such clauses that the penalties are only to apply if the builder has, as far as the building owner is concerned and his conduct is concerned, that time accorded to him for the execution of the works which the contract contemplates he should have.[4]

*Planning work within the contract period*    The contractor has freedom to plan his work within any time constraints.[5] He is not generally obliged to proceed at any particular rate,[6] but it is thought that he cannot allow progress to be so slow that it would be impossible for him to complete by dates specified in the contract. Progress too slow to catch up may, if shown to be wilful, amount, in extreme cases, to an anticipatory breach; the breach does not consist of failing to meet time targets, but of the expressed intention not to comply with the contract time provisions.

*Express provisions as to progress*    Employers are becoming more aware that it is in their interests to take more control over the timing of the works; interim milestones and completion of sections at different times are becoming common. The drafting of such provisions requires caution in a number of respects. Any requirement that the contractor must complete in a given sequence is, unless the contract states otherwise, coupled with a corresponding entitlement on the contractor's part.[7] Furthermore, where failure to meet time targets entitles the employer to agreed (liquidated) damages for breach, these damages must not amount to penalty payments.[8]

## 3. Extensions of time

Where a completion time is agreed, but the contractor is delayed by some matter at the employer's risk, the completion time reverts to a reasonable time unless there is a term in the agreement enabling the time for completion to be extended. Where the time for completion reverts to a reasonable time, any liquidated damages provisions are rendered uncertain and ineffective.[9] In most civil engineering contracts, there are elaborate clauses for extensions of time, designed to avoid this consequence. These clauses,

---

4    *Wells v. Army and Navy Co-operative Society* (1902) 2 *Hudson's Building Contracts*, 4th edn, 346 per Vaughan Williams L.J. at p. 354.
5    *Wells v. Army and Navy Co-operative Society* (1902) 86 LT 764.
6    *Greater London Council v. Cleveland Bridge Engineering Co. Ltd* (1984) 34 BLR 50.
7    *Yorkshire Water Authority v. Sir Alfred McAlpine Co.* (1985) 32 BLR 114; *Holland Dredging v. Dredging and Construction* (1987) 37 BLR 1 (CA).
8    See below, Section 4.

however, tend to focus on the circumstances which entitle the contractor to an extension of time, rather than the proper basis on which it might be computed.

*Concurrent delays*   Where the contractor's delay is caused by a number of concurrent factors, for some of which only the employer is responsible, diffi-cult questions arise as to the computation of the correct extension. The dominant cause approach[10] is frequently used to provide assistance; for example, if there are two causes and the employer is responsible for the dominant cause an extension is granted. Otherwise no extension is granted. This test has been doubted, for where both parties are culpable to a degree, it seems just that each should bear an apportionment of the delay.[11]

*Computing extensions of time*   Where computer-based programming tech-niques are used to compute extensions of time, care should be exercised as these models often lack transparency. Care must be taken to ensure that the assumptions take real construction practice into account. For example, the relationship between resource availability, resource costs and the activ-ities which will be scheduled to run is often poor and unrealistic. Delays are frequently calculated on the basis of an addition or a subtraction model. The addition model takes a credible initial programme and adds in the delay events to give a revised completion date; the difference in original and revised completion dates is the computed extension. The subtraction model requires that an as-constructed programme be devised, from which all the delaying events are subtracted to give the date by which the contrac-tor would have completed but for the delays. Neither of these models take into account the fact that the project is administered from day to day, with-out the benefit of hindsight as to what delays will eventually arise and what their effects will be.

*The opportunity to reprogramme*   It is important for the contractor to know what the completion date is to be and hence to be able to programme the remaining    works.    Under    most    civil    engineering    contracts,    the

---

9     *Peak Construction (Liverpool) Ltd* v. *Mc Kinney Foundations Ltd* (1971) 1 BLR 111 (CA); *Fernbrook Trading Co. Ltd* v. *Taggart* [1979] 1 NZLR 556 (New Zealand Supreme Court); *Perini Pacific Ltd* v. *Greater Vancouver Sewerage and Drainage District* (1966) 57 DLR (2d) 307 (British Columbia Court of Appeal).

10    *Leyland Shipping Co.* v. *Norwich Union Fire Insurance Society* [1918] AC 350 (HL).

11    See, for example, the remarks of Judge Fox-Andrews QC in *H Fairweather & Co. Ltd* v. *London Borough of Wandsworth* (1987) 39 BLR 106.

administrator is required to grant extensions of time.[12] The contract may specify the time in which the extension is to be made, but generally this is not so and the administrator must grant extensions within a reasonable time. What amounts to a reasonable time will depend on the assistance provided by the contractor,[13] the complexity of the delays and the extent of the time consequences ascertainable at that stage.[14] Where the employer or his agent fails to grant a proper extension within a reasonable period, this will normally amount to a breach.[15] It may also provide the contractor with an entitlement under the contract. The contractor is entitled to aim to complete in the currently-set completion date. Accordingly, if he is obliged to accelerate his work to complete in the foreshortened period and he thereby suffers loss, he will be able to recover this loss as damages.

*Adjudication and extensions of time*[16]   Where the administrator believes that no extension of time is due, but an adjudicator (e.g. one appointed under the HGCRA 1996) grants an extension, the contractor may reprogramme his work in accordance with the newly extended completion date. It is thought that if the contractor works to this extension, it cannot later be challenged by the employer. Unlike money, time cannot be paid back. If, however, the employer is strongly of the view that the extension was wrongly decided by the adjudicator, he ought to instruct the contractor to accelerate to complete in the original time and pay a higher rate under objection. This will, it is thought, satisfy the adjudication decision and will also, it is thought, preserve the employer's right to challenge the adjudicator's decision later.

## 4. Liquidated damages for late completion

*Liquidated damages: meaning*   Liquidated damages are all those agreed sums in contracts which are payable upon the breach of contract by the party. They may be contrasted with 'unliquidated damages' which have not been reduced to a defined sum. The amount of unliquidated damages must be proved, which is a difficult, expensive and tiresome process. Liquidated damages are used in a variety of situations, but most commonly as a fixed

---

12   For example, under the ICE Conditions of Contract, the Engineer is empowered to grant extensions of time.

13   For example, where the contractor holds relevant records and has access to project programming data he might reasonably be expected to assist the decision-maker.

14   *Amalgamated Building Contractors Ltd* v. *Waltham Holy Cross UDC* [1952] 2 All ER 452 (CA).

15   *Fernbrook Trading Co. Ltd* v. *Taggart* [1979] 1 NZLR 556 (New Zealand Supreme Court).

16   As to adjudication generally, see Chapter 8.

remedy for the contractor's delay. They are generally expressed as a certain amount of money per unit of time.

*Liquidated damages as an exhaustive agreement*   Liquidated damages fix the amount which may be claimed. An employer cannot claim more because in the particular circumstances his loss is greater. Conversely, substantial liquidated damages are payable even where the innocent party in fact suffers or no loss or even gets a benefit.[17] The arrangement works to the contractor's benefit also; his liability is capped at the agreed damages. In one case the word 'nil' had been inserted as the rate of liquidated damages in a standard printed form of contract. The Court of Appeal decided that this meant that the rate was £0 and that this was

> An exhaustive agreement as to the damages which are or are not to be payable by the contractor in the event of his failure to complete the works on time.[18]

*Defences to the payment of liquidated damages*   The defences include:

(1)   The liquidated damages provision is a penalty.
(2)   There is a delay caused by the employer and there is no extension of time provision to cover that delay.[19]
(3)   There is no date stated in the contract.[20]

Where the defence is made out, the damages will revert to unliquidated damages and the innocent party will have to prove his loss.

*A penalty will not be enforced*   A penalty clause is one where the breach of a term in the contract entitles the innocent party to recover, from the party in default, a sum which does not represent a genuine pre-estimate of his loss and which substantially exceeds it.[21] The law will not enforce a penalty and

---

17   *Clydebank Engineering and Shipbuilding Co.* v. *Casteneda* [1905] AC 6 (HL). The shipbuilders argued unsuccessfully that the late delivery of warships saved them from being sunk by the US Navy along with the rest of the Spanish Armada and that, therefore, they should not have to pay the liquidated damages for delay.
18   *Temloc* v. *Errill Properties Ltd* (1988) 39 BLR 30 (CA), per Lord Justice Nourse.
19   *Peak Construction (Liverpool) Ltd* v. *McKinney Foundations Ltd* (1971) 1 BLR 111 (CA).
20   *Kemp* v. *Rose* (1858) 1 Giff. 258 at 266; but see also *Bruno Zornow (Builders) Ltd* v. *Beechcroft Developments Ltd* (1989) 51 BLR 16 where a completion date linked to a liquidated damages provision was implied.
21   This principle is of general application and is not limited to forfeits for delay. The one exception by custom is a 10% deposit paid upon exchange of contracts for the purchase of land, when the purchaser fails to complete: *Workers Trust and Merchant Bank Ltd* v. *Dojap Investments Ltd* [1993] 2 All ER 370 (PC).

when a clause purporting to be a liquidated damages clause is declared a penalty, the party who sustains a loss as a result will have to prove its amount. The fact that the term 'penalty' or 'liquidated damages' is used is not conclusive. The test relates to the substance of the arrangement rather than its description.[22]

*Challenging a clause as a penalty*   Where there is provision in a contract stipulating a sum to be paid to the innocent party upon a breach, the presumption is that it is not a penalty and hence is enforceable. The burden of proof lies upon the party who seeks to show that the amount is a penalty[23] and the burden is a significant one.[24]

*Sectional completion and liquidated damages*   Difficult problems may arise in relation to sectional completion provisions in the contract. A sectional completion requirement obliges the contractor not only to complete the works as a whole by a certain date, but also requires him to complete specified elements of the work at dates before the date for completion of the overall works. A sectional completion obligation supplemented by liquidated damages provisions related to each section which remains

---

22   The leading case is *Dunlop Pneumatic Tyre Ltd* v. *New Garage and Motor Co. Ltd* [1915] AC 79 (HL). Lord Dunedin, at 86, laid down a series of tests, including

The essence of a penalty is a payment of money stipulated as *in terrorem* of the offending party; the essence of liquidated damages is a genuine covenanted pre-estimate of damages.

23   The test is given by Diplock L.J. in *Robophone Facilities Ltd* v. *Blank* [1966] 1 WLR 1428 at 1447 (CA):

The court should not be astute to descry a 'penalty clause' in every provision of a contract which stipulates a sum to be payable by one party to the other in the event of a breach by the former ... The onus of showing that such a stipulation is a 'penalty clause' lies upon the party who is sued upon it.

His Lordship then appears to use a test by which the evidential burden shifts to the person suing on the 'penalty clause', namely where the amount stipulated is 'extravagantly greater' than the loss which is liable to result from the breach.

24   *Philips* v. *AG of Hong Kong* (1993) 61 BLR 41 at 60 (PC). In *Robophone Facilities Ltd* v.*Blank* [1966] 1 WLR 1428 at 1447 (CA), Diplock L.J. suggests that the evidential burden shifts to the person suing on the 'penalty clause' only when the amount stipulated for is 'extravagantly greater' than the loss which is liable to result from the breach. In *Dunlop Pneumatic Tyre Ltd* v. *New Garage and Motor Co. Ltd* [1915] AC 79 (HL) the contract provided that New Garage were obliged to sell at the list price; they broke this agreement. The contract provided for liquidated damages of £5 for each tyre sold in breach of the agreement. Although the sum of £5 per tyre seemed quite extravagant, the House of Lord dismissed New Garage's contention that the sum was a penalty because these breaches undermined Dunlop's relations with all its other suppliers.

uncompleted at the due date is readily achieved by clear wording in the contract. On occasions, however, parties attempt to create such an obligation simply by inserting words into the liquidated damages clause which indicate a rate per section or element of the work which remains incomplete. In many instances this creates a ambiguity when read with other terms in the contract; the requirement that the liquidated damages provisions of the contract be construed *contra proferentem* then results in the employer's intentions being defeated.[25]

## 5. Alternatives to liquidated damages for delay

*Incentives to complete and the penalty trap*   Employers naturally wish contractors to complete on time. Where the employer is unlikely to suffer major losses if the contractor completes late, the liquidated damages route is not satisfactory. Moreover, if the amount of liquidated damages reflects a genuine pre-estimate of the likely loss, this will provide no incentive. If the amount provides the appropriate incentive, the clause will be struck down as a penalty.

*Optional progress and completion obligations*   There is no contractual objection to an employer offering the contractor a series of optional completion dates, with later dates attracting a reduction in overall price payable.[26] Where the contractor 'opts' for a later date, the price is significantly less than completion at the 'preferred date'. The same approach can be extended to interim progress obligations. It is thought that this device will make liquidated damages provisions obsolete in the near future.

*Lane and site rental*   On highways projects, the closure of lanes causes major inconvenience and disruption. Unfortunately, this is suffered principally by motorists who are not parties to the contract. Liquidated damages for delay cannot, therefore, properly compensate for their inconvenience. Many highways contracts seek to overcome this using the principle of 'lane rental', by which the contractor pays a rental charge for possession of a lane

---

25   *Bramall & Ogden v. Sheffield City Council* (1983) 29 BLR 73.
26   By extension of the reasoning in *Export Credits Guarantee Department v. Universal Oil Products Co.* [1983] 2 All ER 205 (HL).

for construction or maintenance. The rental charge can be designed as an appropriate incentive because it does not represent damages for breach and is not, therefore, subject to the rule as to penalties. The principle can be extended to an ordinary civil engineering project using site rental, but the optional completion date device generally offers more flexibility.

# 6

# The contract and third parties

The doctrine of privity holds that only the parties to a contract can sue or be sued upon it. There are, however, a number of situations where third parties can become involved, either by taking an interest in the subject matter of the contract or performing work under it. In this chapter, the following situations are dealt with:

(1) third party claims under the Contract (Rights of Third Parties) Act 1999[1]
(2) collateral and direct warranties
(3) assignment of benefits under the contract
(4) novation of contract
(5) sub-contracting and vicarious performance.

## 1. The doctrine of privity

The doctrine of privity holds that only parties to a contract have rights and/ or liabilities under it.[2] The doctrine is of great practical importance in the context of civil engineering projects. Contractual responsibility exists only where there is a contractual relationship. For instance, where an employer contracts with a contractor and the contractor sub-lets work to a sub-contractor, the employer may not ordinarily bring a contractual action against the sub-contractor for defective work;[3] nor may the sub-contractors

---

1   At the date of going to press this Bill had not received Royal Assent. It is assumed that it will be enacted in its present form.
2   *Tweddle* v. *Atkinson* (1861) 1 B& S 393.
3   *Dunlop Pneumatic Tyre Co. Ltd* v. *Selfridge & Co. Ltd* [1915] AC 847 (HL).

bring an action in contract against the employer for payment. Contractual relationships frequently differ from the management and administrative relationships. For example, under a traditional contract, where the employer's administrator is an independent engineer, the engineer and contractor will have no contractual connection; thus the contractor will not be able to sue the engineer for pure economic losses caused by negligent failure to certify since such losses are recoverable only under a contract.[4] In addition, the entity named as the 'employer' may be acting in the stead of the substantive employer; the latter will then be unable to enforce the contract.[5]

## 2. Contract (Rights of Third Parties) Act 1999

*The Contract (Rights of Third Parties) Act 1999*   This modifies the strict privity principle. Section 1 provides, *inter alia*:

> 1 – (1)   Subject to the provisions of this Act, a person who is not a party to a contract (a 'third party') may in his own right enforce a term of the contract if—
> (a)  the contract expressly provides that he may, or
> (b)  subject to subsection (2), the term purports to confer a benefit on him.
>
> (2)   Subsection 1(b) does not apply if on a proper construction of the contract it appears that the parties did not intend the term to be enforceable by the third party.
>
> (3)   The third party must be expressly identified in the contract by name, as a member of a class or as answering a particular description but need not be in existence when the contract is entered into.

*Application*   There are a number of excluded contract types, including contracts of employment and some aspects of contracts for the carriage of goods.[6]

*Terminology*   The parties to the contract are described as the 'promisor' and 'promisee'. The promisor promises to do that thing which the third party may enforce. The promisee is the other party to the contract.[7]

---

4    *Pacific Associates* v. *Baxter* [1990] 1 QB 993 (CA). Note that in some Commonwealth jurisdictions, engineers may find themselves liable in negligence: see *Edgeworth Construction Ltd* v. *ND Lea & Associates Ltd* [1993] 3 SCR 206.
5    *Darlington Borough Council* v. *Wiltshier Northern Ltd* [1995] 1 WLR 68 (CA).
6    Section 6 of the Contracts (Rights of Third Parties) Act 1999.
7    Section 1(7) of the Contracts (Rights of Third Parties) Act 1999.

*Defences*   Where the third party seeks to enforce the term, the promisor may (subject to express terms in the contract) set up any defence available under the contract, including rights of set-off against either the promisee or counterclaims against the third party.[8] It is thought that all contractual provisions as to conditions precedent will apply. The third party may not set up the requirement in the Unfair Contract Terms Act 1977 that terms excluding liability for negligence meet the test of reasonableness.[9]

*Prejudicing the rights of the third party*   Where a contract is made for the benefit of a third party, the parties to the contract may not (unless there is an express reservation in the contract, or the third party agrees) rescind or vary the contract to the prejudice of the third party if the third party has relied on it.[10]

*Saving of the promisee's rights*   Rights provided to the third party under Section 1 do not affect any right of the promisee to enforce any term.[11] Where, however, both third party and promisee may both enforce the same term, the Act stipulates that there shall be no double recovery.[12]

*Arbitration*   Where a term in the contract provides that disputes between the third party and the promisor shall be resolved by arbitration, then that suffices as a written agreement and the Arbitration Act 1996 applies. The Contracts (Rights of Third Parties) Act 1999 is unclear on the point, but it is thought that a promisor may stay a dispute under the contract to arbitration; also the arbitrator, it is thought, would, in the normal way, assume jurisdiction over all defences, including counterclaims arising under another contract.[13]

*The impact of the Act on current contracts*   Few clauses are currently drawn so as expressly to confer rights on third parties. In future, many parties may attempt to ensure that the Act has no effect for their contracts by using clear words of exclusion.

---

8   Section 3 of the Contracts (Rights of Third Parties) Act 1999.
9   Section 7(4) of the Contracts (Rights of Third Parties) Act 1999. Note, however, that this is one of the issues still being debated, and Section 7(4) may not be enacted.
10  Section 2 of the Contracts (Rights of Third Parties) Act 1999.
11  Section 4 of the Contracts (Rights of Third Parties) Act 1999.
12  Section 5 of the Contracts (Rights of Third Parties) Act 1999.
13  The most significant matter remaining in issue in the Bill at the time of going to press is the effect on an arbitration clause in the contract. Readers should inform themselves of the final outcome of the debate.

*Potential use to obviate the need for collateral warranties*   Parties in the civil engineering industry frequently use collateral warranties to give rights to third parties.[14] These are often made in favour of financiers or future users of facilities. In addition, employers may take direct warranties against sub-contractors, suppliers, etc. The Contracts (Rights of Third Parties) Act 1999 provides an elegant means of achieving the same effect without the multiplicity of documents. It may be, therefore, that the Act will be widely employed instead of collateral warranties.[15]

## 3. Collateral and direct warranties

Express supplementary agreements (often termed 'direct warranties' or, where they are collateral to another contract, 'collateral warranties') may be used to forge contractual links between parties who would not otherwise be in such a relationship. An example is where financiers wish to have recourse against the contractor if, for any reason, the employer is unable to enforce the contract.[16] Frequently, civil engineering contracts state that the contractor may only sub-let work if the sub-contractor agrees to enter into a supplementary agreement with the employer. Supplementary agreements may also arise by implication; where a prospective sub-contractor makes assurances to the employer to secure the employer's agreement to the sub-contract, a contract may arise by implication.[17]

*Creating a direct contract between the employer and sub-contractors/ suppliers*   The main contract may include a requirement that no work be sub-let unless the sub-contractor agrees to execute a direct warranty in specified terms in the employer's favour. By the warranty, the sub-contractor normally agrees that he will perform the works in such a manner that the main contract will not be broken and where it is broken, the employer may sue him as if he, the sub-contractor, were the main contractor. A sub-contract entered into in breach of this requirement will not be void, but will be a breach (by the main contractor) of the main contract and the

---

14   See Section 3 below.
15   See, for example, Needham-Laing M, Death to collateral warranties? *Construction Law Review*, 1999, 86.
16   An example of a duty of care deed is to found in *Alfred McAlpine Construction Ltd v. Panatown Ltd* (1998) 88 BLR 67 (CA).
17   See *Shanklin Pier Ltd v. Detel Products* [1951] 2 KB 854 where an agreement was created between the employer and the sub-contractor. See also *Independent Broadcasting Authority v. EMI Electronics Ltd and BICC Construction Ltd* (1980) 14 BLR 1 (HL) where no agreement arose.

employer may refuse to allow any performance of that sub-contractor on the works. The direct warranty will normally be made for nominal consideration or as a deed. Supply contracts may be dealt with in the same way.

*Essential suppliers who refuse to execute a direct warranty*   Where an essential sub-contractor or supplier refuses to agree to enter into a direct warranty, the employer's insistence upon it may amount to prevention. But this will be so in the most extreme cases only, since the sub-contractor or supplier can ordinarily be indemnified by the contractor. Where the supplier is adamant that he will not enter into the warranty, then it may be that the employer must mitigate his loss by accepting the supply without the benefit of the warranty.[18]

*Where the sub-contractor warrants the quality of his work as an inducement to the employer*   A warranty of quality may arises where a sub-contractor gives to the employer an undertaking as to quality supported by sufficient consideration. In *Shanklin Pier Ltd* v. *Detel Products*[19] the employer was the owner of a pier which was to be repainted. The manufacturer of a brand of bituminous paint represented to the employer the suitability of their paint. The employer instructed the contractor to place an order with the paint manufacturer. The paint failed and the employer sued the paint manufacturer. It was held that there was a contractual warranty. The consideration supplied by the employer was its promise to instruct its contractor to place a contract with the manufacturer.[20]

*Additional work undertaken by the sub-contractor at the employer's request*   Ordinarily, requests by an engineer for the sub-contractor to do the extra work will be made within the terms of the variations clause in the main contract and with the concurrence of the main contractor. They are properly ordered variations. Where, however, the employer directly

---

18   See *Payzu Ltd* v. *Saunders* [1919] 2 KB 581 (CA) for a case where mitigation required accepting performance from a contractor who was unwilling to abide strictly by the contract.

19   [1951] 2 KB 854.

20   For the limits to this principle, see *Independent Broadcasting Authority* v. *EMI Electronics Ltd and BICC Construction Ltd* (1980) 14 BLR 1. EMI was the contractor and BICC were nominated sub-contractors. IBA claimed that a representation made by BICC during the project constituted a warranty. Viscount Dilhorne said

If this is right, then it would seem to me to follow that any representation, whether made innocently, negligently or fraudulently, which is intended to be acted on and which is acted on creates a contractual relationship. I do not think that this can be right.

requests a sub-contractor to perform work which is genuinely outwith the main contract, a separate agreement will ordinarily come into existence between the employer and sub-contractor. Note the need for authority: a direct request by the engineer to the sub-contractor will not normally bind the employer as the engineer lacks authority unless expressly clothed with it.

## 4. Assignments

*Terminology*   An assignment is the transfer by a party to a contract of some or all of his benefits under that contract. The party who transfers the contractual benefit is known as the assignor, while the person who receives the benefit is known as the assignee. The person who has an obligation to provide the benefit is known as the debtor.

*Examples of assignments in relation to civil engineering contracts*   A contractor may assign his entitlement to retention monies; a concession contractor may assign future incomes from the project. An employer who is selling or transferring an asset may assign the benefit of the performance of the civil engineering contract to the transferee.

*Only benefits may be assigned*   The burden of a contract may not be assigned. In other words, once a party has taken on a contractual obligation, that obligation cannot be discharged by getting another person to promise to do it. Benefits arising under a contract may be assigned provided there is no prohibition in the contract and provided, further, that the benefit is not personal to the original assignor. Subject to contractual prohibitions, the right to require the performance of a civil engineering contract will ordinarily be assignable.[21]

*Prohibitions against assignment*   Prohibitions against assignment are generally effective[22] and bind both parties as against the other; it is thought that a prohibition cannot be set up by the assignor against the assignee. The

---

21   *Linden Gardens Trust Ltd* v. *Linesta Sludge Disposals Ltd* and *St Martins Property Corporation Ltd* v. *Sir Robert MacAlpine & Sons Ltd* [1994] AC 85 (HL).

22   *Helstan Securities Ltd* v. *Hertfordshire County Council* [1978] 3 All ER 262. *Linden Gardens Trust Ltd* v. *Linesta Sludge Disposals Ltd* and *St Martins Property Corporation Ltd* v. *Sir Robert MacAlpine & Sons Ltd* [1994] AC 85 (HL).

precise extent and effect of any prohibition is determined by interpreting the relevant clause; the clause may effectively prohibit the assignment of benefits of any description, including money sums (e.g. retention monies).[23]

*What the assignee can recover*   The assignee, under a successful assignment, cannot recover more than the assignor could. Furthermore, the assignee is subject to any defences which the debtor could set up against the assignor.[24] In practice, a right to have a civil engineering contract performed normally leads to damages which are similar for both assignor and assignee, namely the cost of remedying defects. Should a question arise as to whether the reasonable damages are computed as the (large) cost of remedying defects or the (smaller) diminution in value, then insofar as the personal characteristics of claimant are relevant, it is thought that the debtor cannot be disadvantaged by the assignment, unless the contract clearly states otherwise.

*Where the assignment is ineffective*   Where a purported assignment fails as a result of a prohibition, property in the subject matter of the assignment remains with the assignor. Where the assigned property consists of the right to have the contract performed, the 'assignor' can recover substantial damages on behalf of the 'assignee', even where he no longer has an direct interest in the building and hence has suffered no loss.[25]

*Legal and equitable assignments*   A legal assignment is one which complies fully with the requirements of Section 136 of the Law of Property Act 1925. A legal assignment must be (*a*) in writing, (*b*) absolute (that is, the whole of the relevant debt or benefit must be transferred with no rights remaining to the assignor in the benefit assigned) and (*c*) written notice

---

23   *Helstan Securities Ltd v. Hertfordshire County Council* [1978] 3 All ER 262.

24   *Roxburghe v. Cox* (1881) 17 ChD 520 at 526 (CA).

25   *Linden Gardens Trust Ltd v. Linesta Sludge Disposals Ltd* and *St Martins Property Corporation Ltd v. Sir Robert MacAlpine & Sons Ltd* [1994] AC 85 (HL). St Martins Property Corporation ('Corporation') had attempted to assign the right to the performance of a building contract by MacAlpine to its new owner but this had been defeated by the contractual prohibition on assignment. Lord Browne-Wilkinson said that

> MacAlpine had specifically contracted that the rights of action under the building contract could not without MacAlpine's consent be transferred to third parties ... In such a case, it seems to me proper ... to treat the parties as having entered into the contract on the footing that Corporation would be entitled to enforce contractual rights for the benefit of those who suffered from defective performance but who, under the terms of the contract, could not acquire any right to hold MacAlpine liable for breach.

must be given to the debtor. A legal assignment entitles the assignee to sue in his own right and transfers not only the right to sue for the benefit but also any arbitration clause by which it may be enforced.[26] An equitable assignment does not comply with Section 136, but creates an assignment which the courts will enforce under the principles of equity. An equitable assignee must sue in the name of the assignor or, if the assignor refuses to cooperate, the assignor must be made a defendant in the action. Where a sub-contract term states

> If and to the extent that the amount retained by the employer in accordance with the main contract includes any retention money the contractor's interest in such money is fiduciary as trustee for the sub-contractor

this was held to operate as a valid equitable assignment to the sub-contractor of the relevant portion of the retention monies, which created a trust in favour of the sub-contractor.[27] Although an equitable assignment does not require notice to the debtor, the assignee is at risk if the debtor is not notified because a dishonest assignor may subsequently assign the same property to another assignee; priority as against other assignees depends on the order in which the debtor receives notice of assignment, not upon the dates of assignment.

## 5. Novation of contract

Novation is

> the process by which a contract between A and B is transformed into a contract between A and C. It can only be achieved by agreement between all three of them, A, B and C. Unless there is such an agreement, and therefore a novation, neither A nor B can rid himself of any obligation which he owes under the contract. This is commonly expressed by the proposition that the burden of a contract cannot be assigned unilaterally. If A is entitled to look to B for payment under the contract, he cannot be compelled to look to C instead, unless there is a novation. Otherwise B remains liable, even though he has assigned his rights [i.e. benefits] under the contract.[28]

---

26   *Herkules Piling and Hercules Piling v. Tilbury Construction* (1992) CILL 770. In this case, it was held that the assignment was an equitable assignment and that the assignee could not take advantage of the arbitration clause.

27   *Rayack Construction v. Lampeter Meat* (1979) 12 BLR 30, *Re Arthur Sanders Ltd* (1981) 17 BLR 125 and *Wates Construction (London) Ltd v. Franthom Property Ltd* (1991) 53 BLR 23.

28   *Linden Gardens Trust Ltd v. Linesta Sludge Disposals Ltd and others* (1993) 57 BLR 57 (CA) per Staughton L.J. at 76.

A debtor cannot relieve himself of his liability to his creditor by assigning the burden of the obligation to someone else. This can only be bought about by the consent of all three, and involves the release of all three.[29]

*Discharge and reformation*    Novation involves no more than a contractual discharge and formation. First, the two contracting parties agree to dissolve their contract. Then one of those parties enters into a new agreement with a third party. Each of these steps generally requires consideration; in the case of the discharge, where both parties have unperformed obligations the consideration provided is the release of the other and in the case of reformation consideration is readily found. Novations are frequently undertaken by deed and consideration is not required.

*Examples*    Where a party changes identity, as where there is a reorganisation in a group of companies, novations may take place with the consent of all parties. Commonly also, where an employer engages a designer, that designer's contract may be 'novated' to the contractor at a later date.[30]

# 6. Sub-contracting and vicarious performance

Vicarious performance is where a party's contractual obligations are undertaken by another on that party's behalf. The most common form of vicarious performance is sub-contracting.

### The right to discharge obligations by vicarious performance

Whether a given contract requires personal performance by A, or whether (and, if so, to what extent) A may perform his contractual obligations vicariously, is in my opinion a question of contractual construction. That does not mean that the court is confined to a semantic analysis of the written record of the parties' contract, if there is one. Such is not the modern approach of construction to a commercial contract. It means that the court must do its best, by reference to all admissible materials, to make an objective judgment of what A and B intended in this regard.[31]

In practice most civil engineering contracts provide that a contractor may not sub-contract work unless approval is obtained from the employer or his

---

29  *Tolhurst v. Associated Portland Cement* [1902] 2 KB 660 per Collins L.J. at 668.
30  This topic is dealt with in Chapter 12.
31  *Southway Group Ltd v. Wolff and Wolff* (1991) 57 BLR 33 per Bingham L.J.

agent. There is a general presumption that work of a personal nature, such as architectural conceptual design will not be vicariously performed.[32]

*Terms of the sub-contract*   The same rules of contract apply to sub-contracts as they do to main contractors. Sub-contracts often expressly incorporate 'the terms of the main contract insofar as these are consistent with the sub-contract'. Often the payment provisions and claims procedures are closely linked. Previously, 'pay–when–paid clauses' were common, but they are now generally of no effect in contracts to which the Housing Grants, Construction and Regeneration Act 1996 applies.[33]

*Employer's control over sub-contracted work*   Where work is sub-contracted, the doctrine of privity prevents the employer from having direct contractual control over the sub-contractor. Nevertheless, there are means by which the employer can exercise control. These include:

(1)   a right to object the presence of those on site or on the works
(2)   a right to object to the appointment of proposed sub-contractors
(3)   a right to issue instructions to the contractor—and if work is sub-let, to the sub-contractor—on the mode of performance of the contract.

*Interpretation of powers of approval*   Where a contract clause requires the contractor to obtain prior approval before sub-letting work, the effect of that clause depends upon its true interpretation at law. Often, such clauses provide that the employer's approval shall not unreasonably be withheld. Otherwise, there may be no obligation for the withholding of approval to be reasonable.[34]

*Nominated sub-contractors*   Contracts may contain provisions entitling the employer to instruct the contractor to enter into a sub-contract with a named (or 'nominated') sub-contractor or supplier.

> The scheme for nominated sub-contractors is an ingenious method of achieving two objects which at first sight might seem incompatible. The employer wants to choose who is doing the prime cost work and to settle the terms on which it is to be done, and at the same time to avoid the hazards and

---

32   *Moresk Cleaners v. Hicks* [1996] 2 Lloyd's Rep. 338.
33   One exception is where the employer is insolvent and has not paid the main contractor. Here, a pay–when–paid clause may be effective.
34   *Leedsford Ltd v. Bradford Corporation* (1956) 24 BLR 45 (CA).

difficulties which might arise if he entered into a contract with the person whom he has chosen to do the work.[35]

There are a number of problems associated with nomination. In particular, the courts have been unwilling to make a contractor liable for the work of a nominated subcontractor when the contractor has little control over it. Thus, where there are latent defects in the product supplied by a nominated supplier, the contractor may not be responsible.[36] Where the nominated sub-contractor becomes bankrupt, the employer will be responsible for a timely re-nomination and the contractor may be entitled to damages for delay.[37] An examination of the reported cases[38] shows that each turns on its own contract terms. For example, where a main contractor is obliged to take on a particular designer but there is no restriction on the input the main contractor might have in the design, there is no objection to the contractor taking full responsibility for the design of the sub-contractor.[39]

---

35  *Bickerton v. North West Metropolitan Regional Hospital Board* [1970] 1 WLR 607 per Lord Reid at 611.

36  *Gloucestershire County Council v. Richardson* [1969] 1 AC 480 (HL).

37  *Bickerton v. North West Metropolitan Regional Hospital Board* [1970].

38  *Fairclough Building v. Rhuddlan Borough Council* (1985) 30 BLR 26 (CA); *Percy Bilton v. Greater London Council* [1982] 1 WLR 794 (HL); *Norta Wallpapers (Ireland) Ltd v. John Sisk & Sons (Dublin) Ltd* (1978) 14 BLR 99.

39  *IBA v. EMI and BICC* (1980) 14 BLR 1; Lord Fraser of Tullybelton observed at page 46:

> In the present case, although EMI had no option but to appoint BICC as sub-contractor for the mast, they were not bound to accept any particular design at any particular price. If they had checked BICC's design and had considered it unsatisfactory they would have been entitled to insist on its being improved.

# 7

# Miscellaneous matters associated with civil engineering contracts

In this chapter, a number of issues which may have an important bearing on civil engineering contracts are considered. These are:

(1) access, permits and cooperation
(2) plant and materials
(3) termination clauses and suspension of performance
(4) certificates and contract administration
(5) bonds and guarantees.

## 1. Access, permits and cooperation

*Employer's obligation not to interfere with the contractor's performance*

> There is an implied contract by each party that he will not do anything to prevent the other party from performing the contract or to delay him in performing it. I agree that such a term is by law imported into every contract.[1]

Where the employer, through acquiescence, or by acting in accordance with some entitlement he has under the contract, hinders the contractor, this is not considered a direct prevention. Rather it is an occupational hazard which is part of the contractor's risk burden under the contract.

---

1    *Barque Quilpué Ltd* v. *Brown* [1904] 2 KB 261 (CA) per Vaughan Williams L.J. See also *William Cory & Son Ltd* v. *London Corporation* [1951] 2 KB 476 at 484 (CA), where Lord Asquith made a similar statement but qualified it to exclude situations where the performance was illegal or *ultra vires*.

*Access*   There will be an implied term in every civil engineering contract that the employer will allow the contractor such possession as will reasonably enable him to complete the work in accordance with the contract.[2] The contract often expresses in some detail the degree and extent of the possession of the site to be granted to the contractor. If it does not, the courts will consider what terms relating to access should be read into the agreement. For example, the contractor on a large scale muck shifting project may require complete possession of large tracts of the site. The installation of works in specific locations will require much less complete possession of the site.

*The time of access*   Where no date for possession is agreed, it will be implied that access must be given to the contractor within a reasonable time of making the contract.[3] Losses associated with any delay which, for example, pushes the work into rainy season working (which may have profound effects for earthworks etc.) will be recoverable.[4]

*Where it is, or should be know, that the site will also be occupied by others*   Standard form contracts generally make provision for the situation where a contractor is hindered in his performance by others on the site. Otherwise, the question of the contractor's entitlement to damages for breach depends on the extent of any implied term. The question is one of reasonableness, given what the parties knew or ought reasonably to have known at the time of making the contract.

*The employer's best endeavours to give access*   Whether the obligation to provide the site is absolute or whether the employer's best endeavour will suffice depends on all the circumstances. Where the express terms provide that the employer is to provide space 'so far as he is able' the the obligation is not absolute.[5] Where the employer holds the site open, but third parties prevent access, this will not generally be a breach by the employer.[6]

*Drawings and information*   Detailed drawings and information will often be required as the work progresses and there will normally be an implied obligation upon the employer to supply this material in sufficient detail and in

---

2   *The Queen in Right of Canada* v. *Walter Cabott Construction Ltd* (1977) 21 BLR 46.
3   *The Queen in Right of Canada* v. *Walter Cabott Construction Ltd* (1977) 21 BLR 46.
4   *Freeman* v. *Hensler* (1900) *Hudson's Building Contracts*, 4th edn, Vol 2, 292 (CA).
5   *Kitsons Sheet Metal* v. *Matthew Hall* (1989) 47 BLR 90 at p. 108.
6   *LRE Engineering* v. *Otto Simon Carves* (1989) 24 BLR 131.

sufficient time to enable the contractor to build the works.[7] What is a reasonable time does not depend solely upon the convenience of the contractor.[8] In particular, the contractor is not entitled to require the employer to expedite production of drawings and other information in order to enable the contractor to complete earlier than the contractual completion time.[9] When the contract itself states the timing when information shall be delivered that overrides any term which would otherwise be implied.

*Preparatory and associated works by others*   Often the contractor is held up by the failures of others to carry out preparatory or associated work. Ordinarily, standard form contracts provide that the contractor may claim in such circumstances. Where there is no such express provision, an implied term may arise. In one case, however, a contractor, who was continually impeded, sought the implication of a term that

> The employer would make sufficient work available to the contractor to enable him to maintain reasonable progress and to execute his work in an efficient and economic manner. [10]

The Court of Appeal declined to find such a term.

*Permits*   Permits are frequently required. Standard form contracts generally provide which party is to obtain permits. Where there is no express provision, the party responsible for obtaining any permit is determined by interpreting the contract. In many cases, the interpretation is assisted by asking: 'who is best placed to procure the permit?'.[11] For instance, permits relating to the movements or use of specific items of plant to be used on site are clearly to be obtained by the contractor who supplies the plant unless the contract states otherwise. If, for any reason, the employer must cooperate with the contractor in obtaining the permit it is submitted that the employer will be obliged to cooperate reasonably. Often, however, the employer is best placed to obtain the permit in question and, here, the obligation will fall upon him, unless the contractor has expressly undertaken to obtain it.

---

7   *Merton( London Borough of)* v. *Hugh Stanley Leach* (1985) 32 BLR 51.
8   *Neodex* v. *Borough of Swinton and Pendlebury* (1958) 5 BLR 38.
9   *Glenlion Construction* v. *Guinness Trust* (1988) 39 BLR 89.
10   *Martin Grant & Co. Ltd* v. *Sir Lindsay Parkinson & Co. Ltd* (1984) 29 BLR 31 (CA).
11   *Ellis-Don Ltd* v. *The Parking Authority of Toronto* (1978) 28 BLR 98 (Supreme Court of Ontario).

*Appointment of contract administrators, nominated sub-contractors, etc.*   Where the administration of the contract requires an administrator to be appointed or where works require the nomination of a sub-contractor, the employer is obliged to ensure that the relevant appointments and nominations are made in a reasonable time.

## 2. Plant and materials

*In the absence of any contract provision*   Materials brought onto site by the contractor remain his property until they are fixed into the construction; at that point ownership passes to the owner of the land. Accordingly, even where the materials are subsequently broken out from the construction, they remain the property of the owner of the land. Plant, because of its temporary nature, is deemed not to be affixed and so no property passes.

*Materials paid for by the employer before fixing*   It may that where the contract administrator certifies payment in respect of materials on site, then property may pass at that stage. This will depend on the presumed intention of the parties.

*Vesting clauses*   Civil engineering contracts sometimes provide that property in material and plant passes to (vests in) the employer when it arrives on site in order to give some security in the event of the contractor's default. Such provisions may be effective as between the contractor and employer, but tend to be construed against the vesting so that clear words are required. There will be an implied term that the contractor may use the plant and materials for the purpose of the construction and that unused materials and plant will re-vest at the date they properly leave the site.

*Retention of title clauses*   Ordinarily a supplier's title in materials subsists until delivery, at which point it passes to the contractor. Where the parties to the supply contract intend, expressly or impliedly, that property is to pass at some other time, this provision will prevail.[12] Thus, the parties may agree that property is not to pass to the contractor until full payment is made. As a result, materials owned by the supplier may be fixed into a structure when the employer believes that property has passed to him. At that point, it is not clear whether the ownership passes to the employer or remains with the supplier. The supplier will, of course, be able to recover a reasonable value under the principles of restitution.

---

12   *Aluminium Industrie Vaasen v. Romalpa Aluminium* [1976] 1 WLR. 676 (CA).

## 3. Termination clauses and suspension of performance

### Termination clauses

*Position in the absence of express agreement*   Where there is no express provision for termination, the parties are entitled to terminate their performance unfinished only upon the fundamental or repudiatory breach of the other. A party who commences works is generally entitled to complete them,[13] even when the other party has no further interest in the subject matter of the agreement.[14] This does not mean that the contractor can force his performance upon the employer, especially where the work is to be performed on the employer's land.[15] The remedy is in damages for wrongful withdrawal of the right to do the work. Where an employer fails to pay the contractor, this is not a repudiatory breach unless the employer also indicates that he no longer considers himself bound by the contract.[16] Employers generally preserve their position carefully by citing a contractual reason for withholding money; this makes it difficult for the contractor to claim that the employer has repudiated the contract.[17] However, where the contractor withdraws his labour and plant, this may well amount to a repudiation of the agreement, entitling the employer to treat the contract as at a end and to claim damages for breach; it is, in practice, more difficult to demonstrate that such a withdrawal could reasonably have been consistent with rights under the contract. The HGCRA 1996 now provides that a contractor may suspend performance in the event that the employer fails to make payments without first issuing the requisite notices.[18]

*Express provisions agreed in the contract*   Civil engineering contracts frequently contain determination clauses (sometimes loosely described as 'forfeiture clauses') entitling the employer—and sometimes the contractor also—to terminate performance of the agreement when a specified event occurs. The events specified may be breaches (e.g. not proceeding with due

---

13   *Thomas* v. *Hammersmith Borough Council* [1938] 3 All ER 201; *Edwin Hill & Partners* v. *Leakcliffe Properties Ltd* (1938) 29 BLR 43.

14   *White & Carter (Councils) Ltd* v. *McGregor* [1962] AC 413 (HL).

15   *London Borough of Hounslow* v. *Twickenham Garden Developments* [1971] Ch 233; *Mayfield Holdings Ltd* v. *Moana Reef Ltd* [1973] 1 NZLR 309 (New Zealand Supreme Court); *Tara Civil Engineering Ltd* v. *Moorfield Developments Ltd* (1989) 46 BLR 72.

16   *Mersey Steel and Iron Co.* v. *Naylor, Benzon & Co.* (1884) 9 App Cas 434 (HL).

17   *Woodar Investment Development Ltd* v. *Wimpey Construction UK Ltd* [1980] 1 WLR 571 (HL).

18   See below.

diligence in breach of the agreement) or they may be other commercial events (e.g. the insolvency of the contractor).

*Interpreting determination clauses and wrongful determination*    Clauses entitling one party to terminate performance are generally interpreted strictly; all notice provisions must be complied with and all necessary events must occur.[19] Where, however, the notices served are defective in such a minor degree that the recipient is not prejudiced, the courts may take a business common sense approach, without undue reliance on technicalities; thus

> If a notice unambiguously conveys a decision to determine, a court may nowadays ignore immaterial errors which would not have misled a reasonable recipient.[20]

Where, however, the employer 'jumps the gun' by issuing his notice early, his notice will not be validated by an argument that the contractor has not in fact been prejudiced.[21] Where the breach of the employer or his agent has caused the event,[22] the determination is wrongful and is ordinarily a repudiation.

*Licence to occupy the site following a termination*    Even where the termination is wrongful, a contractor who remains on site after his licence to do so has been revoked is a trespasser.[23] The court will not order the employer to allow the contractor to resume work. Damages are usually an adequate remedy.[24] However, where the contract entitles the employer to use the contractor's plant and materials following a proper determination and the

---

19    *Mardorf Peach & Co.* Ltd v. *African Sea Carriers Corporation of Liberia* [1977] AC 850 (HL). This involved determination of a charterparty. Lord Wilberforce said at 870:

> I would certainly go so far as to agree that the owner has to show that the conditions necessary to entitle him to withdraw have been strictly complied with.

20    *Mannai Investment Co. Ltd* v. *Eagle Star Life Assurance Co. Ltd* [1997] AC 749 (HL) per Lord Steyn at 768.

21    *Afovos Shipping Co.* v. *Pagan* [1983] 1 WLR. 195 (HL). Here, the notice was served by the owner towards the end of banking hours on the last day. This was too early. The charterer had until midnight to make his payment. The notice failed.

22    For example, where delay in issuing plans has delayed the contractor: *Roberts* v. *Bury Commissioners* (1870) LR 5 CP.

23    *Chermar Productions Pty Ltd* v. *Pretest Pty Ltd* (1991) 8 Constr. LJ 44 (Supreme Court of Victoria).

24    In *London Borough of Hounslow* v. *Twickenham Garden Developments* [1971] Ch 233 the employer was refused an injunction to remove the contractor from site. This decision has not been followed. See *Tara Civil Engineering* v. *Moorfield Developments Ltd* (1989) 46 BLR 72 which is to be preferred.

employer is doing so,[25] it is thought that an injunction may be issued to restrain such use where the termination was wrongful.

*The event defined*   When drawing up forfeiture clauses, there should be great clarity as to what amounts to an event which triggers the right to determine. Where, for example, the contractor's insolvency is to trigger a right to determine, this cannot be achieved using the words 'the contractor's default' since insolvency does not normally amount to a breach of contract.[26]

*Termination clauses as a penalty*   Where the consequences of the determination of the contract are excessive, it may be a penalty and hence unenforceable. Where a contract provided that upon termination, the contractor was not to be paid for work already done and all his equipment was to pass into the employer's ownership, this was held to be a penalty. The employer was thus entitled to recover his actual loss and no more.[27]

## Suspension of performance

The HGCRA 1996 entitles a contractor to suspend performance where certain conditions are satisfied. Since a wrongful act of suspension can be a serious breach, often leading to a claim for very substantial damages and since a lawful suspension can be a serious set-back to the employer, causing major consequential losses, it is important that both the employer and the contractor ensure that they have protected their own positions.

Section 112 provides, *inter alia*:

(1)   Where a sum due under a construction contract is not paid in full by the final date for payment and no effective notice to withhold payment has been given, the person to whom the sum is due has the right (without prejudice to any other right or remedy) to suspend performance of his obligations under the contract to the party by whom payment ought to have been made (the party in default)

(2)   The right may not be exercised without first giving the party in default at least seven day's notice of intention to suspend performance, stating the grounds …

---

25   As in *Ranger v. Great Western Railway* (1854) 5 HLC (HL).
26   *Perar BV v. General Surety and Guarantee Co. Ltd* (1994) 66 BLR 72 (CA).
27   *Ranger v. Great Western Railway* (1854) 5 HLC (HL).

*The preconditions for suspension*   The preconditions, which must all be satisfied, are: (*a*) 'a sum due under [the contract] is not paid in full by the final date for payment'; and (*b*) no effective notice to withhold payment has been given; and (*c*) the contractor must give the employer at least seven days' notice of intention to suspend performance.

*Difficulties in being certain that the right to suspend has accrued*   Although this provision appears to give the contractor a real sanction, the unclear language means that it is a power which cannot be used with full confidence until the courts provide judicial guidance. Consider where the contractor submits a detailed account for payment and the employer responds in a perfunctory manner; no proper notices are issued and the contractor suspends performance. The employer refers the matter to adjudication. What is to happen if the adjudicator decides that the sum claimed by the contractor was not in fact due at all because of a proper right of set-off to which the employer was entitled? There are three possible answers: the first is that the right to suspend arises as a result of the employer failing to operate the contract; the second is that any suspension is at the contractor's risk and if it later transpires that he was not due any sum or for some other reason he was not entitled to suspend, then he will pay damages for breach of contract; third, some combination of the two. It is submitted that the third approach is the correct one. Where the employer fails to give any indication of the grounds upon which a claimed payment is being withheld, he must be responsible for suggesting to the contractor that there is no good reason; but where he does give a proper indication (even short of issuing a Section 111 notice), then the contractor will suspend performance at his own risk.

*The content of the notice of intention to suspend performance*   Section 112 (2) provides:

> (2)   The right [to suspend performance] may not be exercised without first giving to the party in default at least seven days' notice of intention to suspend performance, stating the grounds on which it is intended to suspend performance.

Although, the section does not explicitly require it, it seems prudent also to state the date of the notice and indicate the date on or after which the suspension will take place. In addition, a prudent contractor should set out the 'grounds' in detail and should clearly: (*a*) recite the contractor's belief that a sum is due and has not been paid; (*b*) recite the contractual provisions relating to the due date and final date; and (*c*) clearly recite the contractor's belief that no notice of intention to withhold payment has been issued by the employer.

*When can the notice of intention to suspend performance be served?* It may be that the notice can be given in anticipation that the other conditions of a right to suspend may be satisfied at the appropriate date. The following chronology gives an example which appears to meet the requirements of the Act. It assumes that the Scheme applies, so that the period between the due date and the final date is 17 days and the prescribed period for the service of a notice of intention to withhold payment is 7 days:

| | |
|---|---|
| Day 1 | due date for payment of instalment |
| Day 12 | issue of notice of intention to suspend performance |
| Day 18 | final date for payment, payment not received |
| Day 19 | contractor suspends performance. |

At the time of the start of the suspension, all the criteria appear to have been satisfied.

*Losing the right to suspend performance* Section 112(3) provides that the right to suspend performance is lost when the party in default makes payment in full. A cheque may take several days to clear; it submitted that 'payment in full' means the clearance of the cheque or the receipt of some other instrument (e.g. a cheque drawn on a bank's own account) which has no real prospect of being dishonoured.

## 4. Contract administration

*Employer's agent and impartial administrators* As far as the employer is concerned, two types of decision need to be taken in the administration of a civil engineering project:[28]

(1) Decisions about what is in the employer's best interests. For example, where the employer can exercise an option under the contract, he will wish to exercise it to his best advantage.
(2) Decisions about the the rights of the parties under the contract. For example, it has to be decided from time to time whether materials or workmanship comply with the contract, how much the employer is

---

28   Vinelott J. discussed this distinction in *London Borough of Merton v. Hugh Stanley Leach* (1985) 32 BLR 51.

obliged to pay the contractor and whether or not claims for time and money are valid.

*Decisions of the second type require a fair assessment*   In the absence of any agreement as to who should carry out this function, the assessment will of necessity be carried out by the employer or an agent acting on his behalf. On occasions, contracts expressly provide that this role is reserved to the employer.[29] In most commercial civil engineering contracts, however, the interests of the parties are more evenly balanced if the decision is given to a designated professional administrator, usually either an engineer or project manager.[30] The terms of the contract provide the form in which such decisions are to be taken. Traditionally, decisions about payments have been made in 'certificates' or 'decisions' or 'determinations'.

*The effect of certificates and other decisions to be determined by a true interpretation of the contract*   The effect of any certificate, determination, etc., is contractual. A certificate may be final and conclusive or it may be subject to review. There may be pre-conditions to its being effective. All is decided by interpreting the contract in the individual case.

*General principles as to certificates*   Although the effect of any certificate is determined by interpreting the contract in that individual case, various themes have developed.

(1)  *The form of the certificate.* Unless the contract provides otherwise, there is no need for the certificate to be in any particular form; an oral certificate may suffice.[31] Where the word 'certificate' is used, the document (or oral statement) must, however, clearly be the result of a certifying process, though the word 'certificate' (or any derivative) need not, it seems, be used.[32]

(2)  *The administrator need do no more than exercise independent judgment.* An administrator under a construction contract

> must throughout retain his independence in exercising that judgment;
> but provided he does this, I do not think that, unless the contract so

---

29   For example in the contract in *Balfour Beatty Civil Engineering Ltd* v. *Docklands Light Railway Ltd* (1996) 78 BLR 42.

30   For example, the ICE Conditions of Contract provide that 'the Engineer' carries out these impartial functions; the EEC provides that they are carried out by 'the Project Manager'.

31   *Elmes* v. *Burgh Market Co.* (1891) *Hudson's Building Contracts*, 4th edn, Vol. 2, p. 170.

32   *Token Construction* v. *Charlton Estates* (1973) 1 BLR 48 (CA).

provides, he need go further and observe the rules of natural justice, giving due notice of all complaints and affording both parties a hearing.[33]

(3) *Certificate as a condition precedent to payment.* It is clear that under some forms of contract, the issue of a certificate is a condition precedent to payment.[34] Where the contract shows that the administrator is to take an impartial decision, any pressure brought to bear on the administrator by the employer will prevent the employer relying upon a lack of certificate as a defence to the contractor's claim for payment.[35] Likewise, where the administrator wrongly insists on unreasonable pre-conditions before issuing his certificate, the contractor will be excused the requirement to have a certificate.[36]

(4) *The right to set-off.* A certificate issued for payment does not prevent the employer exercising a right of set-off.[37]

*Certificates and adjudication*   Where a party challenges the correctness of a certificate, he may do so by adjudication where the HGCRA 1996 applies or where adjudication is otherwise provided for in the contract. This new legislative framework has rendered the question of certificates less important than hitherto. The adjudicator may decide any 'dispute arising under the contract'. Where certificates are not conclusive, such disputes will include whether a certificate should have been issued and what it should certify. Where the contract provides that a certificate is to be conclusive, and a certificate has been issued, it is submitted that the adjudicator's role is not to decide whether or not he would have made the same certificate, but whether it was properly and impartially made; if he substitutes his own decision, he is not making a decision arising under the contract, but in disregard of the contract.

## 5. Bonds and guarantees

*Position in the absence of express agreement*   Unless the contract provides otherwise, an employer may not insist that a contractor provides any

---

33   Megarry J. in *London Borough of Hounslow* v. *Twickenham Garden Developments Ltd* [1971] Ch 233.
34   See *Lubenham Fidelities* v. *South Pembrokeshire District Council* (1986) 33 BLR 39 (CA), where this was expressly stated to be the case.
35   *Hickman* v. *Roberts* [1913] AC 229 (HL).
36   *Panamena Europa Navegacion* v. *Frederick Leyland & Co.* [1947] AC 428 (HL).
37   *Modern Engineering (Bristol) Ltd* v. *Gilbert-Ash* [1974] AC 689 (HL).

security to safeguard the employer against the contractor's possible default or inability to complete the work.

*Express provisions agreed in the contract*   In practice, most construction contracts provide that the contractor shall provide, at his own expense, a guarantee. The instrument required by the contract is frequently called a bond, but the requirement is generally for a contract of indemnity which entitles the surety to set up any defence against the creditor which is available to the debtor. A bond is a more specific type of instrument, namely a deed whereby the bondsman promises to pay the employer a sum when he calls for it and satisfies any conditions stipulated in the bond, including a bare demand. Bare demand bonds are more commonly used in international projects. Although a guarantee or bond is an autonomous agreement, where the employer calls a bond fraudulently, he may be restrained by injunction.[38] Conduct on the employer's behalf—such as collusion in calling the bond—will release the surety.[39] The meaning is determined by a true interpretation of the instrument.[40] Unfortunately, this is made more difficult since bonds and guarantees tend to be written in archaic language. This is certainly not necessary and obscures the parties' understanding of the nature of the obligations into which they have entered.[41]

---

38   *Themehelp Ltd.* v. *West and others* [1995] 3 WLR 751 (CA).
39   *Bank of India* v. *Patel* [1983] 2 Lloyd's Rep. 298 (CA).
40   *Trafalgar House Construction* v. *General Surety & Guarantee Co.* [1996] 1 AC 199 (HL).
41   See the comments made in *Paddington Churches Housing Association* v. *Technical and General Guarantee Co. Ltd* [1999] BLR 244 at 249:

   This case provides yet another example of the failure of a person or body for whose protection a bond was given to understand the nature of the protection provided ... It should not be difficult to draft a form of bond setting out the bondsman's undertakings in simple positive terms capable of being understood by a building contractor or employer who does not have the benefit of a legal education embracing a form which has its roots in ancient legal history.

# 8

## Extra-contractual entitlements

### 1. Contractual and extra-contractual bases for payment

Contract provides the principal vehicle for entitlements in relation to a civil engineering project; contractual entitlements may either be agreed under the terms the contract or may arise as a claim for damages for breach of contract. In addition, a number of extra-contractual bases for claim may exist, the most common being

(1)  claims for misrepresentation
(2)  claims for restitution in the absence of contract
(3)  claims under a collateral or supplementary contract
(4)  tortious liability for loss caused by a contractor.

These bases of entitlement are explored in this chapter.

### 2. Misrepresentation

During contract negotiations, statements may be made by one party to induce the other to enter the contract on the terms beneficial to the person making the statement. Such statements may become contract terms; where they do not, they are said to be 'mere representations'.[1] If a representation of fact[2] is inaccurate and was a material factor in inducing the representee

---

1   For example, the employer may provide a site investigation report to the contractor; or the contractor may provide information relating to his previous experience.

2   A statement of honest opinion is insufficient: *Bisset* v. *Wilkinson* [1927] AC 177 (PC). Where, however, the opinion is tendered in circumstances where the representor has skill or experience in relation to the subject matter: *Esso Petroleum Co. Ltd* v. *Mardon* [1976] QB 801 (CA).

to enter into the contract, the person to whom the representation is made (the 'representee') may have an action against the person who made the representation (the 'representor'). The action may be framed in one or more of the following ways:

(1) a claim to be entitled to rescind the agreement
(2) an action for fraudulent misrepresentation (deceit)
(3) an action under the Misrepresentation Act 1967.

*Rescission for misrepresentation*   A party who was induced to enter into a agreement by a misrepresentation of a material fact is entitled to rescind the agreement, provided that it is possible to restore any property which had passed in the transaction. Thus, for example, where the contractor has not yet entered onto the site he will be entitled to rescind the agreement. In many practical situations, of course, such as in part-performed civil engineering contracts, it is impossible to restore property passed during the transaction, thus rendering rescission impossible.[3]

*Fraudulent misrepresentation (the tort of deceit)*   A person who is induced to enter into a contract by a fraudulent representation has an action in the tort of deceit against the representor.[4] This action requires a fraudulent misrepresentation, but the word 'fraudulent' has been given wide scope and includes reckless statements.[5] Thus, on a project involving underground work, the engineer's reckless representation that existing protective works were of adequate depth was fraudulent; it defeated a clause in the contract requiring the contractor to be responsible for satisfying himself as to dimensions.[6]

*Misrepresentation Act 1967*   The Misrepresentation Act 1967 provides a number of important rights for a misrepresentee in addition to previously existing causes of action.

---

3   The leading case is *Glasgow and South Western Railway* v. *Boyd & Forrest* [1915] AC 526 (HL). Here the contractor completed a railway with significant groundworks, and then claimed rescission and a *quantum meruit* on the grounds that the employer's engineer had innocently misrepresented the condition of the soil strata. Rescission was refused on the grounds that a restoration of property was impossible.

4   The person making the misrepresentation must, of course, be authorised to do so on behalf of his principal. In the normal course of events, a senior engineer who makes representations of engineering fact during negotiations will be authorised to do so unless such authority is expressly limited.

5   In other words a representation made knowing it to be untrue or reckless as to whether it be true or false: *Derry* v. *Peek* (1889) 14 App Cas 337 (HL).

6   *Pearson* v. *Dublin Corporation* [1907] AC 351 (HL).

Section 2 of the Act provides as follows:

(1) Where a person has entered into a contract after a misrepresentation has been made to him by another party thereto and as a result thereof he has suffered loss, then, if the person making the misrepresentation would be liable to damages in respect thereof had the misrepresentation been made fraudulently, that person shall be so liable notwithstanding that the misrepresentation was not made fraudulently, unless he proves that he had reasonable grounds to believe and did believe up to the time the contract was made that the facts represented were true.

(2) Where a person has entered into a contract after a misrepresentation has been made to him otherwise than fraudulently, and he would be entitled, by reason of the misrepresentation, to rescind the contract, then, if it is claimed, in any proceedings arising out of the contract, that the contract ought to be or has been rescinded, the court or arbitrator may declare the contract subsisting and award damages in lieu of rescission, if of opinion that it would be equitable to do so, having regard to the nature of the misrepresentation and the loss that would be caused by it if the contract were upheld, as well as to the loss that rescission would cause to the other party.

*Section 2(1)—negligent misrepresentation* Section 2(1) provides a new and stronger remedy for negligent misrepresentation by entitling a misrepresentee to recover as if the negligent misrepresentation had been fraudulent. Apart from the element of fraud, the same ingredients are required for a successful Section 2(1) action as for an action in fraudulent misrepresentation. Accordingly, there must be a representation of fact which induces the other to enter into the contract and which causes the loss. The remedy is in damages but this does not deprive the misrepresentee of a right to rescission where it is available. The representor has a defence, namely that 'he had reasonable grounds to believe and did believe up to the time the contract was made that the facts represented were true'. The burden of proof is on the representor to prove his innocence.

*Section 2(2)—innocent misrepresentation where rescission is not feasible* Section 2(2) catches the one situation which is not dealt with by either the tort of deceit or Section 2(1), namely a wholly innocent misrepresentation. There was no common law right to damages for an innocent misrepresentation;[7] the misrepresentee's remedy was in rescission. In most civil engineering cases where works have commenced restitution is

---

7    *Heilbut, Symons & Co.* v. *Buckleton* [1913] AC 30.

impracticable.[8] Section 2(2) provides that the court or arbitrator may award damages in lieu of misrepresentation if it would be equitable to do so.

*Where the misrepresentation has become a term*   Where, for instance, an inaccurate site survey issued at tender stage becomes a term of the agreement, the claimant may elect to advance a claim for misrepresentation or for breach of contract.

*Excluding liability for misrepresentation*   A term which purports to exclude liability for misrepresentation is of no effect except in so far as it satisfies the test of reasonableness.[9]

# 3. Restitution

A restitutionary claim is one where A claims that B has received a benefit from A and it would be inequitable for B not to restore it to A either by the return of goods or by paying for services rendered. A claim for restitution is necessarily made in the absence of a contract governing the situation.

   In the context of civil engineering, restitutionary claims are most frequently advanced in respect of services rendered where there is no contract.[10] This may be because (*a*) A commences work for B in the expectation that a contract will eventuate; or (*b*) A supplies work to B outside the scope of an existing contract; or (*c*) the contract between A and B ceases to exist; or (*d*) A supplies services to B in an emergency. In each of these cases, A claims a reasonable amount for services rendered. This is frequently termed a *quantum meruit*.[11] This expression is used also to refer also to payments under a contract where no payment is fixed, or where the measure of payment is agreed to be a 'reasonable amount'.

*Where one person commences work in the expectation that a contract will eventuate*   It is not uncommon for a professional or contractor to commence work before the contract is finally agreed. Where no contract eventuates, the professional or contractor is entitled to be paid a reasonable amount by the employer provided that the work which was carried out was

---

8    See, for example, *Glasgow and South Western Railway* v. *Boyd & Forrest* [1915] AC 526 (HL). referred to above.
9    Misrepresentation Act 1967, Section 3 as substituted by the Unfair Contract Terms Act 1977.
10   Such a claim is frequently termed a claim in quasi-contract.
11   As much it is worth. See Levine MF and Williams JH, Restitutionary *quantum meruit*—the cross roads, (1992) 8 Constr. L.J. 244.

done for the benefit of the employer and at his request, implied or express.[12] Whether or not there is any such request is determined from an analysis of all the circumstances. Where no request is to be found, the contractor who commences work will do so at his own risk; if no contract eventuates, he will not be not entitled to any payment.[13] Where the party supplying work is entitled to be paid for it on the basis of restitution, the other party cannot counterclaim for late delivery, since there is not a contractually agreed delivery date.[14] Any abatement of the value only applies if the thing itself is worth less,[15] so it seems that an abatement route to reducing payment for late delivery is not open. If a contract is eventually agreed, work provided prior to its agreement will be covered by it, unless otherwise stated in the contract.[16]

**Where one person supplies work to another beyond the scope of an existing contract**   Where a contract exists, the employer frequently requests the contractor to do additional work. Where this is covered by the terms of a variations clause, the additional work will form part of the contractual work and will be paid in accordance with the terms of the contract. However, where the work is beyond the contract scope, the contractor is entitled to be paid a reasonable amount for that work.[17]

**Where the contract ceases to exist**   A contract may be set aside for mutual mistake or duress; it may be rescinded following a misrepresentation. A person who has performed work in pursuance of a non-existent contract at the request of the other party will generally be paid a reasonable amount unless the contract states what is to happen in such an event.[18]

**Where the plaintiff supplies services in an emergency**   Work may be supplied in an emergency in order to preserve the integrity of the work, other property, or the safety of those in the vicinity. Immediate action is often called for and the person specifically responsible for the work is not always available, or in a position to carry out the preventative work himself. An intervener may perform the emergency operations. It may be that where the

---

12   *William Lacey (Hounslow) Ltd v. Davis* [1957] 1 WLR 932.
13   *Regalian Properties plc v. London Docklands Development Corporation* [1995] 1 WLR 212.
14   *British Steel Corporation v. Cleveland Bridge and Engineering Co. Ltd* [1984] 1 All ER 504.
15   *Mellowes Archital Ltd v. Bell Projects Ltd* (1997) 87 BLR 26 (CA).
16   *Trollope & Colls Ltd v. Atomic Power Construction Ltd* [1963] 1 WLR 333 (HL).
17   *Sir Lindsay Parkinson & Co. Ltd v. Commissioner of Works* [1949] 2 KB 632 (CA).
18   For example, Clause 64 of the ICE Conditions of Contract provides rules for payment in the event of frustration.

intervener acts wholly officiously he will not be entitled to be paid;[19] where however he acts in pursuit of some interest he has (e.g. if the collapse would damage his own works) he will be entitled to a reasonable amount.

*The quantification*   A *quantum meruit* means a reasonable sum in all the circumstances. It is not simply the cost of the work plus a reasonable addition for 'profit'. The computation may be informed by the nature and terms of an intended agreement.[20]

## 4. Entitlement under a collateral or additional contract

Several possibilities exist for action under a contract which is collateral to or substitutes for the primary contract.

*Collateral contract*   During negotiations one party may agree to enter into the contract in return for some advantage. For example, a prospective house purchaser may agree to sign the agreement provided some undertaking is given as to the state of the drains.[21] In some cases, this collateral undertaking can form the substance of a contract in its own right, and provides a right of action.

*Compromise agreement*   Where the parties agree to compromise their disputes, a new agreement comes into existence. Thus, where a contractor accepts a promise of £x in full and final satisfaction of all claims, and the employer reneges, the contractor's action is under the compromise agreement.[22]

## 5. Tortious liability for losses caused by a contractor

A tort is a wrong which is actionable at law and which does not depend on agreement. Examples where a party can successfully sue in tort include:

---

19   *Falcke v. Scottish Imperial Insurance Co.* (1886) 34 ChD 235 at 248–249 per Bowen L.J. Where, however, the person who benefits from his intervention knows of it and does nothing to indicate that the intervener will not be paid, it seems that the intervener may claim: Lord Cranworth in *Ramsden v. Dyson* (1866) LR 1 HL 129 140–141 cited in *Proctor v. Bennis* [1887] ChD 740 by Cotton L.J.

20   *Way v. Latilla* [1937] 3 All ER 759.

21   *De Lassalle v. Guildford* [1901] 2 KB 215 (CA).

22   It will be a question of interpretation whether or not the arbitration provisions (if any) in the primary contract will be incorporated into the accord and satisfaction.

(1)  where an engineer or contractor trespasses on another's land including where, for example, a contractor's tower crane oversails land
owned by another.[23]

(2)  where an engineer designs a structure or facility which then collapses
or explodes and people are injured or property is damaged.[24]

(3)  where an engineer writes a report for a client, knowing that it will be
shown to a third party who will rely upon it; where the report is negligent and as a result of relying upon it that third party suffers a loss, the
report writer may be liable,[25] the more so if the third party has indirectly contributed to the report writer's fee.[26]

These matters are not considered here in any detail.

## A contractor's liability in the tort of negligence for defects in structures

One matter which has exercised the courts considerably in the 1980s and
early 1990s is the question of liability in the tort of negligence for economic
loss. This question has now largely been resolved as follows:[27] where a contractor provides defective work in a structure, the defects are properly
characterised as 'pure economic loss' because defects do not cause damage
to anything else; they merely reduce the benefit which the employer
acquires. Pure economic losses are not generally recoverable except under a
contract and, hence, the employer may not recover for defects through the
tort of negligence. The net result is that if, for example, a sub-contractor
performs defective work, the employer cannot sue the sub-contractor
directly in the tort of negligence for the defects. This is crucial where, for
example, the main contractor is in liquidation.

The law remains uncertain in one important area concerning what is
sometimes known as the 'complex structure theory'.[28] Where a contractor

---

23  *Anchor Brewhouse Developments Ltd* v. *Berkeley House (Docklands Developments) Ltd* (1987) 284 EG 625.

24  *Sharpe* v. *ET Sweeting & Son Ltd* [1963] 1 WLR 665; *Eckersley* v. *Binnie and Partners* (1988) CILL 388 (CA).

25  *Hedley Byrne* v. *Heller and Partners* [1964] AC 465 (HL); *Caparo Industries plc* v. *Dickman* [1990] 2 AC 605 (HL).

26  *Smith* v. *Bush* [1990] AC 831 (HL). Here a surveyor provided a negligent house survey to its client, a building society, knowing that the prospective purchaser would also rely on it, having indirectly paid the survey fee.

27  *Murphy* v. *Brentwood District Council* [1991] 1 AC 398 (HL).

28  See, for example, *Nitrigin Eireann Teoranta* v. *Inco Alloys Ltd* [1992] 1 WLR 498.

installs a component of a structure, which, due to the negligent installation, malperforms, causing damage to other parts of the structure, it is not clear whether (and, if so, in what circumstances) such damage is properly described as property damage (which is recoverable) or pure economic loss (which is irrecoverable).[29]

---

29    See *Tunnel Refineries* v. *Bryan Donkin Ltd*, TCC, 8 May 1998, for a recent review of the arguments; here a component of a machine, defective because of negligence of the supplier, broke and wrecked the rest of the machine. The rest of the machine was held not to be 'other property'. Hence, the owner could not recover in tort.

# 9

# Civil engineering professional services contracts

In this chapter, the term 'client' will be used, rather than 'employer' since the client may be a contractor.

## 1. The application of the HGCRA 1996

Professional services agreements for civil engineering works will normally be regulated by the Housing Grants, Construction and Regeneration Act 1996[1] where the works are in Britain and the contract is made on or after 1 May 1998. Section 104(2) provides:

> (2) References in this Part to a construction contract include an agreement—
> (a) to do architectural, design or surveying work
> (b) to provide advice on building, engineering, interior or exterior decoration or on the laying-out of landscape, in relation to construction operations.

Payment schemes must, therefore, comply with the Act.[2] There must be a compliant adjudication scheme.[3] Where the contract does not comply, the Scheme for Construction Contracts applies.

---

1    One of the first decided cases on adjudication under the Act—*Project Consultancy Group* v. *Trustees of the Gray Trust*, TCC, 16 July 1999—involved a professional services agreement.
2    See Chapter 4.
3    See Chapter 11.

Where the works are situated abroad, the Act will ordinarily not apply, even though the professional services in respect of those works are provided in Britain.

## 2. Formation of design and professional services contracts

Contracts for professional services are formed in the same way as other contracts.[4] In practice, there tend to be fewer detailed negotiations than for a civil engineering construction contract.[5] The agreement is typically made in, or evidenced by, an exchange of letters referring to a standard form of conditions or a standard scale of fees such as the forms published by the Association of Consulting Engineers. While professionals frequently contract on the standard conditions published by their own professional institutions, this general practice is, in itself, insufficient to make those conditions part of the contract without the agreement of the client, unless that particular client and consultant habitually contract under those conditions.[6]

## 3. The terms of the agreement

Subject to the HGCRA 1996, any agreed terms prevail. Where the contract is made informally, terms will be implied by the Supply of Goods and Services Act 1982 and by the usual operation of law to give the agreement business efficacy.[7]

*The standard of care*    Unless the parties expressly agree otherwise the principal obligations of the consultant will normally be to exercise 'reasonable care and skill'[8] in and about the performance of the work covered by the agreement. The failure to comply with the standard of reasonable skill and

---

4    See Chapter 2.
5    Although where competitive tendering is used, the process may involve qualification stages, tender bids, negotiations, etc.
6    *Sidney Kaye, Eric Firmin & Partners* v. *Leon Joseph Bronesky* (1973) 4 BLR 1.
7    See Chapter 2.
8    Section 13 of the Supply of Goods and Services Act 1982. The words 'care and skill' are often transposed to read 'reasonable skill and care'.

care is commonly termed 'professional negligence'. Where the agreement stipulates a higher standard, this will be enforced; thus, where the circumstances clearly indicate that the parties intended that the design must be fit for its purpose, such a term will be taken to apply.[9]

*Time*  Unless the agreement states otherwise, the professional will be obliged to complete his work within a reasonable time.[10] This time will be assessed in accordance with all the circumstances, including the known facts at the time of the consultant's appointment. Thus, for instance, if it is known that a structure is to be completed by a particular date, the designer's performance must, insofar as it is reasonably practicable, aim to ensure that this can be achieved. The assessment of a reasonable time also takes into account factors beyond the consultant's control which occur after the appointment. For example, where new design codes are introduced or an independent consultant's report arrives late, these may delay the reasonable time for completion.

*Payment: amount to which the professional is entitled*  Unless the agreement states otherwise, the professional will be entitled to recover a reasonable fee for his work.[11] This will depend on (*a*) the amount, complexity and skill level required in the work, (*b*) the risk undertaken and (*c*) the current market rate for consultants of the relevant experience and eminence. Where the parties cannot agree what is a reasonable rate, expert evidence may be called to provide a reasonable valuation. Fees may be agreed in accordance with the conditions published by professional bodies; these generally provide that the remuneration is tied to the scope and value of the work. Where a lump sum fee is agreed and there is a possibility that the scope of works may increase during the currency of the project, the agreement should say how the remuneration is to be affected. Where this is not done and the professional voluntarily undertakes more work, he will not be entitled to be paid for it[12] unless it was reasonably clear to the client that the professional was expecting to receive additional payment.

---

9   *Greaves & Co. Contractors* v. *Baynham Meikle & Partners* [1975] 1 WLR 1095 (CA). The engineer was employed as sub-contractor to a contractor engaged to design and build a warehouse. A term was implied into the design agreement that the building would be reasonably fit for its purpose as the engineer knew (a) that this was the standard to which the contractor was working and (b) that the contractor intended the engineer's obligation to be on the same basis.
10   Section 14 of the Supply of Goods and Services Act 1982.
11   Section 15 of the Supply of Goods and Services Act 1982.
12   *Gilbert & Partners* v. *Knight* [1968] 2 All ER 248 (CA).

*The mechanics of payment*  Where the Housing Grants, Construction and Regeneration Act 1996 applies, the payee is entitled to be paid by instalments either under an agreement or as provided for by the Scheme for Construction Contracts.[13]

*The extent of the work*  Where the consultant contracts on the basis of a standard form, the scope and extent of the consultant's obligations will be as set out in that form. Where, however, the agreement is informal the parameters of the agreed obligation may be inferred only from the entire factual matrix; for instance, the duty to act as certifier may be inferred from the fact that the construction contract to be used contains certification provisions. Where specialist work is included in the project and that specialist work would normally fall outside the professional's area of competence, it is a question of fact whether or not the professional is responsible for its design. Thus on a project where there is no engineer employed and the architect becomes involved in seeking quotations for specialist engineering work, he may well become responsible for its design.[14]

*The duration of the consultant's obligation*  A professional's design duty continues throughout the design and construction phases. A design which was initially and justifiably thought to have been suitable may in the event turn out to be unsuitable; this will not render the designer negligent. However, the discovery of the design defect prior to the completion of construction reactivates the professional's duty in relation to the design and imposes on him a duty to take such steps as are necessary to correct the results of the inadequate design.[15] It seems, however, that once a project is completed, a consultant need not continue to watch out for it.[16]

*Termination*  Most standard form consultancy contracts contain provisions for terminating the consultancy agreement. Where, however, a consultant has embarked upon a project and there are no agreed termination provisions, the client cannot terminate the consultant's appointment at will. The consultant is to be allowed to complete the project which he

---

13   See Chapter 4.
14   *Richard Roberts Holdings Ltd and Another* v. *Douglas Smith Stimson Partnership* (1988) 46 BLR 50.
15   *London Borough of Merton* v. *Lowe* (1982) 18 BLR 130.
16   It was suggested by the judge at first instance in: *Eckersley and others* v. *Binnie and Partners and others* (1988) CILL 388 that a professional might have a continuing duty. This suggestion was disapproved by Bingham L.J. in the Court of Appeal.

commences. Where he is prevented from so doing this will be a breach of contract entitling him to damages for the loss of profit.[17]

*Reporting deficiencies* Generally speaking, a consultant will not have an obligation to report on deficiencies in the work of other consultants involved in the project unless (*a*) he has agreed to do so, or (*b*) his role in the project is clearly to safeguard the interests of the client in this way.[18] The obligation to report may be more readily demonstrated where the deficiencies can give rise to property or property damage.[19]

## 4. Delegation of design and consultancy services

A civil engineering contractor is normally entitled to sub-let work. Since the employer seeks a defined result, the employer can look to the contractor to perform the contract or to pay for its non-performance in any event. In the case of design or other professional contracts, different considerations may apply. The professional is frequently chosen because of his skill, reputation or flair in a particular field and a client might be rightly aggrieved to find that the work which he had entrusted to a particular person had been delegated or sub-let. In addition, the consultant's obligation is not normally to achieve a specified result but to exercise reasonable skill and care; accordingly, difficult questions of liability may arise where the designer sub-lets the work to an apparently competent person. For these reasons, it has been established that a consultant may not ordinarily delegate design work,[20] including financial appraisals,[21] unless authorised by his client to do so.[22]

## 5. Professional negligence: a failure to exercise reasonable skill and care

The term 'negligence' is ordinarily used to refer to the tort of negligence. However, the composite form 'professional negligence' is used to indicate either a breach of a term in a contract requiring the professional to exercise

---

17  *Thomas v. Hammersmith Borough Council* [1938] 3 All ER 201; *Edwin Hill & Partners v. Leakcliffe Properties Ltd* (1938) 29 BLR 43.
18  *Chesham Properties Ltd v. Bucknall Austin Project Management Services Ltd* (1996) 82 BLR 92.
19  *The Zinnia* [1984] 2 Lloyd's Rep. 211.
20  *Moresk Cleaners v. Hicks* [1966] 2 Lloyd's Rep. 338.
21  *Nye Saunders & Partners v. Alan E Bristow* (1987) 37 BLR 92 (CA).
22  *Investors in Industry v. South Bedfordshire District Council* [1986] QB 1034.

reasonable skill and care, or a breach of a duty owed by a professional in tort. This dual use derives from the fact that the duties owed in the tort of negligence and under a professional contract are identical.[23]

A professional does not necessarily breach his duty by making a design error or by supplying bad advice. He will be in breach only if he fails to use reasonable skill and care.[24] The question whether or not he has met this standard must be answered in the light of all the circumstances existing at the time of the alleged breach. Caution must be exercised when attempting to make assessments after the event.[25]

*Common professional practice*   A person who holds himself out to be skilled in a particular profession must exercise

> the standard of the ordinary skilled man exercising and professing to have that special skill. A man need not possess the highest expert skill; it is well established law that it is sufficient if he exercises the ordinary skill of the ordinary competent man exercising that particular art.[26]

Likewise,

> a professional man should command the corpus of knowledge which forms part of the professional equipment of the ordinary member of his profession. He should not lag behind other ordinarily assiduous members of his profession in knowledge of new advances, discoveries and developments in his field. He should have such awareness as an ordinarily competent practitioner would have of the deficiencies in his knowledge and the limitations in his skill. He should be alert to the hazards and risks inherent in the professional task he undertakes to the extent that other ordinarily competent members of his profession would be alert. He must bring to any professional task he undertakes no less expertise, skill and care than other ordinarily competent members of his profession would bring, but need bring no more. The standard is that of the reasonable average. The law does not require of a professional man that he be a paragon, combining the qualities of polymath and prophet. … In deciding whether a professional man had fallen short of the standards observed by ordinarily skilled and competent members of his profession, it is the standard prevailing at the time of the acts or omissions which provided the relevant yardstick. He is not to be judged by the wisdom of hindsight.[27]

---

23   *North West Water Authority* v. *Binnie & Partners* (1989) QBD Unreported, December 1989.
24   Unless of course the consultant has agreed to supply a design etc. which is fit for its purpose.
25   *Eckersley* v. *Binnie & Partners* (1988) CILL 388 (CA).
26   McNair J. in *Bolam* v. *Friern Hospital Management Committee* [1957] 1 WL 582. This test was expressly approved for construction professionals by Stephen Brown L.J. in *Nye Saunders & Partners* v. *Alan E Bristow* (1987) 37 BLR 92 (CA) at 103.
27   *Eckersley* v. *Binnie & Partners* (1988) CILL 388 (CA) per Bingham L.J. (dissenting).

The standard of care in the context of professional practice is not to be judged according to the standard that a reasonable man would expect from such a professional but from the standard that other members of that profession would consider appropriate.[28] The claimant must therefore establish that what the consultant has done, or his failure to do it, falls below this standard.[29] Inexperienced professionals are required to work to the same standard as experienced professionals doing the same work.[30] Where there are several schools of thought and the defendant can show that he had followed a recognised school he will not be in breach.[31] Where professional work requires an extension to existing practice, professionals discharge their burden by paying consideration to the potential problems as other members of their profession would have considered to be reasonably sufficient[32] and where there are risks, they should ensure that their clients are advised of and are prepared to carry those risks.[33]

Specialist skills are to be judged against the standards in that specialism.[34] The standard of 'common professional practice' is problematical where unique work is being undertaken or where it is so specialist that no body of practice exists; nevertheless the court will determine the relevant standard.

A professional should keep reasonably abreast of new developments.[35] He need not read every article appearing in the professional literature, and

---

28  Unless the standard which other professionals apply is clearly unreasonable: *Lloyds Bank* v. *EB Savory & Co.* [1933] AC 201 (HL).

29  *McLaren Maycroft Co.* v. *Fletcher Development Co. Ltd* [1973] 2 NZLR 100 (New Zealand Court of Appeal).

30  *Wilsher* v. *Essex Area Health Authority* [1987] 2 WLR. 425 (CA): 'this notion of a duty tailored to the actor, rather than to the act which he elects to perform, has no place in the law of tort', per Mustill L.J. at 440. It is submitted that the same applies to contractual liability.

31  *Maynard* v. *West Midlands Regional Health Authority* [1984] 1 WLR. 634 (HL). In *Nye Saunders & Partners* v. *Alan E Bristow* (1987) 37 BLR 92 (CA), Stephen Brown L.J. said at 103 of an architect:

Where there is a conflict as whether he has discharged that duty [to use reasonable skill and care], the courts approach the matter upon the basis of considering whether there was evidence that at the time a responsible body of architects would have taken the view that the way in which the subject of enquiry had carried out his duties was an appropriate way of carrying out the duty, and would not hold him guilty of professional negligence merely because there was a body of competent professional opinion which held that he was at fault.

32  *Independent Broadcasting Authority* v. *EMI Electronics Ltd and BICC Construction Ltd* (1980) 14 BLR 1 (HL).

33  *Victoria University of Manchester* v. *Hugh Wilson* (1984) 2 Con. LR 43.

34  *Maynard* v. *West Midlands Regional Area Health Authority* [1984] 1 WLR 634 (HL).

35  *Eckersley* v. *Binnie & Partners* (1988) CILL 388 (CA).

need not adopt techniques advocated in those articles until such techniques become accepted practice in the relevant discipline.[36]

## 6. Personal liability of consultants acting for a company

Where an engineer, as an employee, issues designs and makes representations, he may—as an individual—be relied upon by his employer's clients. The question arises whether he may attract personal liability. This will necessarily be in tort, for the consultancy contract will be between his employer and the client. In a Canadian case,[37] in a passage approved by the House of Lords[38] it was said of an engineer who had 'sealed' a drawing (i.e. indicated that he, as a qualified engineer, approved it):

> The situation of the individual engineers is quite different. While they may, in one sense, have expected that persons in the position of the appellant would rely on their work, they would expect that the appellant would place reliance on their firm's pocketbook and not theirs for indemnification. Looked at the other way, the appellant could not reasonably rely for indemnification on the individual engineers. It would have to show that it was relying on the particular expertise of an individual engineer without regard to the corporate character of the engineering firm. It would seem quite unrealistic ... to hold that the mere presence of an individual engineer's seal was sufficient indication of personal reliance.

## 7. Examples of professional service obligations

### General design obligations

*The general standard*   A designer must carry out his designs with reasonable skill and care. The designs should be satisfactory when constructed.

### A buildable and supervisable design

> [I]f implementation of part of a design requires work to be carried out on site, the design should ensure that the work can be performed by those likely to be employed to do it, in the conditions which can be foreseen, by the exercise of the care and skill ordinarily to be expected of them. If the work would

---

36   *Crawford v. Charing Cross Hospital* (1953) *The Times*, 8 December.
37   *Edgeworth Construction Ltd v. MD Lea & Associates Ltd* [1993] 3 SCR 206, per La Forest J. at 212, Supreme Court of Canada.
38   *Williams v. Natural Life Health Foods Ltd*, 30 April 1998 (HL).

demand exceptional skill, and particularly if it would have to be performed partly from scaffolding and often in windy conditions, then the design will lack ... 'buildability'. Similarly ... if a design requires work to be carried out on site in such a way that those whose duty it is to supervise it and/or check that it has been done will encounter great difficulty in doing so, then the design will again be defective. It may perhaps be described as lacking supervisability.[39]

These obligations will now be influenced, and presumably enhanced, by the requirements of the Construction (Design and Management) Regulations 1994. Although these regulations impose no direct civil liability on him, the designer must ensure that any designs he prepares pay proper regard to risks during and after construction and gives priority to measures which will protect people.[40]

*Advice from other professionals about the design*   A designer may seek the advice of other professionals, unless prohibited by his terms of engagement. A surveyor must investigate the reliability of any information which is supplied to him.[41] He must not simply rely on the claims of a manufacturer of proprietary products.[42] Where, however, an architect defers to the opinions of a structural engineer retained also by the employer on the suitability as to the adequacy of sub-soils to carry foundations, this will not normally cause the architect to be liable.[43] But, even here,

> if any danger or problem arises in connection with the work allotted to the expert, of which an architect of ordinary competence reasonably ought to have been aware and reasonably could be expected to warn the client, despite the employment of an expert and despite what the expert says, it is in our judgment the duty of the architect to warn the client. [44]

*Extensions to current practice*   Where the designer proposes to extend current practice, he must give proper consideration to any potential problems or risks as other members of their profession would have considered to be reasonably sufficient.[45]

---

39   *Equitable Debenture Assets Corporation Ltd* v. *William Moss Group Ltd* (1984) Con. LR 1, per Judge Newey at 21.
40   Regulation 13 of the CDM Regulations.
41   *Moneypenny* v. *Hartland* (1824) 2 C. & P. 378.
42   *Sealand of the Pacific* v. *Robert C McHaffie Ltd* (1974) 51 DLR (3d) 702 (British Columbia Court of Appeal).
43   *Investors in Industry* v. *South Bedfordshire District Council* [1986] QB 1034.
44   *Investors in Industry* v. *South Bedfordshire District Council* [1986] QB 1034.
45   *Independent Broadcasting Authority* v. *EMI Electronics Ltd and BICC Construction Ltd* (1980) 14 BLR 1 (HL).

For architects [and engineers] to use untried, or relatively untried materials or techniques cannot in itself be wrong, as otherwise the construction industry can never make any progress. I think, however, that architects who are venturing into the untried or little tried would be wise to warn their clients specifically of what they are doing and to obtain their express approval.[46]

*Codes of practice*   Codes of practice, whether drawn up under the direction of British, European or International bodies, may be considered to comprise formal statements of 'good practice'. A failure to undertake work in accordance with a design standard may be prima facie evidence of breach unless it can be demonstrated that the design otherwise conforms with accepted engineering practice by rational analysis.[47]

## Surveys and reports

*Collecting information*   A person engaged to undertake a survey must consider what information is required. Thus a person undertaking a site survey for a development may be in breach if he fails to investigate the history of the site[48] or to investigate adjacent sites for evidence of geological faults which may affect his development.[49] A builder was held to be negligent for failing to take additional precautionary measures (such as obtaining expert advice) when the soil conditions which manifested themselves should have alerted him to the fact that his initial design was inadequate;[50] accordingly if, as they so often do, ground conditions on site indicate that additional site investigation or a revised design is required, the engineer should be prepared to recommend additional necessary work or investigations.

*The preparation of reports*   Where a professional prepares an inaccurate report as a result of his failure to use reasonable skill and care, he will be liable to his client for consequential loss. When deciding whether or not the inaccurate report caused the loss, it is important to distinguish between a duty to provide information for the purpose of enabling someone else to decide upon a course of action and a duty to advise someone as to what

---

46   *Victoria University of Manchester v. Hugh Wilson* (1984) 2 Con. LR 43 per Judge Newey at 74.
47   *Bevan Investments Ltd v. Brackhall & Struthers* (No. 2) [1973] 2 NZLR 45 (New Zealand Supreme Court), per Beattie J. at 66.
48   *Balcomb v. Wards Construction (Medway) Ltd* (1980) 259 EG 765.
49   *Batty v. Metropolitan Property Realisations Ltd* [1978] QB 554 (CA).
50   *Bowen v. Paramount Builders (Hamilton) Ltd* [1977] 1 NZLR 394 (New Zealand Court of Appeal).
51   *South Australian Asset Management Corporation v. York Montague Ltd* (1996) 80 BLR 1 (HL), per Lord Hoffmann at 13.

course of action they should take.[51] Where matters touched upon in the report genuinely cannot be forecast (e.g. future movement of foundations) the engineer must make this clear to the client.[52] In some cases a professional may owe a duty to third parties who rely upon his report. Thus, where accounts were prepared for a company in order to attract investors and were negligently prepared, investors who relied on them and suffered a loss as a result were able to recover from the accountant. They owed a duty

> to any third person to whom they themselves show the accounts, or to whom they know their employer is going to show the accounts, so as to induce him to invest money or take some other action on them. But I do not think the duty can be extended still further so as to include strangers of whom they have heard nothing and to whom their employer without their knowledge may choose to show their accounts.[53]

Professionals frequently attempt to restrict their liability to third parties by the use of disclaimers. Such disclaimers are effective only in so far as they satisfy the reasonableness tests laid down in the Unfair Contract Terms Act 1977.

## Administration and supervision of work

*Contract and financial advice*   An engineer or other professional who strays outside his area of professional competence does so at his own risk. In practice, his advice to an employer about appropriate contract forms should be limited to general indications, suggesting where appropriate that specialist advice be sought. On financial matters, it is often part of the professionals task to give estimates of likely costs. The professional should certainly warn about possible cost increases and this is all the more important where the employer has a clearly limited budget.[54] Where the professional assumes the role of project manager, he should ensure, for example, that relevant insurances are in place.[55]

*Supervision/inspection of work*   An engineer's duty to supervise and/or inspect work in progress will depend on his contract with his client.

---

52   *Matto v. Rodney Brown Associates* [1994] 41 EG 152 (2 EGLR 163).
53   *Candler v. Crane, Christmas and Co.* [1951] 2 KB 164 (CA); see also *Caparo Industries plc v. Dickman* [1990] 2 AC 605 (HL).
54   *Nye Saunders & Partners v. Alan E Bristow* (1987) 37 BLR 92 (CA).
55   *Pozzolanic Lytag Ltd v. Bryan Hobson Associates* [1999] BLR 267.

> The extent of the duty of an architect or engineer to inspect has been dis-
> cussed in may cases. ... But in all the cases, the extent of the duty depends on
> the terms of the contract and the circumstances of the project, which are, of
> course, infinitely variable.[56]

An engineer engaged to supervise the work is not ordinarily obliged to
watch every detail of the work undertaken by the contractor.[57] He is entitled to consider how best to employ the resources which are allocated to the
project; thus where he has no reason to suspect that there is anything amiss,
he need not spend a great deal of resources in supervising the contractor.[58]

56   *Department of National Heritage* v. *Steensen Varming Mulcahy*, ORB, 30 July 1998, per Judge
     Bowsher.
57   *McLaren Maycroft Co.* v. *Fletcher Development Co. Ltd* [1973] 2 NZLR 100 (New Zealand Court of
     Appeal).
58   *East Ham Corporation* v. *Bernard Sunley & Sons Ltd* [1966] AC 406 (HL).

# 10

# Civil engineering claims: entitlements and evaluation

## 1. Formulating claims

Claims in relation to civil engineering contracts may be formulated either as contractual or non-contractual claim.

A contractual claim may be formulated as:

(1) A claim for an entitlement provided for under terms of the contract. For example, Clause 12 of the ICE Conditions of Contract provides that where physical conditions are encountered, which could not reasonably have been foreseen by an experienced contractor, the contractor is entitled to be paid his additional cost for dealing with them. A Clause 12 claim is thus a claim for an entitlement under the contract.

or

(2) A claim for breach of contract. For example, where the contract requires the employer to provide design information at a particular time, but it is not so provided and the contractor thereby suffers delay and loss, he may frame his claim as a breach of the employer's obligation to perform the contract.

Furthermore, a claim may be advanced as a claim related to, but outside, the contract.[1] For example, the claim may be made:

---

1     See Chapter 8, where the general principles of extra-contractual claims are discussed.

(3)   As a claim for misrepresentation. Where, for example, a misleading site investigation report is provided with the tender documents and the report was expressly excluded from the contract, it cannot be a breach of contract. It may help to found an 'unforeseen conditions' claim if there is such a provision in the contract. But if there is not, the contractor may run his claim as a claim for misrepresentation. Of course, the success of this claim depends upon the contractor demonstrating that the employer actually represented that those were the conditions and that the contractor was thereby induced to enter into the contract.

or

(4)   as a claim in restitution where the work claimed for does not form part of the contract work. For example, if the employer asks the contractor to carry out works similar but additional to those specified in the contract, it may be that the contractor decides to run his claim for these additional works as a claim for a reasonable value, especially where the contract rates are low. Such a claim can only succeed where the new works are not covered by a proper variation under the existing contract. Furthermore, if the conduct of the parties indicated that the work was either to be covered by the existing contract or was to form the subject of a new contract governed by the existing rates, then a claim for a *quantum meruit* is not available.

**Claims may be run on alternative bases**   A claimant can claim for the same item of loss as an entitlement under the contract or as a breach or as damages for misrepresentation or as a *quantum meruit*. Of course, if more than one avenue is successful, the claimant can only recover once for the loss. Multiple track claiming often suggests a lack of confidence and claimants invariably opt to claim on the strongest bases only. The preconditions for a successful claim for either misrepresentation or *quantum meruit* mean that a claimant's best avenue is almost always a claim under or for breach of the contract, where this is possible.

**The basis of valuation permitted**   Frequently the claimant's entitlement under the contract will be more generous than the damages which he would recover for breach of contract.[2] Accordingly, given the choice, he may

---

2   For the measure of damages see below at Section 6.

prefer to run his case primarily on the basis of an entitlement under the contract.

*The dispute resolution mechanisms provided for in the contract* The applicability of such mechanisms to the dispute will depend upon the true construction of the contract provisions[3] and the factual basis alleged. For instance, a claim for restitution on the grounds that no contract eventuated deprives the claimant of the benefit of any dispute resolution mechanisms in the contract.[4] A claim for misrepresentation may fall within the terms of the dispute resolution clause, although this will depend on the construction of the contract.

*Notices* Civil engineering contracts frequently contain provisions which purport to deprive or limit[5] the right to claim unless proper notices are served. While a failure to comply with notice provisions may prevent the contractor from receiving benefits under the contract, breaches of contract or extra-contractual remedies will not normally be affected.

## 2. Statements of claim and response

### General introduction

A party is entitled to submit a claim for the consideration of the other party at any time. Likewise, a party to a contract to which the Housing Grants, Construction and Regeneration Act 1996 applies may refer a dispute to adjudication at any time. The HGCRA 1996 stipulates no format for the claim. The Scheme for Construction Contracts simply requires a Notice of Adjudication to set out briefly

> (a) the nature and a brief description of the dispute and of the parties involved; (b) details of where and when the dispute has arisen; (c) the nature of the redress which is sought ...[6]

### The Referral Notice

---

3   *Ashville Investments* v. *Elmer Contractors* [1989] QB 488 (CA); *Strachan & Henshaw Ltd* v. *Stein Industrie (UK) Ltd and GEC Alsthom Ltd* (1997) 87 BLR 52 (CA).

4   Including statutory adjudication, which is limited to disputes arising 'under the contract'— Housing Grants, Construction and Regeneration Act 1996, Section 108(1).

5   For example, under the ICE Conditions of Contract, 7th Edition, notices are required in a number of situations. They rarely operate as conditions precedent, but the claimant's right to claim is limited if his failure to serve a notice has prejudiced the Engineer: see Clause 53(5).

6   Scheme for Construction Contracts, Part I, Paragraph 1(3).

shall be accompanied by copies of, or relevant extracts from, the construction contract and such other documents as the referring party intends to rely upon.[7]

Despite this permissible informality, the presentation of a well-structured claim is extremely desirable both in the case of a simple claim for consideration by the other party and in adjudication proceedings. In the case of arbitration proceedings it is imperative that the claim is set out properly. The 'fair resolution of disputes' requires that each party has a fair opportunity to deal with his opponent's case.[8] This is only possible where the case is clear and the issues can be properly defined. In the case of litigation proceedings, the court will insist that the Statement of Case complies with these same principles.

### The content of the statement

Subject to any requirements in the contract, the parties are entitled to formulate and set out their claims as they wish.[9] All claims, however, should contain a clear statement of (*a*) the nature of the claim, (*b*) the basis of the alleged entitlement and (*c*) the sums claimed and their computation.

*A clear statement of the nature of the claim*   The claim should state whether it is a claim for breach of contract or, alternatively, a claim for an entitlement under the contract or some other type of claim. A party is entitled to advance alternative claims even if they are inconsistent providing one claim does not undermine another.

*A clear statement of the basis of the entitlement*   Where a breach of contract is alleged, the term which is allegedly breached should be set out (or clearly referred to) and the facts of the alleged breach should be set out succinctly. The claim should include all material facts so that the other party is alerted to the case being advanced.[10] The case must be pleaded with sufficient particularity and the causation between any breaches of contract alleged and the consequences which it is alleged flow from them must be outlined.[11]

---

7    Scheme for Construction Contracts, Part I, Paragraph 7(2).
8    Arbitration Act 1996, Sections 1 and 33. See Chapter 11.
9    GMTC *Tools and Equipment Ltd* v. *Yuasa Warwick Machinery Ltd* (1994) 73 BLR 102 (CA).
10   *Philipps* v. *Philipps* (1878) 4 QBD 127 at 139 per Cotton L.J.
11   *Wharf Properties* v. *Eric Cumine Associates* (1991) 52 BLR 1 (PC).

*A clear statement of the amount and calculation of the sums claimed*   For example, where a breach of contract is alleged, the claim will be for damages and a calculation giving a proper breakdown of how the damages have been calculated must be submitted.

*Complex interactions*   Generally speaking, the causal link between any breach and damage[12] must be pleaded and demonstrated. In some civil engineering situations, however, the interactions between individual items are complex and the precise cause and effect for each individual loss is difficult to establish. Whilst there can be no excuse for failing to state in proper particularity the basis of the claim,[13] the complexity which arises in civil engineering cases has been judicially recognised; a party is not, it seems, obliged to show the individual effect of each and every minor factor where the degree of interaction is great.[14]

*Special claims*   A claim for interest should normally be specifically pleaded. Any matter which the party pleading will allege constitutes a release from the claim of the other party should specifically be pleaded.[15] Normally no special allowance should be made for tax payable.[16]

## 3. Set-off and counterclaim

### Set-off

Where A claims against B, B may reply that the sums claimed by A are to be diminished or extinguished because A owes B money. Here, a potential set-off arises. A set-off may be claimed at law or in equity. Furthermore, the contract may provide its own set-off provisions.

---

12   Or, as the case may be, between the term giving an entitlement and the sums claimed.
13   *Wharf Properties v. Eric Cumine Associates* (1991) 52 BLR 1 (PC).
14   J. *Crosby & Sons v. Portland Urban District Council* (1967) 5 BLR 121 (DC); *London Borough of Merton v. Hugh Stanley Leach* (1985) 32 BLR 51; *Mid Glamorgan County Council v. J Devonald Williams & Partners* (1991) 8 Constr. L.J. 61; *Bernhard's Rugby Landscapes Ltd v. Stockley Park Consortium Ltd* (1997) 82 BLR 39; *Naura Phosphate Royalties Trust v. Matthew Hall Mechanical & Electrical Engineering Pty Ltd* [1994] 2 VR 386 (Supreme Court of Victoria); *John Holland Construction & Engineering Pty Ltd v. Kvaerner RJ Brown Pty Ltd* (196) 82 BLR 81 (Supreme Court of Victoria). See also Byrne D, Total costs and global claims [1995] ICLR 531; Wilson M, Global claims at the cross roads (1995) 11 Constr. L.J. 15.
15   For example, an accord and satisfaction or a time bar under the Limitation Acts.
16   *British Transport Commission v. Gourley* [1956] AC 185; *Amstrad plc v. Seagate Technology Inc.* (1997) 86 BLR 34.

*Set-off at law*   This species of set-off is rare in civil engineering cases. A prerequisite is that the sum to be set-off has been ascertained with certainty.[17] A surveyor's estimate does not render a claim ascertained.[18]

*Set-off in equity*   B may set-off unliquidated cross-claims against A's claim providing they arise out of or in connection with the same contract as A's claim against B.[19] Equitable set-offs are generally effective, even where the contractor is in possession of a certificate ostensibly entitling him to immediate payment.[20] There are two distinct types of equitable set-off: (*a*) a cross-claim against the claimant arising out of or in connection with the same contract as the claim; (*b*) a claim by the respondent to be entitled to abate the claim because the work which the claimant claims to be paid for is defective and/or incomplete and hence has a value less than that claimed. Claims properly framed as (*a*) are true equitable set-offs while those properly framed as (*b*) are known as abatement defences.[21]

*Contractual set-off*   Where the parties agree upon a term relating to set-off this will be enforced according to its construction. Thus, rights of set-off in respect of other contracts between the two parties may be generated. Terms purporting to exclude rights of set-off are construed strictly, so that a term excluding set-offs may not exclude abatement defences.[22]

## Counterclaims

Where A claims against B and B believes that he has a claim against A, he may counterclaim, providing the relevant tribunal has jurisdiction to entertain it. The tribunal will ordinarily have jurisdiction to hear any cross-claim which amounts to a set-off, since a set-off is a defence.

---

17    *Axel Johnson Petroleum AB* v. *MG Mineral Group AG* [1992] 1 WLR 270 (CA).
18    *B Hargreaves Ltd* v. *Action 2000* (1992) 62 BLR 72 (CA).
19    *Hanak* v. *Green* [1958] 2 QB 9 (CA).
20    *Modern Engineering (Bristol) Ltd* v. *Gilbert-Ash* [1974] AC 689 (HL).
21    *Mondel* v. *Steel* (1841) 8 M & W 858; *Modern Engineering (Bristol) Ltd* v. *Gilbert-Ash* [1974] AC 689 (HL). In *Mellowes Archital Ltd* v. *Bell Projects Ltd* (1997) 87 BLR 26 (CA) it was said:

> The difference between abatement and set-off is only of significance in very particular situations, namely special issues of limitation … or where, as in our case, a contractual limitation on remedies confines itself to 'set-off'.

22    *Acsim (Southern)* v. *Danish Contracting and Development Co. Ltd* (1989) 47 BLR 55 (CA).

## 4. The evaluation of claims

Where a claim is made, and the respondent's liability is demonstrated, the amount due must be evaluated. The amount due may be different depending on how the claim is framed. The following may be recovered

(1) contractual entitlement: sums due in accordance with the contract terms[23]
(2) damages: compensation due as a result of the respondent's breach of duty[24]
(3) *quantum meruit*: a reasonable sum payable where the employer has requested the contractor to perform the work but no contract exists between them.
(4) indemnity: a sum payable in order to reimburse a party for some payment which he is obliged to make.[25]
(5) contribution: a party who has reasonably settled a claim, may sue others who are liable for the same damage for a contribution to the amount paid in settlement.[26]

### Damages

Damages are compensatory and are awarded to place the innocent party in the position in which he would have been had the breach not occurred.[27] Damages in contract are valued as the loss which the claimant has suffered as a result of the breach of contract, namely the difference between what he has received and what he would have received without the breach. Damages for misrepresentation are valued as the loss experienced by the

---

23  For example, under Clause 12 of the ICE Conditions of Contract. Where adverse ground conditions are encountered which could not reasonably have been foreseen by an experienced contractor, this does not cause the employer to be in breach of contract. Nevertheless, the contractor is entitled to his cost plus an element for profit as an entitlement under Clause 12.
24  For example, breach of contract, breach of duty not to misrepresent, breach of a tortious duty of care.
25  For example, where a surety guarantees to pay over a sum to an employer in the event of a contractor's default, he is entitled to an indemnity from the contractor.
26  Civil Liability (Contribution) Act 1978. Section 1(4). See Chapter 11, Section 1, and the cases there referred to.
27  In *Robinson* v. *Harman* (1848) 1 Exch. 850 Parke B. said

> The rule of common law is, that where a party sustains a loss by reason of a breach of contract, he is, so far as money can do it, to be placed in the same situation, with respect to damages, as if the contract had been performed.

claimant as a result of the misrepresentation.[28] Damages for breach of a tortious duty are valued as the loss suffered by the claimant as a result of the breach of duty to take care.[29]

The claimant must demonstrate that he has suffered the loss[30] and that the damage is attributable to the respondent's breach. This involves demonstrating that: (*a*) the breach caused the loss; (*b*) the loss was reasonably foreseeable; (*c*) the evaluation proposed is reasonable in all the circumstances; and (*d*) the claimant is not claiming for any loss which he could have avoided.

### Causation

> If the damage would not have happened but for a particular fault, then that fault is the cause of the damage; if it would have happened just the same, fault or no fault, the fault is not the cause of the damage.[31]

> Causation is to be understood as the man in the street, and not as either the scientist or the metaphysician, would understand it.[32]

In instances where more than one cause may have contributed to the claimant's damage the position can be complex.

> [T]here is no abstract proposition, the application of which will provide the answer in every case, except this: one has to ask oneself what was the effective and predominant cause of the accident that happened, whatever the nature of that accident may be.[33]

**Reasonable foreseeability**    The test of reasonable foreseeability is set out in *Hadley* v. *Baxendale*.[34]

---

28  See *East* v. *Maurer* [1991] 1 WLR 461 (CA) (Fraudulent misrepresentation); on the measure of damages recoverable under Sections 2(1),(2) of the Misrepresentation Act 1967 see *Royscot Trust Ltd* v. *Rogerson* [1991] 2 QB 297; *Gran Gelato* v. *Richcliff* [1992] Ch. 560; *William Sindall plc* v. *Cambridgeshire County Council* [1994] 1 WLR 1016 (CA).

29  See *South Australian Asset Management Corporation* v. *York Montague Ltd* (1996) 80 BLR 1 (HL) for the meaning of 'as a result of' the breach.

30  Where a claimant's damage is made good by a third party (e.g. a purchaser who buys a defective building at the full market rate) complications may arise. Where the loss is made good by an insurer, the insurance company is entitled to take over the claimant's claim under the principle of subrogation.

31  *Cork* v. *Kirby MacLean Ltd* [1952] 2 All ER 402 (CA); see also *Barnett* v. *Chelsea and Kensington Hospital Management Committee* [1969] 1 QB 428.

32  *Yorkshire Dale Steamship Co. Ltd* v. *Minister of War Transport* [1942] AC 691 (HL) per Lord Wright at 706.

33  *Yorkshire Dale Steamship Co. Ltd* v. *Minister of War Transport* [1942] AC 691 (HL) per Viscount Simon L.C. at 698.

34  (1854) 9 Ex 341. Strictly speaking, this case relates to breaches of contract only. However, in tort, fairly similar principles apply.

> Where two parties have made a contract which one of them has broken, the damage which the other party ought to receive in respect of such breach of contract should be [1] such as may fairly and reasonably be considered as either arising naturally i.e. according to the usual course of things, from such breach of contract itself, or [2] such as may reasonably be supposed to have been in the contemplation of both parties at the time they made the contract as the probable result of such breach of it. [Enumeration added.]

This passage refers to two distinct limbs as indicated. The first relates to the damage which may be recovered in the general case; while the second, often termed 'special damages', is recoverable where the claimant notifies the respondent at the time of making the contract that these losses are likely in the event of a breach.

*Reasonable evaluation*   The decided cases show that there are no hard and fast rules.[35] The measure of damage cannot be determined in the abstract. The law does not disregard the hopes and aspirations or the individual pre-dilections of the particular claimant in applying the basic principle.[36]

*Where a substantial loss is suffered by someone other than the claimant*   The doctrine of privity provides that only a party to a contract may sue upon it. Frequently, for reasons of financial convenience, the 'employer' under a contract is not the owner of the works. Here, defective work causes loss to the person whose property it is. Since the employer suffers no loss, what damages can he claim? The answer, technically, is that he can recover no damages. However, the law makes every effort to ensure that a proper claim is not lost in a 'black hole'. A number of devices are used to allow the employer to recover for the owner's loss or to allow the owner to sue directly.[37]

---

35   *British Westinghouse Electric and Manufacturing Co.* v. *Underground Electric Railways of London* [1912] AC 673 (HL) (at page 688):

> The quantum of damages is a question of fact, and the only guidance which the law can give is to lay down general principles which afford at times but scanty assistance.

See *Forsyth* v. *Ruxley Electronics Ltd* [1995] 3 WLR 118 (HL).

36   *Radford* v. *De Froberville* [1977] 1 WLR 1262.

37   See, for example, *Alfred McAlpine Construction Ltd* v. *Panatown Ltd* (1998) 88 BLR 67 (CA); *Darlington Borough Council* v. *Wiltshire Northern Ltd* [1995] 1 WLR. 68 (CA); *Linden Gardens Trust Ltd* v. *Linesta Sludge Disposals Ltd* and *St Martins Property Corporation Ltd* v. *Sir Robert MacAlpine & Sons Ltd* [1994] AC 85 (HL). See the discussion on the doctrine of privity in Chapter 6, Section 1, and assignment in Chapter 6, Section 4.

*Mitigation of loss*   The claimant may not claim for any loss which he could reasonably have avoided. This is commonly termed 'the duty to mitigate'.[38] This terminology is slightly misleading as the claimant is under no duty to mitigate the loss. However, his entitlement to damages is limited to losses which could not reasonably be prevented by mitigation.[39]

*Liquidated damages*   Liquidated damages are prospective damages which have been reduced to a fixed sum agreed in the contract. They may be contrasted with 'unliquidated damages' which have not been reduced to a defined sum and which are the ordinary damages recoverable upon a breach of contract. Liquidated damages are dealt with in Chapter 5, section 4.

### Entitlements under the contract

Parties often include terms in their agreement designed to regulate entitlement in a variety of eventualities. These include payment for variations to work ordered in accordance with the contract, the occurrence of stated risks, etc. The entitlement which follows from such clauses is a question of interpreting the agreement. It it suggested that the following rules of interpretation are appropriate in the absence of anything to the contrary. First, where no method of computation is provided, the claimant must demonstrate that the sums claimed flow from the relevant event and that he has mitigated his loss. Second, where the claimant agrees to supply the respondent with a service under the agreement, he may claim a reasonable sum for profit in all the circumstances, unless the agreement provides otherwise.

### Quantum meruit

A claim for *quantum meruit* is a claim for a reasonable sum in all the circumstances.[40]

## 5. Interest and finance charges claims

The resolution of civil engineering claims often takes a considerable time. Therefore, interest frequently constitutes a major element of claims. Interest may be claimed as:

---

38   *Payzu Ltd* v. *Saunders* [1919] 2 K.B. 581; Lord Haldane in *British Westinghouse Electric and Manufacturing Co.* v. *Underground Electric Railways of London* [1912] AC 673 (HL).

39   *The Solholt* [1983] 1 Lloyd's Rep. 605.

40   See Chapter 8.

(1) an entitlement under the contract
(2) damages flowing from breach, or
(3) under statute.

*Interest as an entitlement under the contract*   The parties may agree that late payment of sums or entitlements attracts interest, including compound interest.[41] Terms entitling the contractor to, for example, direct loss and expense may be interpreted to include the recovery of 'financing charges'.[42] For some classes of contract, terms as to interest may be implied by statute.[43]

*Interest as damages flowing from breach*   The common law maintains a fiction that a person suffers no loss by the late payment of damages owed to him; hence he cannot claim interest as damages for the delayed payment of those damages under the first limb of *Hadley* v. *Baxendale*.[44]

> There is no such thing as a cause of action in damages for late payment of damages. The only remedy which the law affords for delay in paying damages is the discretionary award of interest pursuant to statute.[45]

Any claim for interest as damages must, therefore, fall within the second limb of *Hadley* v. *Baxendale*. This requires[46] that the claimant must have advised the respondent at the date of making the contract that he would incur 'special damages' (e.g. interest) if the claimant is not paid promptly.

*Interest under statute*   The court/arbitrator have discretionary powers under the Supreme Court Act 1981 and the Arbitration Act 1996 to award interest from the date when a cause of action arose.

## 6. Prolongation, delay and disruption claims

Contractors frequently claim that their work has been delayed, prolonged or disrupted. They will be entitled to additional costs or damages only if

---

41   Civil engineering contract provisions have not always been well-drawn and difficulties have arisen with their interpretation. See, for example: *Secretary of State for Transport* v. *Birse-Farr Joint Venture* (1993) 62 BLR 36.

42   For example, in *FG Minter Ltd* v. *WHTSO* (1980) 13 BLR 1 (CA) the 'direct loss and/or expense' provisions of the 1963 Standard Form Building Contract were construed to include financing charges. See also *Rees & Kirby* v. *Swansea City Council* (1985) 30 BLR 1 (CA); *Ogilvie Builders Ltd* v. *Glasgow District Council* (1994) 68 BLR 122 (Scottish Court of Session).

43   Late Payment of Commercial Debts (Interest) Act 1998. See Chapter 4, Section 7, for a more detailed discussion.

44   *London, Chatham and Dover Railways* v. *South Eastern Railway* [1893] AC 429 (HL).

45   *President of India* v. *Lips Maritime* [1988] AC 395 (HL) per Lord Brandon at 425.

46   *Wadsworth* v. *Lydall* [1981] 1 WLR 598 (CA).

they can bring their claim within one of the usual heads of claim. Thus, if they wish to claim for breach of contract, they must identify the term,[47] show that the employer is in breach of it and that his loss is caused by the breach. Alternatively, the employer may have misrepresented the state of facilities or quality of access or possession to be provided, in which case a claim for misrepresentation may lie. Alternatively (and most commonly) there will be provisions in the contract entitling the contractor to claim in a variety of circumstances.[48]

## *Distinction between extensions of time and prolongation and delay claims*

A claim for an extension of time[49] is essentially a claim that no damages for delay may be levied by the employer until some later date than that originally stated. A claim for prolongation bears a resemblance to an extension of time claim; for example, project programming techniques are frequently employed to evaluate both and indeed it is rare that a prolongation claim will succeed where an extension of time claim does not.[50] However, a claim for prolongation etc. is directed at the loss which the contractor suffers as a result of disruptions for which the employer is responsible. In short, a claim for an extension of time is a defensive claim while a claim for delay, prolongation, etc., is an active claim.

## *Prolongation, delay and disruption: heads of claim*

Claims for prolongation etc. frequently arise from reduced site access, craneage, storage facilities, etc., interference from other contractors or personnel, delays in issue of information, free-issue materials, preceding work

---

47    This may be an express or an implied term; see Chapter 3 on implied terms as to 'cooperation between the employer and contractor'.

48    In the ICE Condition of Contract, 7th Edition, for instance, the Contractor is entitled to claim for interference from other contractors etc. where such interference could not reasonably have been foreseen. In addition Clause 52(4) provides for the revision of contract rates as a result of variations which render the original rates unreasonable or inapplicable.

49    For 'extensions of time', see Chapter 5.

50    Although the delay may come within the extension of time clause, that of itself is not sufficient to allow the contractor to claim for prolongation. Thus, under the ICE Conditions of Contract, the contractor is entitled to an extension of time for exceptionally inclement weather; however, the effect of this is simply to stop the clock running on the contract period and does not rank as a matter for which the employer assumes any other responsibility. See also *Henry Boot Construction Ltd* v. *Central Lancashire Development Corporation* (1980) 15 BLR 8.

or necessary reprogramming causing inefficient working (e.g. winter working). The heads of claim are frequently as follows. Note that these heads may overlap and it is important to ensure that no 'double recovery' is sought.

*On-site resources being employed for a longer period than was reasonably anticipated*  This may be due to the site facilities, such as offices, power, etc., being required for a longer period;[51] or specified items of plant being required for longer as where delay pushes work into the winter period.[52] In some civil engineering contracts, there are provisions to bill some items as time-related[53] so that the effects of prolongation associated with these items may readily be computed.

*Increased allocation of on-site resources*  Where the efficiency of plant or operatives is reduced, additional machinery or personnel may be needed to achieve the same rates of production. Furthermore, additional or different measures may be required, such as additional cranes, safety measures, etc. In addition, disruption may cause wastage of or damage to materials, as where concrete or asphalt deliveries are held waiting or non-productivity of personnel or plant where work is delayed. The costs of additional resources on site are readily computed, and it is a question of fact and reasonableness whether any item of site resource is allowable for any period. Consequential effects may involve complex interactions, and the contractor must demonstrate reasonably that the loss claimed did occur as a result of the delay of which he complains.[54]

*Increased 'main-office overheads'*  The costs of head office overheads are usually claimed as a supplement to the direct on-site costs. Their computation is frequently controversial.[55] It is suggested that the factors which

---

51    These items are frequently referred to as 'preliminary items' as they are usually placed first in a traditional bill of quantities.

52    *Freeman* v. *Hensler* (1900) *Hudson's Building Contracts*, 4th edn, Vol. 2, 292 (CA).

53    For example, where the contract incorporates CESMM 3, section 7 of which deals with time- and other method-related charges.

54    See the discussion above at Section 2 above and the cases cited there.

55    A number of formulae have been advocated, including the Hudson Formula, the Emden Formula and the Eichlay Formula (USA). The use of formulae has received some qualified support e.g. *Finnegan* v. *Sheffield City Council* (1988) 43 BLR 124. But they are based on unstated assumptions about the state of the market, the uniformity of application of overheads to site work, etc. It is submitted that they provide no advantage over a traditional calculation in which the overheads which are being claimed are set out with reasonable particularity. In this regard see also *Alfred McAlpine Homes (North) Ltd* v. *Property and Land Contractors Ltd* (1995) CILL 1130.

might properly be taken into account include (*a*) whether the proportion claimed for overheads is within the normal range for that type of contractor;[56] (*b*) whether the overhead percentage claimed is supported by historical accounts; and (*c*) the relationship between site work and the correlated head-office input.[57]

*Profit*   Unless the contract provides otherwise,[58] profit may only be recovered, in principle, to the extent that it represents a loss. In a buoyant market it will not be difficult to show that the resources used on the project could have been profitably employed elsewhere. Where the market is slower it is suggested that there should still be a presumption that a modest level of profit be recoverable.[59]

*Interest or financing charges*   These are dealt with at Section 5 above.

## 7. Claims for defective work

A contractor whose work does not comply with the specification is in breach of contract. His employer is entitled to recover damages. Various bases for computing the quantum of damages have been suggested, including the diminution in value of the building and the cost of repair. Frequently the building experiences only a small diminution in value despite the defective work whereas the cost of remedying that work may be significantly higher. In such a case the contractor will attempt to have damages calculated on a diminution basis while the employer will seek to recover the cost of the remedial work. It is clear that there are no hard and fast rules and that the court will take all the circumstances into account before deciding on the appropriate basis of computation.[60] Non-compliance

---

56   Some small specialist contractors may run with high head-office overheads due to the small scale of their operations, the need to supply engineering support on all projects, etc. Larger contractors working in areas involving highly repetitive low-technology work may run with significantly lower overheads.

57   For example, for some low-technology projects, such as straightforward muck-haulage, a major increase in work on a well-established site may entail little head-office input. However, an apparently minor change which requires careful planning may result in significantly more than a proportionate increase in head-office input.

58   For example, the ICE Conditions of Contract provide for profit for additional permanent or temporary work where there is interference from other contractors: Clause 31(2).

59   The levels of profit margin in the construction tend to seem small, since they are expressed on turnover. However, a contractor's turnover is usually several times the capital he employs, the profit margin expressed as a proportion of capital employed is significantly higher.

60   See, for example, *East Ham Borough Council* v. *Bernard Sunley Ltd* [1966] AC 406 (HL).

with the specification may have a number of practical effects and it is suggested that it is not possible to decide how damages should be calculated without a consideration of these effects. Where reinstatement is the appropriate basis of calculation of quantum, the time at which those damages are to be calculated is usually at the time when those defects could reasonably have been detected.[61]

Due to the delay in resolving defects claims, an employer who finds defects in his structure frequently calls upon consultants to advise him as to remedial works. Where consultants advise remedial works the employer is, generally, entitled to rely upon their advice and recover the cost from the party or parties liable for the defects. Where, however, the consultant gives advice which is itself negligent, the contractor is not required to pay for any additional or excessive work.[62] The employer must give credit for any improvement over the original specification.[63] Where, however, the fact employer gets new for old, this is not normally betterment if the employer had no option.[64]

## 8. Limitation

An aggrieved party may take action to recover under a contract or other entitlement during the 'limitation period'.[65] Upon the expiry of the limitation period, any potential action is said to be 'time barred'. The limitation period in most civil actions related to civil engineering is six years and runs from the time at which the 'cause of action' arises; the exception is an action under a contract 'under seal' where the limitation period is twelve years. The appropriate proceedings must be commenced before the limitation period expires. In litigation, the proceedings begin when the claim is taken out; arbitration proceedings commence with the issue of the relevant notice.

---

61   *East Ham Corporation* v. *Bernard Sunley & Sons Ltd* [1966] AC 406 (HL).

62   *Frost* v. *Moody Homes Ltd & others* (1989) CILL 504; *The Board of Governors of the Hospital for Sick Children* v. *McLaughlin & Harvey plc* (1987) 6 Constr. LJ 245.

63   This is termed 'betterment'. See for example, *Richard Roberts Holdings Ltd and Another* v. *Douglas Smith Stimson Partnership* (1988) 46 BLR 50.

64   *Harbutt's 'Plasticine' Ltd* v. *Wayne Tank and Pump Co. Ltd* [1970] 1 WLR (CA) per Cross L.J.:

> They replaced it in the only possible way, without adding any extras. I think that they should be allowed the cost of replacement. True it is they got new for old; but I do not think that the wrongdoer can diminish the claim on that account. If they added extra accommodation or made extra improvements, they would have to give credit. But that is not this case.

65   Limitation Act 1980: see Sections 5 and 8 respectively for simple contracts and deeds.

*Contractual causes of action*   In contract cases, the cause of action arises at the time of the breach of contract. For civil engineering design and/or construction works this is usually the date of completion on site, since a contractor who creates a defect is entitled to remedy it before completion[66] and a designer has a continuing duty in respect of the design until the works are complete.[67] Breaches by the employer of contractual obligations to provide access, produce drawings, information, etc., occur at the time when the employer failed to provide or produce them; the cause of action for claims which arise out of a failure to certify sums probably arises at the date of the final certificate.

*Causes of action based on tortious negligence*   In cases based in negligence, the cause of action arises at the date of damage, which frequently leads to a longer limitation period. Employers who wish to sue engineers frequently do so in the tort of negligence to take advantage of this extended period.[68]

*Concealment*   Section 32 of the Limitation Act 1980 provides that where the defendant has been fraudulent, and this has affected the claimant's knowledge of his right to make a claim against the defendant, the period of limitation does not run until the claimant has, or could reasonably have had, knowledge of his cause of action[69] and may be reactivated by a subsequent act of concealment.[70]

---

66   *P and M Kaye v. Hosier & Dickinson Ltd* [1972] 1 All ER 121 (HL) where Lord Diplock said at 139:

> Provided that the contractor puts it right timeously I do not think that the parties intended that any temporary disconformity should of itself amount to a breach of contract by the contractor.

67   *London Borough of Merton v. Lowe* (1982) 18 BLR 130.

68   At one stage it was thought that where the employer and engineer were in a contractual relationship, the limitation period would be that prescribed by the contract: *Tai Hing Cotton Mill v. Liu Chong Haing Bank* [1986] AC 80 (PC). However, it now seems clear that a claimant may take advantage of a tortious remedy even where a contractual remedy exists: see *Henderson v. Merrett* [1994] 3 WLR 761 (HL). Note also that the Latent Damage Act 1986 provides that no negligence claim may be brought more than 15 years after the breach of duty. The protection offered by this rule is likely, on occasions, to be of value to professionals.

69   See, for example, *Gray v. TP Bennett & Son* (1987) 43 BLR 63.

70   *Sheldon v. RHM Outhwaite Ltd* (Underwriting Agencies) Ltd [1996] 1 AC 102 (HL).

# 11

# Dispute resolution

In this chapter, the general principles of dispute resolution will be described. In Chapter 15, there is a more detailed examination of some of the civil engineering industry standard procedures for dispute resolution. Although each dispute resolution technique is examined in isolation, it is common for several tiers of dispute resolution to be agreed.[1]

## 1. Negotiated settlement

*General*   The vast majority of disputes and differences are settled by amicable negotiation leading to a binding agreement. Such agreements are frequently made at the conclusion of a project and include a full and final settlement of all claims by both parties.[2] Such settlements are variations to the existing contractual obligations and hence need to be agreed by properly authorised agents of the negotiating parties; likewise, the agreement must be entered into freely[3] and supported by consideration.[4]

---

1   See, for example, Lewis D, *Dispute resolution in the new Hong Kong International Airport core programme projects*, ICLR (1993) 76. Here the dispute process contained four tiers: (1) a decision of the engineer; (2) a mediation; (3) an adjudication; and (4) an arbitration.
2   Such an agreement is commonly referred to as an 'accord and satisfaction'. See *Hamlin v. Edwin Evans (a firm)* (1996) 80 BLR 85 (CA) for an example of a party prevented from suing for major defects because of a previous compromise agreement.
3   *North Ocean Shipping Co. Ltd v. Hyundai Construction Co. Ltd* [1979] QB 705.
4   See *D & C Builders v. Rees* [1966] 2 QB 617.

*Formality*   In principle, there is normally no need for a negotiated settlement to be put into writing or indeed to be evidenced in any way whatever.[5]
It is enforceable if made orally at a meeting.

*Failure to honour a settlement agreement*   A subsequent failure on the part
of one of the parties to honour the settlement will not revive the earlier
contract; the appropriate cause of action is for breach of the settlement
agreement.

*Effect of settlement on other contract relationships*   The complex contractual arrangements in civil engineering practice mean that one dispute may
span across several contracts. For example, alleged defects in work caused
by a sub-contractor may be in issue both under the sub-contract and under
the main contract. Where a contractor makes a settlement with the
employer, this will not fix his rights against the sub-contractor. He will still
need to prove against the sub-contractor that the sub-contractor is, in principle, liable and, if so, that the amount of the settlement was reasonable.
Where appropriate, a party who has reasonably settled a claim, may sue
others who are liable for the same damage for a contribution to the amount
paid in settlement.[6] In an action for damages based on a settlement with
another party, the court is not required to approach the question of the
amount of the settlement as if the matter were being tried from the start.

> If, upon the evidence, the Judge is satisfied that the damages would have
> been somewhere around the figure at which the Plaintiff had settled he
> would be justified in awarding the settlement figure. I do not consider that it

---

5   If the settlement includes a transfer of an interest in land then evidence in writing is required. Also
    note that Section 53(1) of the 1925 Law of Property Act requires that a 'disposition of an equi-
    table interest or trust subsisting at the time of the disposition must be in writing [and] signed …'.
    This technical provision may be important when the contents of funds held by one party on behalf
    of another—e.g. the retention fund held by an employer pending the successful conclusion of the
    maintenance period—are included in a settlement. If such a fund is to form part of the agreement,
    the settlement should always be in writing and signed to avoid a later claim that it is unenforceable
    for lack of formality.
6   Civil Liability (Contribution) Act 1978. Section 1(4) provides:

    A person who has made or agreed to make any payment in bona fide settlement or compro-
    mise of any claim made against his in respect of any damage … shall be entitled to recover
    contribution in accordance with this section without regard to whether or not he himself is or
    ever was liable in respect of the damage, provided, however, that he would have been liable
    assuming that the factual basis of the claim against him could be established.

    For cases on the interpretation of this Act see *Friends Provident Life Office* v. *Hillier Parker May &
    Rowden* [1996] 4 All ER 260 (CA) and *Birse Construction* v. *Haiste Ltd* [1996] 1 WLR 675 (CA).

is part of his duty to examine every item in these circumstances. … The Plaintiffs must establish a prima-facie case that the settlement was a reasonable one. If the Defendants fail to shake that case, the amount of the settlement can properly be awarded as damages.[7]

## 2. Adjudication

### Introduction

Adjudication, in its most general sense, is the presentation of a dispute by the parties to a third party neutral, who then proceeds to make a decision. This decision can either be (*a*) advisory, (*b*) binding on the parties until it is reversed by an arbitrator or the court, or (*c*) final and binding. Since the coming into force of the Housing Grants, Construction and Regeneration Act (HGCRA) 1996, the term 'adjudication' has acquired, in the UK at least, a more specific meaning: adjudication which complies with the provisions of that Act. Such adjudications are discussed in detail in this section because of their importance for dispute resolution in the UK. However, it must not be forgotten that construction is an international business and the Act only applies to British contracts; accordingly, it is fitting that some attention is also paid to forms of adjudication commonly used around the world, including Dispute Review Boards, and these will also be dealt with.

### Statutory Adjudication

*The Housing Grants, Construction and Regeneration Act 1996*[8]    By Section 108 of the HGCRA, a party to a construction contract has the right to refer a dispute or difference for adjudication in accordance with an adjudication procedure which complies with a number of criteria set out in that section. The section does not invalidate procedures which are non-compliant, but a party who wishes to take advantage of a compliant procedure is entitled to do so. By regulations made under the Act, the 'default' procedure is set out in the Scheme for Construction Contracts 1998. The courts have enforced a number of adjudication decisions with the clear intention of giving full effect to the will of Parliament.[9] Adjudicator's decisions are enforced

---

7  *Biggin & Co. Ltd* v. *Permanite Ltd* [1951] QB 314 per Singleton L.J. at 317. He also called attention to the fact that, in that case, 'the settlement was arrived at under advice …'.

8  The full text of Part II of the Act is set out in Appendix 2, together with the text of the Scheme for Construction Contracts (1998).

9  *Macob Civil Engineering Ltd* v. *Morrison Construction Limited* [1999] BLR 93.

quickly[10] by summary judgment[11] unless there is a arguable doubt as to whether the adjudicator has jurisdiction.[12]

*The application of the HGCRA 1996*   The Act applies to all contracts in writing,[13] entered into on or after 1 May 1998, which relate to[14] 'construction operations' in England, Wales or Scotland irrespective of whether or not the law of England and Wales or Scotland otherwise applies.[15] 'Construction operations' are given an extended meaning which includes not only contracts for the carrying out of construction operations, but also contracts for the arrangement and design of construction contracts.[16] Accordingly, the HGCRA applies to most civil engineering management, design and construction contracts for work in Britain. There are a number of specific exceptions which are important for civil engineering. These include drilling or tunnelling for the purpose of extracting minerals and access and machinery for works relating to the basic utilities and to the chemical, food and pharmaceutical industries.[17] Furthermore, Private Finance Initiative projects are excluded.[18] Where only part of the contract relates to construction operations, that part only is subject to the Act.

*The adjudicator's decision, the engineer's control and the employer's rights*   The adjudicator's decision should not impinge upon the proper management of the project. Although the adjudicator may make decisions about the acceptability of materials or methods under the contract or extensions of time in accordance with the contract, such decisions are designed to relate to payment only. They do not, or at least should not, deprive the engineer of his authority to control the project; nor do they deprive the employer of the right to have the product he wants. The principal function of the adjudicator's decision is to regulate the agreement and to maintain proper cash flow rather than to direct the project.

---

10   *Outwing Construction Limited* v. *H Randell & Son Limited* [1999] BLR 93.
11   *Macob Civil Engineering Ltd* v. *Morrison Construction Limited* [1999] BLR 93.
12   *Project Consultancy Group* v. *Trustees of the Gray Trust*, TCC (Dyson J.), 16 July 1999.
13   Section 107 of the Housing Grants, Construction and Regeneration Act 1996. Note that the expression 'in writing' is given an extended meaning and includes, for example, contracts evidenced in writing: Section 107(2)(c).
14   The question arises as to whether the Act applies to collateral warranties, settlements, etc. See Macaulay ML, 14 Constr. L.J. 318—321 (1998).
15   Section 104(6), (7) of the Housing Grants, Construction and Regeneration Act 1996.
16   Section 104(1), (2) and Section 105 of the Housing Grants, Construction and Regeneration Act 1996.
17   Section 105(2) and Section 105 of the Housing Grants, Construction and Regeneration Act 1996.
18   Construction Contracts (England and Wales) Exclusion Order 1998, SI 648.

Consider the following examples:[19]

(1)   The engineer condemns materials placed by the contractor as being unsuitable. The adjudicator decides that those materials, in his opinion, comply with the terms of the contract. This does not mean that those materials must stay in place; rather, it means that if the engineer considers them to be unsuitable and wishes them to be removed, he must instruct the contractor to remove them. This will amount to a variation order entitling the contractor to additional money and an extension of time.

(2)   The contractor claims an extension of time. The engineer rejects the claim. The adjudicator, however, grants an extension. This does not mean that the contractor necessarily has an extended time to complete the works. If the engineer wants the works completed in the original time, he must issue an instruction accelerating the works, which will entitle the contractor to additional money.

In each case, the adjudicator simply makes a decision about the meaning of the contract in relation to the facts on site. But it is the engineer/employer who decides how the decision is to be implemented in practice. It is they who decide ultimately whether the materials are to be replaced and/or whether time for completion is to be extended. Each of the decisions is later reviewable. If the court/arbitrator finds that the materials did not comply or that an extension was not due, then the variation order in each case were unnecessary and the additional money paid to the contractor can be recovered.

**Procedures which comply with the Act**   A compliant procedure meets the basic tests laid down in Section 108 of the Act. These tests require that the contract shall:

(1)   enable a party to give notice at any time of his intention to refer a dispute to adjudication

(2)   provide a timetable with the object of securing the appointment of the adjudicator and referral of the dispute to him within 7 days of such notice

(3)   require the adjudicator to reach a decision within 28 days of the referral or such longer period as is agreed by the parties after the dispute has been referred

---

19   See, also, Blackler T, Statutory adjudication under the HGCR Act 1996—some difficulties, in *Contemporary Issues in Construction Law*, Vol. II. *Construction Law Reform: a Plea for Sanity*, Ed. John Uff QC, Construction Law Press, 1997, pp. 1–12.

(4)  allow the adjudicator to extend the period of 28 days by up to 14 days, with the consent of the party by whom the dispute was referred
(5)  impose a duty on the adjudicator to act impartially
(6)  enable the adjudicator to take the initiative in ascertaining the facts and the law
(7)  provide that the decision of the adjudicator is to be binding until the dispute is finally determined by legal proceedings
(8)  provide that the adjudicator is not liable for anything done or omitted in the discharge or purported discharge of his functions as adjudicator unless the act or omission is in bad faith, and that any employee or agent of the adjudicator is similarly protected from liability.

Many contract drafting bodies have drawn up adjudication procedures which purport to comply with Section 108.[20] Any such procedural rules must be read together with the other terms of the contract; the entire procedure must be compliant.

*Where the contractual procedure is (or may be) non-compliant*   Where an adjudicator makes a decision under a non-compliant procedure, that decision will remain contractually binding and hence immediately enforceable provided there is a term in the procedure to that effect. Even where there is no such express term, and the adjudicator decided in good faith but wrongly that the contractual procedure complied with the Act, his decision will probably still be immediately enforceable under the Act. Following *Macob*,[21] it seems that a bona fide statutory adjudication may not be impeached on the grounds that it is technically defective, unless the adjudicator lacks basic jurisdiction. Where the converse situation applies (i.e. the adjudicator declines to act under a compliant procedure but instead makes his decision under the Scheme) this decision will also be immediately enforceable for the same reason. Thus, arguments about whether or not the various contractual schemes comply with Section 108 are rendered largely academic.

### Adjudication under the Scheme for Construction Contracts

*General*   The adjudication procedure in the Scheme for Construction Contracts is not easy to follow.

---

20   See, for example, the ICE Adjudication Procedure in Chapter 15.
21   *Macob Civil Engineering Ltd* v. *Morrison Construction Limited* [1999] BLR 93 per Dyson J. at 99:

I would hold, therefore, that a decision whose validity is challenged is nevertheless a decision within the meaning of the Act … .

*Adjudication in accordance with the law* The Scheme requires the adjudicator to 'reach his decision in accordance with the applicable law in relation to the contract ...'.[22] An adjudicator does not have to wait for legal submissions or to restrict his consideration to such submissions as he receives. Many adjudicators will not be lawyers and so the power enabling him 'provided he has notified the parties of his intention, [to] appoint ... legal advisers'[23] will be useful where difficult points arise. It must be emphasised, however, that the adjudicator must not delegate his decision making power; he must assess any advice he receives and must take his own decision.

*Terminology* A party who refers a dispute is called the 'referring party'.[24] The other party is not specially named. An 'adjudicator nominating body' is

> a body ... which holds itself out publicly as a body which will select an adjudicator when requested to do so by a referring party.[25]

*Notices* There are two defined notices under the Scheme:

(1) the 'notice of adjudication'[26] commences the adjudication and gives notice that the referring party intends to refer the matters stated therein to adjudication;

and not more than seven days later, the referring party serves

(2) the 'referral notice'[27] which is issued to the adjudicator referring the dispute to him.

*Notice of intention to seek adjudication* An adjudication begins by the issue of a 'notice of adjudication' which is technically a notice of intention to refer a dispute to arbitration. Copies are sent to every other party to the contract. The notice should set out basic information such as the nature and brief description of the dispute and other similar material details.[28]

*Appointment of an adjudicator* The Scheme contains fairly detailed provisions for appointment, in order to cater for the large range of possibilities,

---

22    Scheme for Construction Contracts 1998, Part I, Paragraph 12(1).
23    Scheme for Construction Contracts 1998, Part I, Paragraph 13(f).
24    Scheme for Construction Contracts 1998, Part I, Paragraph 1(1).
25    Scheme for Construction Contracts 1998, Part I, Paragraph 2(3).
26    Scheme for Construction Contracts 1998, Part I, Paragraph 1(1).
27    Scheme for Construction Contracts 1998, Part I, Paragraph 7(1).
28    Scheme for Construction Contracts 1998, Part I, Paragraph 1(3).

including that the adjudicator may decline to act or resigns.[29] The basic principles are that:[30]

(1) where an adjudicator is named in the contract, the referring party serves the notice of adjudication on the adjudicator, requesting him to act as adjudicator

(2) where no adjudicator is named (or the named adjudicator will not act) and a nominating body is named, the referring party sends the notice of adjudication to the named body with a request to select a person to act as adjudicator

(3) where neither adjudicator not nominating body is named in the contract, the referring party sends the notice of adjudication to an adjudicator nominating body together with a request to select a person to act as adjudicator.

*The referral notice*    Within seven days of the adjudication notice, the referring party serves a Referral Notice[31] on the other parties and the adjudicator. The referral notice is accompanied by 'copies of, or relevant extracts from, the construction contract and such other documents as the referring party intends to rely upon'.[32]

*Procedure following the referral notice*    The Scheme provides no details as to how the adjudication is to be conducted from this point on, save that it requires the adjudicator to reach his decision within 28 days of the notice of referral. It does, however, empower the adjudicator to make his own enquiries, to meet the parties and to give directions as to the timetable.[33] It does not provide specifically for the other party to have an opportunity to serve a response to the material served by the referring party, although the adjudicator's duty of impartiality necessarily requires this. The key moderators in any procedure are the timescales laid down; see below.

*Timescales*    One of the remarkable features of adjudication under the HGCRA 1996 is the short timescales. The following outline schedule shows the full allowable time interval for each major phase in the proceedings:

---

29   Scheme for Construction Contracts 1998, Part I, Paragraph 9(3).
30   Scheme for Construction Contracts 1998, Part I, Paragraph 2(1)(a), (b) and (c).
31   Scheme for Construction Contracts 1998, Part I, Paragraph 7(1).
32   Scheme for Construction Contracts 1998, Part I, Paragraph 7(2).
33   Scheme for Construction Contracts 1998, Part I, Paragraph 13(g).

Day 1            notice of adjudication
                 [7 days]
Day 8            referral notice
                 [28 days]
Day 36           the adjudicator makes his decision.

The 7 day interval between the notice of adjudication and referral notice is not open to extension under the Scheme. The 28 day period between the referral notice and the decision can be extended to 42 days if the referring party consents and longer if all parties consent.[34]

Whilst the above intervals provide the basic time framework, there are also other internal requirements aimed at ensuring that the basic time framework is maintained. For example, a person who receives a request to act as adjudicator must indicate whether or not he is willing to do so within 2 days[35] and a nominating body must communicate the selection of an adjudicator within 5 days.[36]

*The effect of failure to comply with time deadlines*    Given the strictness of the time limits, the question naturally arises as to the effect of any non-compliance with them. Arguably, a missed deadline takes the adjudication outside the terms of the Scheme and the decision will be invalid. However, the courts have normally been prepared, where possible, to interpret time requirements in a way which saves the procedure.[37] It is submitted that especially when the beneficiary of a decision has himself met time limits, the benefit will not be lost because of a dilatory adjudicator or nominating body.

*Where a party wishes to add further disputes*    A party may wish to have additional disputes adjudicated at the same time. For example, a connected dispute may arise after the issue of the notice of adjudication; or the other party's defence may entail setting off his alleged entitlement under some other claim under the contract. Paragraph 8(1) of the Scheme provides that

> The adjudicator may, with the consent of all the parties to those disputes, adjudicate at the same time on one or more disputes under the same contract.

---

34   Scheme for Construction Contracts 1998, Part I, Paragraph 19(1).
35   Scheme for Construction Contracts 1998, Part I, Paragraph 2(2).
36   Scheme for Construction Contracts 1998, Part I, Paragraph 5(1).
37   For example, time terms in a contract are normally assumed not to be of the essence.

This need for consent suggests, at first sight, that the adjudicator cannot hear any such additional matter at the same time without the consent of the other party. However, since adjudication is a legal procedure, it is submitted that an adjudicator is not only entitled, but obliged, to consider any argument which is a defence at law or in equity, whether or not it amounts to a distinct claim. For example, where a contractor claims that he has not been paid upon a certificate and commences adjudication, the employer may wish to claim that he has set-off the cost of defective work under the same contract. Since this, if proven (and subject to terms in the contract as to the set-off and the issue of the relevant notices in Section 111), amounts to a good defence, it seems that the adjudicator may not only consider this matter, but must do so if the employer supplies him with the details and evidence.[38] Although such a set-off appears to fall within the discretionary wording of Paragraph 20,

> The adjudicator … may take into account … matters under the contract which he considers are necessarily connected with the dispute …

it is thought that, on principle, the adjudicator should do so.

*Overlapping adjudications*   Since civil engineering contracts, including those for design or construction services, will contain adjudication rights, there is significant scope for overlap. An employer and main contractor may, for instance, have a dispute which is similar to one between the main contractor and a sub-contractor. If an adjudication decision is given in the first dispute, is the sub-contractor bound by it? Paragraph 9(2) provides

> An adjudicator must resign where the dispute is the same or substantially the same as one which has previously been referred to adjudication, and a decision has been taken in that adjudication.

This gives no indication as to what constitutes a dispute which is 'substantially the same'. It is thought that such a dispute must involve the same parties as in the instant dispute. Clear words would be required to bind the world at large and the Scheme provides only that 'The decision of the adjudicator shall be binding on the parties …'.[39]

*Adjudicator's decision*   When the adjudicator makes his decision it is

---

38   Scheme for Construction Contracts 1998, Part I, Paragraph 17:

> The adjudicator shall consider any relevant information submitted to him … .

39   Scheme for Construction Contracts 1998, Part I, Paragraph 23(2).

binding on the parties, and they shall comply with it, until the dispute is finally determined by legal proceedings.[40]

There is, however, no general sanction for failing to act upon the decision but the beneficiary can have it enforced by the courts by summary judgment.[41]

*Fees and costs*   An adjudicator is entitled to recover reasonable or agreed fees for his work. The parties are jointly and severally liable for these, but there appears to be a power to apportion his costs as between the parties.[42] However, there seems to be no provision empowering an adjudicator to award a successful party his own costs to be paid by another party.

*Challenging a decision*   An adjudicator's decision may be challenged on the grounds that there is no contract or that the contract is not one to which the HGCRA 1996 applies.[43] It may also be challenged to the extent that his decision deals with matters which were not referred to him.[44] Whilst summary judgment is normally available to enforce an adjudicator's decision, it will not be given where there is a strong case that the decision was outside the reference or the HGCRA 1996 does not apply.[45] It is thought, however, that where an adjudicator clearly has jurisdiction in principle, his decision can only be challenged if it is shown that he is not impartial or if the beneficiary of the decision has procured the decision by fraud. There may also be a marginal possibility of challenging a decision on the grounds that it is so perverse that the adjudicator must have exceeded his jurisdiction.

---

40   Scheme for Construction Contracts 1998, Part I, Paragraph 23.
41   *Macob Civil Engineering Ltd* v. *Morrison Construction Limited* [1999] BLR 93. See below at Section 4 for the summary judgment procedure.
42   Scheme for Construction Contracts 1998, Part I, Paragraph 25. This power is suggested in oblique terms rather than clearly spelled out; there are no guidelines as to how the power might be exercised.
43   *Project Consultancy Group* v. *Trustees of the Gray Trust*, TCC (Dyson J.), 16 July 1999. For exclusions see generally Sections 105(2) and 106 of the Act. In addition, the Secretary of State has exercised his power under Section 105(3) to extend the exclusions and has excluded Private Finance Initiative and development contracts: see the Construction Contracts (England and Wales) Exclusion Order 1998, SI 648.
44   *Macob Civil Engineering Ltd* v. *Morrison Construction Ltd* [1999] BLR 93, per Dyson J. at 98:

> If his decision on the issue referred to him is wrong, whether because he erred on the facts or the law, or because in reaching his decision he made a procedural error which invalidates the decision, it is still a decision on the issue. Different considerations may well apply if he purports to decide a dispute which was not referred to him at all.

45   *Project Consultancy Group* v. *Trustees of the Gray Trust*, TCC (Dyson J.), 16 July 1999.

*Adjudication under contractually agreed procedures*

Parties are entitled to agree any dispute resolution procedure they wish, including adjudication procedures. Where the agreed procedure, properly interpreted, meets the compliance requirements of the HGCRA 1996, this procedure applies and the Scheme for Construction Contracts does not. Where the agreed procedure fails to meet the compliance requirements, both it and the Scheme will apply and the referring party may elect which to use.[46]

*Published adjudication procedures*[47]    A number of bodies publish adjudication procedures. The publications include the *ICE Adjudication Procedure*, the *TecSA Adjudication Rules*, the *CEDR Model Rules for Adjudication* and the *CIC Model Adjudication Procedure*.

*Dispute Adjudication Boards*    A commonly used form of adjudication used in international projects is the Dispute Adjudication Board.[48] Such an adjudication is contractual and so its powers, constraints and the effects of its decisions (e.g. whether binding or not) are determined by construing the terms of the agreement which establish the board.

## 3. Conciliation, mediation and similar processes

*The meaning of the terms 'conciliation' and 'mediation'*    The terms 'conciliation' and 'mediation' have no fixed or settled meaning at law. The terms are frequently used interchangeably. Some writers use them to distinguish between different procedures but there is no uniformity in the usage of the terms. Nevertheless, it is commonly understood that both conciliation and mediation are, in essence, assisted negotiation. A third party neutral (a mediator or conciliator) attempts to bring the parties together either in open session or in private meetings (or both) to discuss and to facilitate a resolution. However, the role and powers of the neutral will vary from case to case and are, in every instance, a result of agreement between the parties.

---

46    It is assumed that the referring party must elect to use one or other.
47    See Miles D, A comparison of the available adjudication rules with the Government Scheme, 14 Constr. L J 311—317 (1998). This comparison is set out in a useful tabular format.
48    For example, see the FIDIC contracts. Also termed a Dispute Review Board. The Board will normally have three members, but may have one or five etc. See Pike A, Disputes review boards and adjudicators, ICLR [1993] 157; Jaynes GJ, Disputes review boards—yes! ICLR [1993] 452.

*Facilitative and evaluative conciliation/mediation* The assisted negotiation may be facilitative or evaluative: in the former case, the neutral is not entitled to suggest terms for settlement or to make a recommendation; in the latter the neutral is enjoined to make a recommendation.[49] It is important that the neutral's terms of reference are clearly defined in advance, including his power, if any, to meet the parties in private and any requirement to maintain confidences.[50]

*Other similar processes* A number of other procedures are available which have features in common with conciliation and mediation. An example is the non-binding tribunal process, where a panel is established to hear submissions from the parties. Typically the panel will include an employee of each party who is not directly involved with the dispute. There will also be a neutral on the panel. The parties may present their case in a strictly limited time and the panel then retires to make a recommendation. The parties are not obliged to accept the recommendation.[51]

# 4. Expert determination

The parties may agree that a person or panel is to be empowered to determine any dispute without any appeal.[52] This is commonly known as an 'expert determination'. It has a long history; in the nineteenth century engineers under civil engineering contracts frequently issued unappealable certificates.[53] Where the parties agree to be bound by the expert's decision, it may only be upset by fraud or a manifest departure from the terms of his appointment.[54]

---

49 This is the position under the conciliation provisions of the ICE Conditions of Contract; the conciliator makes a recommendation which becomes binding if neither party issues a notice of intention to proceed to arbitration.
50 See the ICE Conciliation Procedure in Chapter 15 which provides a useful model.
51 This procedure is frequently termed a 'mini-trial'. See Henderson DA, Mini-trial of construction disputes, ICLR (1994) 442.
52 In *Lee* v. *Showmen's Guild* [1952] 2 QB 329 (CA) at 342 Lord Denning suggested that a tribunal from which there was no appeal was bound to conform with the rules of natural justice. It is submitted that Lord Denning's comments are to be viewed against the unusual background facts of that case.
53 For example, *Sharpe* v. *San Paulo Railway Co.* (1873) LR 8 Ch. App 597.
54 *Jones* v. *Sherwood Computer Services* [1989] EGCS 172; *Campbell* v. *Edwards* [1976] 1 All ER 785. See also McGaw MC, Adjudicators, experts and keeping out of court (1992) 8 Constr. L.J. 332.

# 5. Arbitration

Arbitration is frequently used to resolve disputes arising under civil engineering contracts. Arbitrations are normally conducted under agreed rules of procedure, including the ICE Arbitration Procedure or the CIMAR Rules. These rules are discussed in Chapter 15. A number of general issues relating to arbitration are discussed here.

*Applicable law*   The 'seat' of an arbitration is the place that the arbitration proceedings are deemed to be located, either as a matter of fact or agreement between the parties. The Arbitration Act 1996 came into force in February 1997 and applies to all arbitrations which have a 'seat' in England, Wales or Northern Ireland and were commenced after that date. In the main body of this section, it will assumed that the Arbitration Act 1996 applies; the position in international arbitrations will then be discussed.

## The nature of arbitration

*General principles* Arbitration is the referral of a dispute to a person (the arbitrator) for his determination. The agreement to refer the dispute is ordinarily made before the dispute arises. The arbitrator must be impartial and act fairly, giving each party an opportunity to put his case and deal with that of his opponent.[55] The arbitrator's decision is made in accordance with the law[56] and is final,[57] although the courts retain a limited jurisdiction to set aside or vary an award or remit it back to the arbitrator for his further consideration.[58] The procedure is not set by the statute. The parties may, through institutional rules, agree the procedure; or to the extent that this is not done, the procedure is adopted by the arbitrator and is tailored to the circumstances of the case.[59]

*Multi-party arbitrations*[60]   Arbitration relies upon agreement between the parties to a contract. The complexity of contractual relationships in civil

---

55   Arbitration Act 1996, Section 33.
56   Arbitration Act 1996, Section 46(1). The parties may, however, agree that the arbitrator is able to decide the dispute 'in accordance with such other considerations as are agreed'.
57   Arbitration Act 1996, Section 58. Note also that, where the parties have agreed this, the arbitrator may make a 'provisional order' which is binding in the interim, but may be later reviewed by him: Section 39.
58   Arbitration Act 1996, Sections 68 and 69.
59   Arbitration Act 1996, Section 33.
60   See Fox-Andrews J, Construction industry disputes: Official Referee or technical arbitrator—the pros and cons (1992) 8 Constr. L.J. 2. This discusses the benefit of court proceedings where there is likely to be a multi-party situation.

engineering projects means that any one set of facts may lead to contractual disputes between more than two parties. It may be convenient for these to be dealt with in a single arbitration (a 'consolidated arbitration'), or in parallel (arbitrations heard 'concurrently') or serial arbitrations. A multi-party arbitration requires the agreement of all those to be involved. A number of devices attempt to achieve this result:

(1) Where the arbitration clause obliges the parties to join into another arbitration. For example, a sub-contract arbitration clause may provide that if the main contractor and sub-contractor have a dispute which is substantially the same as a dispute between the main contractor and employer, then the main contractor can require that the sub-contract dispute be referred to the main contract arbitrator.[61] The arbitrator, however, cannot be compelled to take on the new arbitration and the employer may have objections.[62] Furthermore, the disputes for both arbitrations may not crystallise at the same time.[63]

(2) Back-to-back contracts with name borrowing. A back-to-back main and sub-contract may entitle the sub-contractor to claim against the employer using the main contractor's name. Questions often arise here concerning the extent to which the main contractor must cooperate with the sub-contractor, the liability of the main contractor for costs and the question of discovery of the sub-contractor's documents.[64]

(3) Agreement in standard rules to empower the arbitrator to consolidate arbitrations or to have hearings in a connected dispute heard concurrently.[65] This is workable provided that all parties have signed up to the same rules.

---

61 CECA Form of Sub-Contract, Clause 18. But there are limits to how long the sub-contractor can be made to wait: *Redland Aggregates Ltd* v. *Shepherd Hill Engineering Ltd* [1999] BLR 252 (CA). Here, the sub-contractor issued an arbitration notice against the main contractor. The main contractor issued a notice requiring that the sub-contract arbitration be heard together with the arbitration under the main contract. In fact, the main contractor took no steps to proceed with the main contract arbitration. The Court of Appeal held that there was an implied term that unless the main contractor took reasonable steps to proceed with the main contract arbitration, the subcontractor was not obliged to wait.
62 *Higgs & Hill Building Ltd* v. *Campbell Davis Ltd* (1982) 28 BLR 47.
63 *Multi-Construction (Southern) Ltd* v. *Stent Foundations Ltd* (1988) 41 BLR 98.
64 *Lorne Stewart Ltd* v. *William Sindall plc and NW Thames Regional Health Authority* (1986) 35 BLR 109.
65 Both the ICE Procedure and the CIMAR Rules provide that the arbitrator may hear connected arbitrations concurrently. They do not allow consolidation without the agreement of the parties. See Chapter 15.

*Binding arbitration agreements*    An 'arbitration agreement means an agreement to submit to arbitration present or future disputes'.[66] The agreement must be in writing;[67] this requirement is satisfied where the contract is made by reference to a written form of arbitration agreement or to a document which has the effect of making that clause part of the agreement.[68] Thus, where a civil engineering contract provides that 'the ICE Conditions of Contract, 7th Edition apply to this contract', the arbitration clause in that agreement is also incorporated.

*The jurisdiction of the arbitrator*    The arbitrator's jurisdiction in respect of any dispute depends upon three key tests each being met:

(1)   The arbitration clause must be wide enough to encompass the dispute. For example, if the clause is confined to disputes 'arising under the contract' then there is no jurisdiction for a claim based on misrepresentation; whereas if the clause is extended to 'disputes arising out or in connection with the contract' then a misrepresentation claim can be heard.[69]

(2)   The dispute must be immediately arbitrable. For example, all conditions precedent to arbitration (e.g. a referral of a matter to the arbitrator under Clause 66 of the ICE Conditions of Contract) must have been met.

(3)   The dispute must fall within the scope of the matters actually referred to that arbitrator. For example, if the dispute referred relates to 'a claim under Clause 12 of the Contract in relation to the tunnel works' the arbitrator will not have jurisdiction in relation to another structure.[70]

---

66   Arbitration Act 1996, Section 6(1).
67   Arbitration Act 1996, Section 5. 'The provisions of this Part [of the Act] apply only where the arbitration agreement is in writing ...'. However, the definition of 'in writing' is extremely broad and includes where the agreement is evidenced in writing.
68   Arbitration Act 1996, Section 6(2). *Black Country Development Corporation* v. *Kier Construction* (1996) 80 BLR 110.
69   *Ashville Investments Ltd* v. *Elmer Contractors Ltd* [1989] QB 488 (CA).
70   Note, however, that Rule 5.2 of the ICE Arbitration Procedure 1997 clothes the arbitrator with jurisdiction over all matters

   connected with and necessary to the determination of any dispute or difference already referred to him whether or not any condition precedent to referring the matter to arbitration has been complied with.

   The CIMAR Rules provide that, where a dispute has already been referred to him, the arbitrator may abrogate the requirement where a subsequent dispute is referred. See the commentary in Chapter 15.

The arbitrator is entitled to 'rule' upon his own jurisdiction, but this ruling may be challenged. The substance of the challenge is that the arbitrator is not empowered to hear a dispute, then provided the challenger raises the issue immediately he may seek to raise the matter in the court, either as a preliminary point of jurisdiction or as a defence to the enforcement of the award.

*The award and reasons* An arbitrator's decision is contained in his award. The award should be in writing and in accordance with the formalities set out in Section 52 of the Arbitration Act 1996. It should be supported by reasons. These must allow the parties broadly to see and understand the arbitrator's reasoning; they must be sufficient to allow the Court to see the legal reasoning involved.[71] The content will include:[72]

(1) recitals as to the arbitration agreement, the mode and date of appointment of the arbitrator, the procedure adopted and so forth
(2) a statement of the issues in the case
(3) the arbitrator's summing up of submissions of law and his decisions on the law[73]
(4) the arbitrator's summing up of the evidence and his findings of fact[74]
(5) a statement of the arbitrator's overall decision on the issues in the light of his findings of law and fact.

---

71  *The Nimenia* [1986] QB 802 per Sir John Donaldson M.R. at 807: a reasoned award is one which 'states the reasons for the award in sufficient detail for the court to consider any question of law arising therefrom'.

72  In *Bremer Handelgesellschaft mbH* v. Westzucker GmbH [1981] 2 Lloyd's Rep. 130 (CA): at 133 Donaldson L.J. made a broad statement as to the matters which a reasoned award should contain. He said that the key was to tell a story 'logically, coherently and accurately' and that an arbitrator's award differed from a judgment in that 'arbitrators will not be expected to analyse the law. It will be quite sufficient that they should explain how they reached their conclusion'. Nevertheless, he went on to say that any reasoning offered would not be unwelcome.

73  *Universal Petroleum Co. Ltd* v. Handels und Transportgesselschaft mbH [1987] 2 All ER 737 (CA) per Kerr L.J. at 748:

A reasoned award is usually requested in order to lay the foundation for a possible application for leave to appeal. An arbitrator should therefore remember to deal in his reasoned award with all issues which may be described as having a 'conclusive' nature, in the sense that he should give reasons for his decisions on all issues which lead to conclusions on liability or other major matters in dispute on which leave to appeal may subsequently be sought.

74  *JH Rayner (Mincing Lane) Ltd* v. Shaher Trading Co. [1982] 1 Lloyd's Rep. 632 per Bingham J. at 636:

arbitrators should set out what, on their view of the evidence, did or did not happen, and should explain succinctly why, in the light of what happened, they reached their decision.

## The role of the courts in supervising arbitrations

*Staying court proceedings*  An agreement to arbitrate does not preclude parties from litigating disputes; indeed any provision which purports to exclude access to the court is void.[75] Where one party to an arbitration agreement commences litigation proceedings, the court will, however, order a stay to arbitration on the application of the other party,[76] provided that application is made before taking a step in the proceedings. In such a case, the only grounds for refusing a stay to arbitration are where 'the arbitration agreement is null and void, inoperative, or incapable of being performed'.[77] The fact that the issue is so obvious that it can hardly be said that there is a dispute at all will not prevent the matter being stayed to arbitration.[78] Nor will the fact that arbitration is not immediately available to the plaintiff because the contract requires prior procedures to be effected be of any weight.[79]

*Removal of an arbitrator*  The court may remove an arbitrator upon the application of a party where: (*a*) there are circumstances which give justifiable doubts as to his impartiality, (*b*) where he does not posses the qualifications required by the arbitration agreement, (*c*) he is physically or mentally unfit or (*d*) he has refused or failed to conduct the proceedings properly with reasonable despatch.[80]

*Challenging the award*  The decision of the arbitrator is recorded in 'the award'. A party who is dissatisfied with the award may challenge it by alleging that:

---

75  However, an agreement to make the arbitrator's award a condition precedent to bringing an action is enforceable: *Scott* v. *Avery* (1856) 5 HLC 811 (HL).

76  Arbitration Act 1996, Section 9.

77  Arbitration Act 1996, Section 9(4).

78  *Halki Shipping Corporation* v. *Sopex Oils Ltd* [1997] 1 WLR 1268; *Hayter* v. *Nelson* [1990] 2 Lloyd's Rep. 265. Note that where a claimant seeks to enforce an adjudicator's decision by summary judgment this will normally form an exception to this rule.

79  Under the ICE Conditions of Contract, a stay may be granted before the matter is referred to the Engineer: *Enco Civil Engineering Ltd* v. *Zeus International Development Ltd* (1991) 8 Constr. L.J. 164.

80  Arbitration Act 1996, Section 24. See also the cases under the earlier legislation: *The Elissar* [1984] 2 Lloyd's Rep. 84 (CA); *Modern Engineering (Bristol) Ltd* v. *C Miskin & Son Ltd* [1981] 1 Lloyd's Rep. 135 (CA); see also *Lovell Partnerships (Northern) Ltd* v. *AW Construction plc* (1996) 81 BLR 83 (Commercial Court) where the arbitrator was removed for failing to crystallise the nature and terms of the issues to be determined so that the parties had no real opportunity to address him as to them.

(1)  The arbitrator has no jurisdiction.[81] The applicant must have raised his objection at the earliest moment or else he will have lost his right to object.[82]

(2)  There has been a 'serious irregularity' in the proceedings. The categories of serious irregularity are set out exhaustively in Section 68 of the Arbitration Act 1996. Again, the applicant must have raised his objection at the earliest moment or else he will lose his right of challenge.[83]

(3)  There is a serious error of law, which substantially affects the rights of a party and is either (*a*) obviously wrong or (*b*) is seriously doubtful and a matter of general public importance.[84] The parties may agree expressly to forego the right to appeal.[85]

In the case of a challenge on the grounds of serious irregularity or an appeal on a point of law, the application must be made within 28 days of the date of the award.[86] Where, for whatever reason, the parties are slow in collecting the award, time continues to run.

Where an award is successfully challenged, and the arbitrator has jurisdiction, the award will ordinarily be remitted back to him for his reconsideration unless the correction is so obvious that the court will make it or where the parties can no longer have faith in the arbitrator.

## The powers and responsibilities of the arbitrator

*Relevant skill and experience*  Where the arbitration agreement requires that the arbitrator must possess stated qualifications, the appointment of a person who does not have such qualifications is potentially voidable[87] unless the parties waive their objections, for example, by failing to make their objection at the earliest opportunity.[88] In any event a prospective arbitrator should consider whether he has the appropriate skills and training before taking an appointment. An arbitrator who fails to demonstrate the elementary skill and knowledge of an arbitrator may be considered incapable of properly conducting the proceedings.[89]

---

81   Arbitration Act 1996, Section 66(3).
82   Arbitration Act 1996, Section 73.
83   Arbitration Act 1996, Section 73.
84   Arbitration Act 1996, Section 69.
85   Arbitration Act 1996, Section 69(1).
86   Arbitration Act 1996, Section 70(3).
87   Arbitration Act 1996, Section 24.
88   Arbitration Act 1996, Section 73.
89   *Pratt v. Swanmore Builders Ltd and Baker* [1980] 2 Lloyd's Rep. 504.

*Impartiality*   An arbitrator must be impartial.[90] He should not have any commercial or other similar connection with either party which may give rise to a suspicion in the mind of a reasonable person that he may be biased.[91]

*General powers and duties*   The prime duty of the arbitrator is given in Section 33 of the Arbitration Act 1996. He shall:

(*a*)   act fairly and impartially as between the parties, giving each party a reasonable opportunity of putting his case and dealing with that of his opponent, and

(*b*)   adopt procedures suitable to the circumstances of the particular case, avoiding unnecessary delay or expense, so as to provide a fair mean for the resolution of the matter falling to be determined.

Subject to this and any express agreement made by the parties, the arbitrator is 'master of the procedure to be followed'.[92] Correspondingly, the parties 'shall do all things necessary for the proper and expeditious conduct of the arbitral proceedings'.[93]

*Civil engineering professionals and the use of specialist knowledge*   Technically skilled and experienced arbitrators are generally entitled to use their own specialist knowledge in making directions as to procedure[94] and reviewing the evidence.[95] However, two types of specialist knowledge— general and particular—must be distinguished.[96] An arbitrator may use his general knowledge freely; particular knowledge acquired by the particular arbitrator and peculiar to him must, however, be disclosed so that the parties may deal with it.[97] Section 34(2)(g) of the Act provides that an

---

90   Arbitration Act 1996, Section 1(a).

91   Specialist barristers frequently practice from the same chambers. Occasionally one barrister may appear as advocate, while another is arbitrator. Likewise, civil engineering arbitrators and advocates frequently engage in joint activities, such as writing books and learned institution committee membership. This should not generally cause any suspicion of lack of independence: See, for example, the Paris Court of Appeal's decision in *Kuwait Foreign Trading* v. *Icori Estero SpA*, reported in [1993] ADRLJ 167 supporting this view.

92   *Bremer Vulcan* v. *South India Shipping* [1981] AC 909, per Lord Diplock at 985 (HL).

93   Arbitration Act 1996, Section 40.

94   *Carlisle Place Investments Ltd* v. *Wimpey Construction (UK) Ltd* (1980) 15 BLR 109.

95   *Jordernson & Co.* v. *Stara, Kopperbergs Bergslag Atkiebolag* [1931] 41 Lloyd's Rep. 201; *Mediterranean & Eastern Export Co. Ltd* v. *Fortress Fabrics (Manchester) Ltd* [1948] 2 All ER 186.

96   *Fox* v. *Wellfair* [1981] 2 Lloyd's Rep. 514 (CA); *F R Waring (UK) Ltd* v. *Administracao Geral do Acucar e do Alcool* [1983] 1 Lloyd's Rep. 45.

97   *Top Shop Estates* v. *Danino* (1984) 273 EG 197 (1 EGLR 9).

arbitrator may, subject to the agreement of the parties, take the initiative in ascertaining the facts; this overturns the rule that an arbitrator must not gather his own evidence.[98] An arbitrator may receive evidence from experts, appointed by himself or the parties; expert evidence should be impartial.[99]

## International arbitration

Any international litigation must take place in national courts, usually in public and in the national language; arbitration is preferred because of the neutrality of the forum, privacy, the avoidance of language problems, the convenience and adaptability of the process.

*Applicable laws*   A number of different laws may be applicable to different facets of the arbitration.

(1)   The law of the contract. This is the law agreed by the parties or, failing agreement, the law which has the closest connection to the contract. This law governs the substantive rights of the parties.

(2)   The law of the arbitration agreement. This is normally the same law as governs the contract, but need not be. This law governs such matters as the validity of the agreement to arbitrate.

(3)   The law of the seat. This is normally the law of the place where the arbitration is held, although it is technically possible for parties to agree that the arbitration has a seat in Country X and to hold it in Country Y. This law governs the procedural aspects of the arbitration, including the rights to make applications to the court for interim or conservatory relief and the right to appeal against the award.

*International procedures*   Most international civil engineering contracts provide for arbitrations to be administered by international bodies, such as the International Chamber of Commerce or the London Court of International Arbitration.[100]

---

98   *Owen v. Nicholl* [1948] 1 All ER 707 (CA); *Fox v. Wellfair* [1981] 2 Lloyd's Rep. 514 (CA); *Top Shop Estates v. Danino* (1984) 273 EG 197 (1 EGLR 9); *Mount Charlotte Investments plc v. Prudential Assurance* [1995] 10 E.G. 129 (1 EGLR 15).

99   *Whitehouse v. Jordan* [1981] 1 WLR 246 (HL); *University of Warwick v. Sir Robert MacAlpine* (1988) 42 BLR 1; *The Ikarian Reefer* [1993] 2 Lloyd's Rep. 68 at 81.

100   A brief exposition of international arbitrations conducted under these rules is given in Chapter 15.

### Postscript: the rise and fall of the Crouch decision

Civil engineering contracts frequently contain an arbitration clause entitling the arbitrator to 'open up, review and revise' certificates and decisions of the Engineer. In *Northern Regional Health Authority* v. *Crouch*[101] it was decided that since (*a*) the parties have agreed to give the arbitrator this special power and (*b*) the function of a court is to enforce the parties' agreement rather than to operate it for them, the court cannot 'open up, review and revise' the engineer's certificates and decisions. Thus, where a party commences litigation and seeks a remedy which necessarily entails the opening up of a decision made under the contract, the court will not have power to do this.[102] The decision was criticised by many[103] but generally followed. It caused considerable difficulty in many cases. The starkest illustration of this came in 1996, in *Balfour Beatty* v. *Docklands Light Railway*.[104] The contract was in the ICE 5th Edition form with two major amendments: the Engineer's functions were carried out by the Employer and the arbitration clause was deleted. Because of *Crouch*, the Employer's opinion became the final word on what was due, since there was no arbitrator and the court could not open up, review or revise that opinion. Now, however, after 14 years, the House of Lords in *Beaufort Developments* v. *Gilbert-Ash*[105] decided that *Crouch* had been wrongly decided. Accordingly, in the absence of express words in the contract, the court will be able to open up, review or revise an opinion, certificate or decision.

## 6. Litigation

Litigation is the term used to describe dispute resolution in the courts. Civil engineering cases are normally heard in the Technology and Construction

---

101  [1984] QB 644 (CA).
102  A party in this situation was said to be 'Crouched'.
103  But by no means all.

> That wording ['open up review and revise any decision ... certificate ...' etc.] appears to me to show clearly that the parties' agreement is not just that they will use the arbitration machinery but that they are giving up the right to use any other means for determining the disputes covered by it. That wording does not exclude all possible resort to the courts; but it is quite inconsistent with allowing the courts an open jurisdiction to do that which an arbitrator is empowered to do

> per Lord McCluskey in *Costain Building and Civil Engineering Ltd* v. *Scottish Rugby Union plc* (1993) 69 BLR 85 (Scottish Court of Session, Inner House) at 107; a case on the 5th Edition of the ICE Conditions of Contract.

104  *Balfour Beatty Civil Engineering Ltd* v. *Docklands Light Railway Ltd* (1996) 78 BLR 42.
105  *Beaufort Developments (NI) Ltd* v. *Gilbert-Ash NI Ltd* (1998) 88 BLR 1 (HL).

Court (TCC)[106], a specialist list of the Queen's Bench Division of the High Court. Disputes are normally heard by a single judge, though they may sit with 'assessors' who are technical experts who advise the judge. Where either party wishes to challenge a judgment, they may appeal to the Court of Appeal. If not satisfied with that judgment, they may seek leave to appeal to the House of Lords; but this is not available as of right.

## Recent changes in the procedures of the court

*Prior to April 1999*   Prior to 1999, the Rules of the Supreme Court applied to all High Court Proceedings. These were published by Sweet & Maxwell together with commentary in a book with white covers and the rules were often known informally as the 'White Book'. The County Court Rules differed in a number of respects and were published by Butterworths together with commentary in a book with green covers and these rules were informally known as the 'Green Book'.

*The new rules*   In 1996, Lord Woolf completed a major review of civil procedure. As a result, Parliament enacted the Civil Procedure Act in 1997. This provided for a committee to be formed to draw up radical new rules of civil procedure which apply both to the County Court and the High Court. The resulting Civil Procedure Rules were published in early 1999 and came into effect on 26 April 1999. The White and Green Books are now obsolete.

*Innovations*   The major innovations of the Civil Procedure Rules include:

(1)   The court's duty to manage cases. Under the old rules, detailed procedures dictated the conduct of the case; now the judge must consider which procedures are appropriate.
(2)   Case allocation. Cases are allocated to a 'track'. There are three tracks—small claims, fast track and multi-track. Cases heard in the TCC are automatically allocated to the multi-track.[107]
(3)   Offers of settlement. While parties could previously make an offer of settlement, the benefits were largely confined to the defendant. Both parties are now encouraged to make offers of settlement.

---

106   Until late 1998 called the 'Official Referees Court'.
107   Part 49, Practice Direction—Technology and Construction Court, Paragraph 4.1.

*Describing and discussing the new Civil Procedure Rules*   The Civil Procedure Rules (CPR) are published in 51 'Parts'. For example, Part 36 describes
the procedure for making an offer of settlement and such an offer is known
as a 'Part 36 Offer'. Within each Part are to be found 'rules' and these may
have sub-rules. For example Rule 36.10(1) provides

> If a person makes an offer to settle before proceedings are begun which com-
> plies with the provisions of this rule [viz Rule 36.10], the court will take that
> offer into account when making any order as to costs.

### Action for damages—a brief outline

*Parties*   The person, firm or company making a claim is called the 'Claimant'.[108] The 'Defendant' is the person, firm or company defending against it.

*Starting proceedings*   CPR Part 7 sets out the procedure for starting a
claim. Rule 7.2(1) provides: 'Proceedings are started when the court issues
a claim form at the request of the claimant'. Civil engineering cases will,
ordinarily, be started in the Technology and Construction Court (TCC).
This is a 'specialist proceeding' within the meaning of Part 49 and, so, it is
necessary to examine the Practice Directions under Part 49 for the procedure to be followed. Paragraph 2.1 of the Practice Direction provides

> Before the issue of a claim form relating to a TCC claim, the claim form ...
> should, if it is intended that the case be allocated to the TCC, be marked in
> the top right hand corner 'Technology and Construction Court'.

Paragraph 2.4 of the Practice Direction says

> Where a claim form marked as mentioned in paragraph 2.1 is issued in the
> Royal Courts of Justice, the case will be Assigned to a named TCC judge (the
> assigned judge) who will have the primary responsibility for the case manage-
> ment of that case. All documents relating to that case should be marked
> under the words 'Technology and Construction Court' in the title, with the
> name of the assigned judge.

Although the Practice Direction does not say so, the same procedure
should also be followed when the claim is issued in a District Registry (i.e.
from the High Court Registry at a major legal centres outside London such
as Leeds, Newcastle, etc.).

*Documents defining the claim and the defence*   The claimant's case is
defined in his statement of case. The defendant's defence is set out in his

---

108   Formerly 'the Plaintiff' under the previous rules.

statement of case, entitled a 'defence'. A third document ('a reply') is usually also served by the claimant. Lawyers still call these documents by the traditional term, 'pleadings', although that term does not appear in the CPR. The CPR provides in Part 16 what must be included in each of these pleadings; for example, a claim for interest must be made in detail. Importantly, each pleading must also contain a 'statement of truth'. The Practice Direction to Part 22 sets out the required form of statement. In the case of a claimant's particulars of claim, this is

> I believe [or, The claimant believes] that the facts stated in the particulars of claim issued on [date] are true.

Where a lawyer signs on a claimant's behalf, he is taken to have explained the claim to the client and to have obtained the client's acceptance that he honestly believes the statement to be true. A similar regime prevails for a defence and a reply.

*Additional claims*    Additional claims may be made; for example the Defendant may make a counterclaim against the Claimant or the Defendant may make a claim against a third party. These are dealt with in Part 20 of the CPR and are called 'Part 20 claims'. The situation can become complicated in multi-party actions. Consider the following example: Wimbledon Water Company claims for defects in a pumping station against ABC Contractors Ltd. ABC Contractors Ltd then claims that any defects are the responsibility of its sub-contracted designer, Water Lines Ltd. Water Lines claims that the fault lies partly with a supplier of components Valve Tech. Inc. and partly also with Wimbledon Water who gave Water Lines incorrect information. When all the parties have been joined in, the names set out in the heading will be:

**Wimbledon Water Company**
> *Claimant and Part 20 Defendant (2nd claim)*

—and—
**ABC Contractors Ltd**
> *Defendant and Part 20 Claimant (1st claim)*

—and—
**Water Lines Limited**
> *Part 20 Defendant (1st Claim) and Part 20 Claimant (2nd and 3rd claims)*

—and—
**Valve Tech. Inc.**
> *Part 20 Defendant (3rd Claim)*

*Management of the case*   Within 14 days of the defendant acknowledging service of the claim or serving his defence, the claimant makes an 'application for directions'. A directions hearing will be fixed. Before this hearing, each party completes a 'case management questionnaire'.[109] These are exchanged and returned in advance. At the directions hearing, directions are given[110] as to disclosure of documents held by the other party, service of a Scott Schedule and the headings to be used, witnesses of fact, expert witnesses, inspections/samples and a pre-trial review.[111] The latter is a short hearing at which directions as to 'trial bundles' (i.e. files of documents to be available at trial), service of documents on computer disc, openings statements, etc., are given.

*Trial*   The hearing takes place before the trial judge. Parties are generally represented by advocates, who are barristers[112] or solicitors with rights of audience.[113] Advocates will open and present their clients' cases and will examine their own witnesses and cross-examine the opponent's witnesses. Proceedings are adversarial in nature; the judge listens to the arguments put and evidence adduced by the parties. After the hearing, the judgment is given. This may be spoken from notes or a typed draft; sometimes a written judgment may simply be handed down.

*Expedited forms of procedures*   In order to expedite the process for parties with an obviously meritorious case, a number of procedures have been devised:

(1)  Part 24 Summary judgment. This is available to a claimant where there is no real defence or to a defendant to have a claim with no merit dismissed. Where the application is made by the claimant the test is whether the defendant has 'no real prospect of successfully defending the claim or issue'. Where the application is made by the defendant, the test is whether the claimant has 'no real prospect of succeeding on the claim or issue'.

---

109  Part 49, Practice Direction—Technology and Construction Court, Appendix 1.
110  Part 49, Practice Direction—Technology and Construction Court, Appendix 2 sets out a template for directions.
111  Part 49, Practice Direction—Technology and Construction Court, Appendix 3 gives a questionnaire and Appendix 4 sets out a template for directions.
112  Also referred to as 'counsel'.
113  See also the Access to Justice Act 1999, Sections 44 to 48.

(2) Part 25, CPR 25.7 Interim payment. When (*a*) a defendant admits a claim or (*b*) the claimant has been given judgment in principle with damages to be assessed or (*c*) the court is satisfied that

> if the claim went to trial, the claimant would 'obtain judgment for a substantial amount of money …' then the court may order an interim payment of 'a reasonable proportion of the likely amount of the final judgment'.

*Arbitration applications*   Parties may under closely defined circumstances make an application to the court in respect of arbitration proceedings. This may include where an award has been made and a party challenges it on the grounds that there has been a serious irregularity[114] or as application for leave to appeal on a point of law.[115] Such applications are made in accordance with Practice Direction 49G.

---

114 Arbitration Act 1996, Section 68.
115 Arbitration Act 1996, Section 69.

# 12

# Particular types of civil engineering contract

In this chapter a number of contract types will be outlined. In particular:

(1)  traditional contracts, with Engineer's design and administration
(2)  contracts with contractor's design
(3)  international turnkey contracts
(4)  contracts with specialist management
(5)  term contracts
(6)  concession (BOT) contracts.

## 1.  Traditional contracts

### Historical introduction

The expression 'traditional contract' is widely used to describe an arrangement where the employer takes on a consulting engineer to advise him on all aspects of the scheme, to design the works and to administer the construction through to completion. The qualifier 'traditional' is somewhat misleading. These contracts originated in the eighteenth century and are a direct product of the separation of design and construction activities and the high status accorded to individual engineers.[1] The contracts were

---

1    Architects, likewise, attained high status and the RIBA Contract (which is now the JCT Contract) developed in much the same way as the ICE Contract described here. The JCT is also a traditional contract, with the Architect occupying a position of supremacy as the Engineer does in civil engineering contracts.

drawn up with the advice of these engineers and reflect the supreme position occupied by 'the Engineer' who was wholly responsible for the design, approval of complete work and certification for payment of that work. The same pattern was seen for both private and public schemes. To begin with contracts were highly individual.[2] Over time, however, more or less uniform features began to emerge. By the end of the nineteenth century, many local authorities (who had, by this time, taken over responsibility for the majority of civil engineering works), began to adopt 'standard forms'. Later, in the late 1930s, these were taken as the basis for an industry wide standard contract, described as the ICE Conditions of Contract.[3] This contract form is now in its seventh edition. The concept has been exported and the general shape of the ICE Conditions of Contract is seen in many contracts used overseas, not least the FIDIC Red Book Contract.

## The features of a traditional contract

*Outline procedure*   An employer appoints an engineer to be the lead consultant on the project. Other consultants may also be appointed directly by the employer, to work in consultation with and under the general coordination of the engineer. The project is developed, surveyed and designed by the engineer's staff, under his direction. When the design is complete, or nearly so, the engineer (on the employer's behalf) invites tenders for the work. The engineer appraises the tenders and selects a contractor. The employer then engages the contractor to do the work to the approval of the Engineer.

*Design responsibility and construction responsibility*   The key feature of a traditional contract is the almost complete separation of design and construction responsibility. The engineer assumes a responsibility to the employer to exercise 'reasonable care and skill'. The contractor agrees to

---

2    Brunel drew up particularly onerous contracts for the contractor: see *Ranger* v. *Great Western Railway* (1854) 5 HLC (HL) discussed in Chapter 7, Section 3, where a Brunel drawn contract contained an unenforceable penalty upon termination.

3    See Rimmer EJ, The conditions of engineering contracts, *Proc. ICE*, February 1939. (Rimmer was a civil engineer, QC and the co-editor of the 8th edition of *Hudson's Building and Engineering Contracts.*) The conditions were originally drawn up by the ICE and the Association of Consulting Engineers. For the second edition, the drafting team was joined by delegates from the Federation of Civil Engineering Contracts (which has reformed as the Civil Engineering Contractor's Association). These three bodies continue to be responsible for the 7th edition.

build what is shown on the drawings and described in the specifications. The contractor's design responsibility extends only to what is implied in the selection of good materials and in the provision of good workmanship.

*The 'Engineer' and the 'engineer'* The employer appoints an engineer (lower case 'e' to denote a general appointment) to carry out designs. When the contract is let, the engineer may become the Engineer (with upper case 'E' as described in the ICE and FIDIC Contracts) with set of clear and distinct roles to perform under the contract. The Engineer is a defined person under the contract and the functions of the Engineer must be undertaken by him in that capacity.

*Computing the price* Traditional civil engineering contracts, such as the the ICE Conditions of Contract, tend to use a measure and value scheme for computing the price. This is not essential to the traditional arrangement. In building work, the main traditional contract is based on a lump sum;[4] and in the process industry, a common traditional form of contract uses cost reimbursement.[5]

*Perceived problems with the traditional arrangement* Four major concerns are expressed over contract types where the Engineer is the administering professional. Firstly, consulting engineers tend to be learned in technical matters, but their political and strategic management skills have been questioned; many employers asks whether it is appropriate for control of financially and politically sensitive projects to be handed over to an independent engineer. Secondly, Engineers under a traditional contract do not set out to 'manage and coordinate' the project; they assume a rather lofty detachment and many employers do not find this acceptable. Thirdly, the Engineer's impartiality is often questioned, rightly or wrongly. It is said that he identifies too closely with the employer's camp and is thus unable to take impartial decisions as between employer and contractor. In any event, it is said, many important decisions he must take (e.g. over the contractor's right to claim additional money) involve an indirect challenge to the Engineer's own competence and that he cannot be trusted to be impartial in such a case. And fourthly, complete separation of design and construction means that opportunities for enhanced efficiency and buildability are lost.

---

4    The JCT Standard Form of Building, 1998.
5    IChemE Green Book.

Many of these concerns are overstated, but they are taken sufficiently seriously for many employers to experiment with new forms of contract.

*The ICE Conditions of Contract, 7th Edition*   This standard form is examined in detail in Chapter 13.

## 2. Contracts with contractor's design

*The separation of design and construction*   Contractors in many industries are familiar with the idea of contractor's design. This was so, even in the consulting engineer's heyday. Consider one example. A nineteenth century carriage was a complex piece of engineering; but a contract to provide a carriage was invariably undertaken on a design and build basis. Of course, carriages had evolved over the centuries, whereas most civil engineering projects of the industrial revolution demanded unprecedented innovation, requiring the application of skills possessed by a mere handful of engineers. There was only one viable pattern of contract for these work, namely contractors working to designs produced by an engineer, constructed under his direction. The situation today is very different. Contracting companies employ graduate engineers and many have specialist design offices. There is, in short, far less need, than hitherto, to maintain the historical distinction between design and construction.

*Benefits of contractor's design*   The benefits of contractor's design depend to a large degree on the specific circumstances of the project and the design arrangements. They may include:

(1)  Single point and clear cut liability. Defects of every character are the contractor's responsibility, unless he can show some special defence. Defects claims are more readily settled and, where resisted, it is easier and cheaper for the employer to succeed. Beneficial side effects include less conflict on site and a reduction of defensive and uncooperative modes of behaviour.
(2)  Enhanced design standard. Under a design appointment, the designer is obliged to do no more than exercise reasonable care and skill. This by no means guarantees that the works he designs will be suitable for their purpose. Under a design and construct arrangement, the completed construction must be, unless the parties agree otherwise,

reasonably suitable for its purpose. In practice many employers agree to a lower standard of design liability for commercial reasons.[6]

(3)  Opportunity for the contractor to enhance to the constructability of the project and enhance both time and cost efficiency. This advantage can be significant where the contractor is engaged as part of the team right at the outset, but may be very much diminished if the contractor is selected after the design is all but fixed.

*Descriptions of contracts with contractor's design*  Contracts with contractor's design are variously described as 'design and build', 'package deal' or 'turnkey'. The true meaning of a contract is not determined from its description, but from its terms, read as a whole.

> There is a tendency to designate certain types of contract by names such as 'design and build', 'turnkey' or 'package deal'. Such names may help to sell the services of those who put them forward. In law, they are, however, unhelpful in ascertaining the rights and obligations of the parties to them. They may, indeed, be misleading.[7]

*Employer's control and contractor's responsibility*  How can the employer retain sufficient control over the design, while at the same time passing full responsibility for it to a contractor? In many instances (e.g. in process plants), the employer's demand for control is answered by the stipulation of carefully defined performance criteria. In other cases, the employer wants a particular aesthetic look or structural arrangement (e.g. a cable stayed structure). Here, he initiates the design himself and then passes it over to the contractor, who assumes responsibility for it *ab initio*. Whilst this procedure can be very effective, it is not without its difficulties. For example, the contract must deal with the possibility of fundamental design defects arising while the design was under the employer's control; for example, if the specified cable stayed arrangement proves impracticable, how can the design be adjusted? Who must approve the new design? Will any necessary re-design be a breach by the contractor? If so, what damages flow? There are also practical shortcomings; indeed, many of the oft-cited benefits of the design–build arrangement may be lost because the project has, to all

---

6   The commercial reasons include the fact that many designers do not carry insurance for a fitness for purpose standard of care and this narrows the employer's market. Under the ICE Design and Construct Contract the standard is explicitly 'all reasonable skill care and diligence'. The GC/Works/1 Single Stage Design and Build (1998) Contract provides that the employer may select in Clause 10(2) from Alternative A, which is essentially reasonable care and skill or Alternative B, which is fitness for purpose.

7   Editors' commentary to *Viking Grain Storage* v. *TH White* (1985) 33 BLR 103 at 104.

intents and purposes, been commenced on a traditional basis and the description 'design and build' is used to describe a legal rather than an engineering concept.

*Contractor's design and contractor's responsibility for design*    The essence of a contract 'with contractor's design' is not that the contractor necessarily does the design, but that he is responsible for it. He may not design the works because he sub-contracts the design to a specialist firm. But, very commonly, the reason is that the employer's own designer has largely completed the design before the contractor ever becomes involved; as part of his tender, the contractor checks the design before agreeing to take over responsibility for it. But his practical (as opposed to legal) involvement in the design is minimal. In principle, the contractor's liability for the design is not affected by whether or not the contractor actually carries out the design. What is important is whether or not he has agreed to take responsibility for it.

*A classification of contracts with contractor's design*    The quality and extent of a contractor's design obligation cannot be expressed as a single measure, but consists of a number of discrete characteristics:

(1)  The scope of the contractor's design responsibility. Some contracts provide that the contractor is to be responsible for the design of the entire works, while others limit his obligation to particular structures or aspects of the work.
(2)  The degree of contractor's involvement in and control over the design. A contractor who has won a competitive design competition may be appointed to undertake the whole of the design, through from the initial concept to the final details. At the other extreme, the contractor may assume responsibility for a design which has been worked up in detail by the employer's designer and in which he has little or no input.
(3)  The standard of obligation. In some contracts, the contractor warrants that the design is done with reasonable care and skill. In other contracts, he warrants that the finished product is reasonably fit for all its known purposes.

## The obligation to design

*Express and presumed obligation to design*    The contractor's obligation for design will normally be set out expressly. Where the circumstances indicate it, the obligation to design may also arise from the presumed intention of the parties. In one case,[8] a shipyard drew up an outline design for a propeller; the thickness of the medial lines was given and the note on the drawing read 'edges to be brought up to fine lines'. The manufacturer cast the propeller. When put into operation, it did not perform satisfactorily and the ship was refused Lloyd's classification. It was held that the manufacturer was obliged to provide a propeller which was fit for its purpose and this required such design as was necessary to achieve this.

*Agreement to design*   Where the contract shows clearly that the contractor is to take full responsibility for the design, then he will be liable even for defects in the design which which arose before he was involved.[9] Where the contract contains contradictory provisions, these may allow the contractor to avoid liability. But where the purpose of the contract is clearly to place the burden for design upon the contractor, then the court will not shrink from enforcing this intended result. In an offshore civil engineering case,[10] for example, the primary terms of the contract clearly stated that the contractor was to take responsibility for all the design *ab initio* and to comply with detailed performance requirements. The contract documentation included a partial design, prepared by the employer's design team. There were defects in that design, which caused the contractor significant technical problems. It was held that the contractor was responsible for those defects and was required to provide a product which met the agreed performance requirements without additional payment. Where, however, a contractor is obliged to adopt designs carried out by a third party, and has no opportunity to check nor right to object to them, clear words indeed will be required to make the contractor liable.[11]

---

8    *Cammell Laird & Co. Ltd* v. *Manganese Bronze and Brass Co. Ltd* [1934] AC 402 (HL). In house building cases, the builder normally assumes responsibility for the design; but see *Lynch* v. *Thorne* [1956] 1 WLR 303 (CA) which is thought to be wrongly decided.

9    *Davy Offshore Ltd* v. *Emerald Field Contracting Ltd* (1991) 55 BLR 1.

10   *Davy Offshore Ltd* v. *Emerald Field Contracting Ltd* (1991) 55 BLR 1.

11   *Norta Wallpapers (Ireland) Ltd* v. *John Sisk & Sons (Dublin) Ltd* (1978) 14 BLR 99 (Irish Supreme Court). In this case, the contractor was obliged to use a 'patent' systems roof to be designed, supplied and installed by a specified company. Since there was no right to object it was held that the presumed intention was that the contractor was not to be liable for the design. This is an unusual case. The opposite, and more usual, result was achieved in *IBA* v. *EMI Electronics Ltd and BICC Construction Ltd* (1980) 14 BLR 1 (HL). Lord Fraser observed at page 46:

   In the present case, although EMI had no option but to appoint BICC as sub-contractor for the mast, they were not bound to accept any particular design at any particular price. If they had checked BICC's design and had considered it unsatisfactory they would have been entitled to insist on its being improved.

   The fact that he does not avail himself of the opportunity to check the design and object will not, it is thought, enable him to avoid liability.

*Express agreement as to defective employer-provide design*   Many contracts provide specifically for where there are defects in the preliminary design supplied by the employer.[12] Agreed provisions normally entitle the contractor to an instruction, which may or may not, depending on the circumstances, carry an entitlement to additional payment.[13]

*Unattainable or impracticable design requirements*   Where the employer stipulates aspects of the design which are unattainable or impracticable, the position (in the absence of express agreement) will turn on the presumed intention of the parties.[14] Generally, where the contractor has taken over responsibility for the design *ab initio*, he will be responsible for the bad design commissioned by the employer.[15] But, in the absence of clear words to the contrary, a number of apparent exceptions may exist:

(1)   Where the defect is not a design defect at all. For example, where the stipulated design capacity of a reservoir is based on incorrect survey data, acceptance of responsibility for 'design' may be interpreted not to include acceptance of liability for preliminary survey work.

(2)   Where the contractor has neither opportunity nor entitlement to object to the design.[16]

(3)   Where the structure is impossible to build as designed. For example, the contractor may assume responsibility for the design of a light

---

12   For a general discussion see Hammond CG, Dealing with defects: defective owner-provided preliminary design in design–build contracting [1998] ICLR 193.

13   Under the ICE Design and Construct Conditions of Contract, for example, defective design in the Employer's Requirements may constitute a 'discrepancy', entitling the contractor to an instruction. Whether or not he is entitled to additional cost and an extension of time depends on whether

> such instructions delay or disrupt his arrangements or methods so as to cause him to incur cost beyond that reasonably to have been foreseen by an experienced contractor at the time of the award of the Contract.

Clause 5(1)(c)(ii).

14   McRae v. *Commonwealth Disposals Commission* (1950) 84 CLR 377 (High Court of Australia) per Dixon and Fullagar J.J.

> The common law has generally been true to its theory of simple contract, and it has always regarded the fundamental question as being: 'what did the promisor really promise?' Did he promise to perform his part in all events … ? So questions of intention or 'presumed intention' arise, and these must be determined in the light of the words used by the parties and reasonable inferences from all the surrounding circumstances … the problem is fundamentally one of construction'.

15   *Davy Offshore Ltd* v. *Emerald Field Contracting Ltd* (1991) 55 BLR 1.

16   *Norta Wallpapers (Ireland) Ltd* v. *John Sisk & Sons (Dublin) Ltd* (1978) 14 BLR 49 (Irish Supreme Court).

weight cable stayed structure, which appears stable at the date of the contract using conventional calculations. If later wind-tunnel tests show the concept to be fundamentally flawed (rather than requiring adjustment) then this may amount to frustration.[17] There can be no frustration if the matter is dealt with by express terms,[18] including where, on its true interpretation, the contract requires the employer to issue instructions as to what is to be done.[19]

## The standard of design responsibility

*The standard of care: general*  Where a contractor agrees to design and construct works, his obligation is (in the absence of express terms) to provide works which are fit for their known purpose.[20]

> In the absence of a clear contractual indication to the contrary, I see no reason why one who in the course of his business contracts to design, supply and erect a television aerial mast is not under an obligation to ensure that it is reasonably fit for the purpose for which he knows it is intended to be used. ... Counsel for the [design–build contractor], however, submitted that, where a design, as in this case, requires the exercise of professional skill, the obligation is no more than to exercise the care and skill of the ordinary competent member of the profession. However, I do not accept that the design obligation of the supplier of an article is to be equated with the obligation of a professional man in the practice of his profession.[21]

*The design gap: express contract terms as to standard of liability*  Many contractors are reluctant to assume a responsibility to provide works which are fit for their purpose, which leaves them exposed to the employer and

---

17  In *Thorn* v. *Mayor and Commonalty of London* [1876] 1 AC 120 (HL) a construction technique suggested by the employer's engineer was shown on contract drawings. The House of Lords held that this was a matter at the contractor's risk. But Lord Cairns suggested if the difference between what was envisaged and what was possible is 'so peculiar, so unexpected, and so different from what any person reckoned or calculated upon, that it is not within the contract at all'.

18  *McAlpine Humberoak* v. *Mc Dermott International* (1992) 58 BLR 1 (CA).

19  In *Davy Offshore Ltd* v. *Emerald Field Contracting Ltd* (1991) 55 BLR 1, the employer's defective partial design was included in the contract. The contractor argued that there was an impasse: if he executed the defective design he was in breach; if he changed it, he was in breach. He claimed to be entitled to a variation (and, of course, to be paid for it). Held that the scheme of this contract showed that the contractor was to take full responsibility for the design and he was not therefore entitled to a variation. But other contracts may produce a different result.

20  *IBA* v. *EMI Electronics Ltd and BICC Construction Ltd* (1980) 14 BLR 1 (HL).

21  *IBA* v. *EMI Electronics Ltd and BICC Construction Ltd* (1980) 14 BLR 1 (HL) per Lord Scarman at 47.

deprives them of a similar cause of action against their own designers.[22] Accordingly, many design and construct standard form contracts specifically limit the contractor's responsibility to the provision of a design which is done with reasonable skill and care.[23]

*Reasonable suitability for purpose*   Where the obligation is to design the works to be suitable for their purposes, those purposes should be made clear in the contract. Where, the works only have one proper purpose,[24] then defects which render the works unsuitable for that purpose are clearly breaches. Where the works may have more than one purpose (e.g. a warehouse which can be used for storing a variety of items) it is thought that, unless the parties agree otherwise, the court will not burden the contractor with having to ensure suitability for unusual purposes.

*Selection of materials*   Where the agreed design standard requires the contractor to exercise reasonable care and skill, he will nevertheless be responsible for selecting suitable materials, unless the contract expressly states otherwise.[25]

> [A] person contracting to do work and supply materials warrants that the materials which he uses will be of good quality and reasonably fit for the purpose for which he is using them unless the circumstances of the contract are such as to exclude any such warranty.[26]

Where the employer specifies particular materials, the contractor will remain responsible for any lack of quality in them, even where there is only one source;[27] in such a case he will not, however, warrant (unless the

---

22   It may be possible to require the designer to contract on a fitness for purpose basis; however, many are unwilling to do this either for professional or insurance reasons. Note also, that in some cases, a designer may owe a fitness for purpose design duty: see *Greaves & Co.* v. *Baynham Meikle* [1975] 1 WLR 1095 (CA).

23   Including the ICE Design and Construct Conditions of Contract. The GC/Works/1 Contract (1998) contains an optional Clause 10(2) which allows the employer to specify either a 'reasonable care and skill' or a 'fitness for purpose' type responsibility.

24   For example, *Viking Grain Storage* v. *TH White* (1985) 33 BLR 103 (grain storage depot); *IBA* v. *EMI Electronics Ltd and BICC Construction Ltd* (1980) 14 BLR 1 (HL) (television aerial mast).

25   The ICE Design and Construct Contract provides that selection of materials is to be done with 'all reasonable skill, care and diligence'—Clause 8(2)(a). The GC/Works/1 Single Stage Design and Build (1998) Contract provides in Clause 31(2) that materials, except for those selected by the employer, will be fit for their intended purposes.

26   *GH Myers & Co.* v. *Brent Cross Service Co.* [1934] 1 KB 46 per du Parcq at 55.

27   *Young & Marten Ltd* v. *McManus Childs Ltd* (1969) 9 BLR 77 (HL). Here, the employer specified a particular brand of roof tiles, available only from one supplier. It was conceded by Counsel and accepted by the House of Lords that this excluded any warranty of fitness for purpose.

contract states otherwise) that the materials are reasonably suitable for their purpose.[28] Where, the provision of materials requires a particular design input, the position is uncertain. For example, a concrete mix is not merely selected, but designed; likewise, many new materials, including plastic composites, require advanced design.

*Construction (Design and Management) Regulations 1994*   Under the CDM Regulations, the designer must ensure that any design he prepares pays proper regard to risks during and after construction and gives priority to measures which will protect people. Where the employer's requirements contains a provision that the design will accord with the Regulations[29] and the works are not safe, e.g. for cleaning, as required by the Regulations, this may amount to a 'defect' entitling the employer to undertake 'remedial works', the cost of which may be claimed from the contractor.

## Novation of designers

Employers frequently appoint designers to initiate the design. At tender stage, the employer may stipulate that the successful tenderer will be obliged to take over the appointment of the designers. This requires the prior approval of the designers, usually obtained at the time of their initial appointment.

*The benefits of novation*   Benefits of novation may include (*a*) the retention of the designer's experience of the scheme; (*b*) avoidance of a 'liability gap' where a new designer blames the first designer for not providing a good basic design and the first blames the new designer for failing to understand and/or implement the design properly; (*c*) added incentive for the designer, who knows that his involvement in the project will be a continuing one.

*The mechanism whereby the contractor takes over the appointment*   The legal device used is novation.[30] This involves the dissolution of the contract between the designer and employer and the creation of a contract between

---

28   In *Young & Marten Ltd* v. *McManus Childs Ltd* (1969) 9 BLR 77 (HL) the employer was able to recover from the contractor for the lack of quality resulting from a latent defect in roof tiles which had been specified by the employer.

29   Note that the breaches of the Regulations, except in two specific cases, do not confer civil liability—Reg. 21. Hence the need for this term to be agreed.

30   See Chapter 6, Section 5, for novation generally. See McNicholas P, Novation of consultants to design build contractors (1993) 9 Constr. L.J. 263.

the designer and contractor. This is usually achieved using deeds to avoid any question of lack of consideration.

*Contractor's concerns as to novation*    As far as the contractor is concerned, the novation should preserve his reasonable rights against the designer. He will wish to ensure:

(1)   That the duration of designer's appointment is coterminous with the duration of design liability under the civil engineering contract. For example, if the contractor is to be responsible for the design *ab initio*, the designer's appointment should also be novated to the contractor *ab initio*. Likewise, if the contractor's contract with the employer is under deed (with a 12 year limitation period) then the agreement between the contractor and designer should also be by deed.

(2)   That, insofar as it is commercially practicable, the designer's liability to the him is coextensive with his liability to the employer. For example, if the designer's appointment is on a reasonable care and skill basis but the contractor's liability is on a fitness for purpose basis, the contractor has taken on an added exposure. This consideration is extremely important where key aspects of the design involve innovation; here the potential gulf between reasonable care and skill and suitability for purpose is greatest. In practice, it will normally be difficult to enhance the designer's standard of obligation at the time of novation.

(3)   That the damages claimable by the contractor against the designer relate specifically to the losses which are suffered by the contractor, including the express likelihood of damages payable to the employer.

(4)   That the designer's insurances are good and secure.

*Employer's concerns as to novation*    As far as the employer is concerned, the novation should not allow any uncovered risk. Considerations include:

(1)   Ensuring that there is a complete and absolute transfer of the designer's obligation to the contractor. It will not readily be presumed that there is any enlargement of the designer's obligation, so any obligations retained may not be transferred as intended.

(2)   Requiring a direct collateral warranty (supplementary to the novation) from the designer in the event of the contractor's inability to meet any claim.

(3)   Ensuring that it is possible to assign the substantial benefit of the construction contract (including rights under the direct warranty).

(4) Ensuring that the designer's insurances are good and secure.

*Novation and the designer's performance* The designer's performance may prove less than satisfactory for a variety of reasons, ranging from insolvency to incompetent work. The contractor's remedies, if any, will depend upon an interpretation of the contract. Where the designer is unable to perform properly, it is thought that, in the absence of agreed terms, the contractor may appoint a substitute designer of his choice; the situation will normally differ from a nomination in which the employer is to appoint a replacement.[31]

## Tendering practice

Under a traditional contract, the employer (or his engineer) will have completed (or virtually completed) the design at tender stage. The employer will be able to approve the final drawings and specifications prior to tender. The drawings and specifications are passed to the tenderers who price the works. Traditionally, a bill of quantities is provided, showing the quantities, or at least their approximate values. Under a design and build contract, the position can be very different.

The tender process can be described in terms of the number and types of stages it involves.

*Single stage tendering* This is where the contractor receives documents and submits his tender based on those documents. Although described as 'single stage' there is often a significant period of negotiation following the submission of tenders. A single stage process is appropriate where the employer's requirements are purely functional (e.g. a temporary haul road) or the employer has already identified those features over which he wants strict control and has commissioned an outline design which fixes them. Tender documents for a single-stage design and construct project will include the 'employer's requirements'. In his submission, the contractor offers the 'contractor's proposals'. The former describes the criteria which the employer required the design to meet and the latter sets out the ways in which the contractor proposes to meet them. Many design and construct contracts[32] have a defined order of precedence in which the employer's requirements take priority over the contractor's proposals.

---

31 See *Bickerton v. North West Metropolitan Regional Hospital Board* [1970] 1 WLR 607 (HL).
32 Including both the GC/Works/1 Single Stage Design and Build (1998) Contract and the ICE Design and Construct Conditions of Contract.

*Multiple stage tendering*   Where the employer wishes the contractor to supply the conceptual as well as the detailed design, a multiple stage approach is inevitable. Such cases include utility or process contracts where the employer relies on the contractor's expertise[33] and design competition projects.[34] Tendering contractors often compete through a number of distinct stages; competitive qualification for each subsequent round may be judged using aesthetic, environmental, safety, likely cost, risk and time criteria. In the final stage, where a significant element of detailed design is required, unsuccessful finalists may in some instances be partly compensated for their efforts during this final stage.

## Payment for contracts with contractor's design

*Payment approaches*   Many design and build contracts use a lump sum payment arrangement. But this is by no means universal. Cost reimbursable contracts may be appropriate where a 'fast track' approach is required. And where the design is essentially fixed before construction, a measure and value approach may be a sensible approach.

*Payments under a lump sum arrangement*   A number of methods of computing interim payments are frequently used, including:[35]

(1) *Milestone payments.* The contract contains a schedule of tasks against which there are agreed payments. The right to payment for each task matures when that task is substantially complete. The contractor may be entitled to be paid when he has completed the task. Alternatively, the contract may set periodic payment dates; payment is made for all tasks completed within that interval.

(2) *Progress schedule.* The contract lists out a schedule of tasks against which there are agreed payments. The contractor is entitled to periodic payments for all tasks in proportion to the degree of completion.

(3) *Budget schedule.* The contractor is paid in accordance with a pre-agreed plan. The payment is thus made without regard to the value of the work actually completed. This approach can, technically speaking, lead to overpayment but the schedule is usually drawn up so

---

33   See section 3 below for international turnkey contracts.
34   These are extremely common in Continental Europe and becoming more so in the UK.
35   See, for example, the GC/Works/1 Single Stage Design and Build (1998) Contract: Clause 48 provides that each of these three methods may be used as an alternative.

that this is unlikely unless progress is so slow that the determination provisions can be activated.

The choice of payment method must take into account the cost of administering it. Method (1) above is extremely simple to operate, provided that a fair definition of 'substantially complete' is used and provided also that the contract makes provision for what is to happen if a task is delayed by a matter for which the employer bears responsibility. Method (2) entails a significant amount of measurement. Method (3) is simple and straightforward. But in order to ensure that the contractor is never overpaid, the payments must be designed so that they do not cover the cost of the projected work; accordingly, the contractor is obliged to run the project in a cash flow deficit. In the case of contracts for work in Britain, the payment provisions must comply with the Housing Grants, Construction and Regeneration Act 1996.

*Additional payments* In addition to the agreed tender sum, the contract must make provision for variations, prolongation and disruption and other events entitling the contractor to additional payments under the contract.

## 3. International turnkey (engineer–procure–construct) contracts

*Terminology* There is no standard definition of a turnkey contract; the origin of the expression is obscure.[36] In international civil engineering practice, it tends to mean a contract where the contractor 'engineers, procures and constructs' the facility in a state ready for operation and to a standard which meets pre-set performance specifications. A turnkey contract is, thus, a species of design and build contract; but in the international arena, the concept goes beyond a mere design and construct contract. It suggests a scheme in which there is significant utilisation of intellectual property (e.g. patented technologies) and 'know-how' by the contractor.[37] Turnkey projects usually combine major civil engineering works with mechanical and

---

36  It seems most likely that it simply refers to what the employer will have to do to get a functioning project, viz 'turn the key'. The expression 'clé-en-main' appears in French which gives credence to this theory. The term 'package deal' is often used to denote the same idea.

37  This aspect is stressed by Schneider ME, Turnkey contracts: concept, liabilities, claims [1986] 3 ICLR 388; Franklin SM, Critical issues in turnkey contracts for heavy plant [1990] 7 ICLR 269; and Wiwen-Nilsson T, Supply of technology and specifications in turnkey heavy plant contracts [1990] 7 ICLR 282.

electrical installations. Frequently, there is also involvement by the contractor in the finance of the project (even if only by deferred payments) and in passing on skills to the employer's workforce during and after completion.

*When are turnkey contracts used?*   Turnkey contracts are widely used for major utilities projects with an international dimension. They are coming to be used more and more for major infrastructural projects. Development agencies, such as the World Bank or the European Bank for Reconstruction and Development, prefer to minimise their risk exposure and, for that reason, prefer turnkey arrangements. But even where governments are able fully to finance major plant or infrastructural projects themselves, they prefer to adopt a turnkey arrangement because of the single-point liability and the opportunity to train local staff.

*Standard form turnkey contracts*   The are several recognised forms of turnkey contract, including those drawn up by:

(1)   The Fédération Internationale des Ingénieurs-Conseils (FIDIC). This is an umbrella body for associations of consulting engineers. It formerly published an 'Orange Book' which was used as the basis of turnkey contracts. In September 1999 it published its Silver Book Contract which is designed specifically for turnkey projects.

(2)   The Engineering Advancement Association of Japan (ENAA). This is an association of Japanese engineering companies. The Japanese were the pioneers of turnkey contracting and ENAA's Model Form of International Contract for Process Plant Construction (Turnkey Lump Sum basis) 1986 was the first internationally recognised contract specifically for turnkey works.[38] This was updated in 1992 and in 1996 a new ENAA Model Form for Power Plant Contracts was published.[39] The ENAA contracts are widely used on World Bank projects.

(3)   European International Contractors (EIC). This is an association of 15 European construction industry federations. In 1995 it published its turnkey contract,[40] which is also widely used.

---

38   Gould G, Some comments on the policies and drafting of the ENAA Model Form Contract for Process Plant Construction [1988] 5 ICLR 205.

39   Hoshi H, ENAA Model Form of Contract for Power Plant Construction [1997] 8 ICLR 61; Wiwen-Nilsson T, The 1996 Edition of the International Contract for Power Plant Construction—a brief review [1997] 8 ICLR 273.

40   Goedel J, The EIC Turnkey Contract—a comparison with the FIDIC Orange Book [1997] 8 ICLR 32.

*Tendering practice* One of the disadvantages of turnkey contracting is that submitting a tender involves very significant cost. The tenderers need to be reassured that there is a 'level playing field'.[41] Furthermore, because each contractor may approach the project in a very different way, it is not easy to compare bids. Unless some pre-qualification occurs, well-experienced and resourced contractors may feel themselves at a disadvantage because their superior designs may lose out to underdesigned bids by others. Furthermore, unless bona fide, but unsuccessful, bidders receive some recompense, the bids will not be properly thought through. In order to resolve these difficulties, a number of tendering strategies have been suggested. The World Bank suggests a two-stage process; the first is prequalification and the second is the submission of detailed bids by a small number (three or less) contractors. Others have suggested more sophisticated systems. For example, one commentator[42] proposes a five stage process:

(1) *Pre-feasibility stage*: here the employer's engineers investigate existing technologies and prepare a report setting out the basic features, costs, risks and rewards of the project
(2) *Feasibility*: the employer invites interested firms to demonstrate their capacity and experience and to comment upon the pre-feasibility report
(3) *Bidding*: three or so contractors are invited to prepare detailed bids on the basis of objectively defined criteria, with each contractor being paid a fixed fee to part-compensate for the costs of bidding
(4) *Evaluation of bids and provisional award*: the most advantageous bidder is provisionally offered the job 'subject to contract'
(5) *Negotiation*: final details are ironed out, including technical clarifications, definition of scope, etc. If a deal can be struck, the contractor is formally awarded the contract.

A potential method of ensuring an objective selection process once a small group of contractors has been pre-qualified is to fix the key variable, whether it be price or output. For example, if the project is for a power plant, the price may be fixed at $x million. A formula is issued to each tenderer by the employer which gives an overall score based on objectively measurable quantities, such as: time to completion (months), power

---

41 Rosenburg G, International construction procurement—the developing regulatory framework [1997] ICLR 168. This deals with the guidelines for procurement under IBRD and IDA credits. The guidelines are not binding, but advisory.
42 Westring G, Turnkey heavy plant contracts—from the owner's point of view [1990] 7 ICLR 234 at 235.

outputs (GW averaged over the first year of operation), fuel consumption (e.g. tonnes of coal measured over the first year of operation), maintenance input (man hours measured over a five year period), etc. The contract contains 'liquidated damages'[43] provisions for failure to meet the bid performance levels. The contractors' bids are assessed against the formula and the winning bid is the one which scores highest.

*The governing law*   When establishing any international contract, it is important that the governing law of the contract is clearly defined.

*Defining the scope*   One of the key features of a turnkey contract is that the contractor is, subject to well-defined risks, liable for all shortcomings in the works. However, this only applies to matters within the project scope. As a result one finds that many disputes in relation to turnkey contracts are, in essence, disputes about what is and what is not within the scope of the project. Examples include: (*a*) where a contractor has to abstract cooling water from a watercourse and that abstraction causes consequential erosion in the existing watercourse, is the contractor responsible for rectifying that problem? (*b*) where the geometry of the site forces the contractor to cut a slope at a particular angle, at which angle it fails due to a deep-seated failure, do the contractor's responsibilities extend to deep-stabilisation of slopes? It is important, therefore, for the parties to agree clearly the extent of the project scope.

## 4. Contracts providing for a specialist management function

*The need for a specialist management function*   Traditional contracts assume a particular, and relatively simple, management and time structure. An engineer considers, surveys and designs the scheme. A single contractor then takes on responsibility for the works, sub-contracting where necessary and always with the approval of the engineer. In the 1960s, employers became frustrated at the time taken by sequential design and construction and by the lack of control they appeared to have during the construction phase of the works.[44] Employers in the oil and gas industries had, for many

---

43   Optional performance provisions are more secure than liquidated damages provisions: see Chapter 5, Section 5.

44   See, for example, *Greater London Council v. Cleveland Bridge Engineering Co. Ltd* (1984) 34 BLR 50 where the question was whether the contractor was obliged to proceed with due speed.

years, used contracts which explicitly enhanced the role of a project manager. These models led traditional construction employers to experiment with contract management arrangements. In the early 1980s, particularly in the building industry, a variety of new management based contract arrangements were developed and used extensively.

*The description of a management contract and its effect in law*   Contracts with a specialist management function often bear names such as 'management contracting' and 'construction management'. These descriptions have, to some limited degree, acquired a recognised meaning. The effect of a contract at law, however, is derived from an analysis of its terms taken as a whole and this is nowhere more true than for a management contract. Descriptions such as 'management contract' are insufficiently fixed or certain to render any assistance whatever in interpreting the contract and can be misleading because they suggest that because two contracts are so described, then the same results and consequences follow. In fact, two documents, both described as 'management contracts' and both fitting the general descriptions of such contracts, can, in the most fundamental respects, produce quite different obligations.[45]

*Classifying management contracts*   The terminology used is by no means uniform. In *Department of National Heritage* v. *Steensen Varming Mulcahy*[46] the following classification was suggested:

There are four main forms of management contract.

1.  *Management Contracting* where all contractors enter into a contract with the Management Contractor. The Managing Contractor is responsible for procurement and financial control and each contractor is paid by the Managing Contractor. The Managing Contractor is paid a fee by the Client together with the costs of all the contractors.
2.  *Project Management Contracting* where the Project Manager provides a technical service to the client and manages work on behalf of the client. The Project Manager has full control of budget and programme and is paid on a staff reimbursement basis.
3.  *Design and Management* is similar to Project Management with the addition of design responsibility.

---

45   *Copthorne Hotel (Newcastle) Ltd* v. *Arup Associates* (1997) 85 BLR 50 (CA) and *Chester Grosvenor* v. *Alfred McAlpine* (1991) 56 BLR 115. Counsel in the former case argued that the liability of the management contractor under each contract should be the same. Judge L.J. at 57 described the resemblance as 'superficial'.
46   Official Referees Court, 30 July 1998, per Judge Bowsher at paragraph 82.

4.  *Construction Management* [where] the client enters directly into separate contracts with the designers, Construction Manager and the the Works or Trade Contractors, and the Construction Manager manages the Works Contractors on behalf of the client.

This classification covers the basic approaches by which a specialist manager may participate formally in a project. It also demonstrates the overlap between the various schemes.

*Management Contracting*  The arrangement known commonly by this description is the most strikingly original form. The other schemes are all based on the traditional contract arrangement with the manager being vested with the supremacy formerly reserved for the engineer.[47] In management contracting, however, there is a profound shift. The contractor becomes the key ally of and adviser to the employer. Unlike a traditional contractor, who may not sub-contract work without consent, here the management contractor is obliged to sub-contract all field work. The basic features of this approach are set out in the following extract from *Copthorne Hotel v. Arup Associates*:

> Under an orthodox contractual regime for construction work, the main contractor is typically responsible to the employer for the due and timely execution of the contract works and liable in damages for breach of contract. If any such breach arises from the act or default of a sub-contractor that is no defence to the employer's claim.
>
> It is not in dispute that the form of management contract in use here modifies that regime and those consequences to some extent. In place of a "main contractor" with the primary responsibility of executing the works and liberty to have part carried out by "sub-contractors" there is a 'management contractor' with the primary obligation of ensuring that the works are executed and a duty to achieve that end by letting out the whole of the works in "packages" to "works contractors." Apart from those differences of function and terminology there a shift in liability. Provided the management contractor complies with certain procedural and other requirements, its liability to the employer for some breaches of contract is limited to the amount which it recovers from the works contractor at fault.[48]

In order to achieve this effect a number of contract provisions are required. The terms of the management contract: (1) require the

---

47    In *Bernhard's Rugby Landscapes Ltd* v. *Stockley Park Consortium Ltd* (1997) the contract was based squarely on the ICE Conditions of Contract, 5th Edition, with the Engineer replaced by a Project Manager; other necessary amendments were also made.

48    *Copthorne Hotel (Newcastle) Ltd* v. *Arup Associates* (1997) 85 BLR 22, per Judge Hicks at 33. This statement was largely repeated in the judgment of Judge L.J. in the Court of Appeal at 85 BLR 50.

management contractor to sub-let the 'field work' in 'works packages'; (2) requires the management contractor to ensure that all sub-contracts contain prescribed terms; (3) entitles the management contractor to recover all costs paid to sub-contractors plus a fee. The prescribed terms of the sub-contracts include that: (1) the sub-contractor is bound by all the obligations owed by the management contractor to the employer; (2) the sub-contractor waives his right to contend that the main contractor has incurred no loss. Where the management contract does not expressly provide for the standard of care which the management contractor must exercise in his management of the project, a term providing that he will exercise reasonable care and skill will normally be implied.[49]

*Project Management, Design and Management and Construction Management*   In each of these approaches, the employer engages the designer, manager and each of the contractors directly. These approaches are, from a contract organisation viewpoint, therefore, essentially the same as a traditional contract and are variants of each other. Indeed, by substituting the word 'Project Manager' for 'Engineer' in a traditional contract, and enhancing his powers to coordinate the contractor's on-site activities, a very substantial part of the effect is achieved immediately. This, indeed, has been the preferred approach of a number of employers.[50] It has the advantage of enabling a familiar set of conditions to be used. In the GC/Wks series of contracts, the administrator formerly known as the 'Supervising Officer' has been replaced by a 'Project Manager' in order to emphasise a new enhanced management role. The Engineering and Construction Contract has adopted the concept of a managing administrator as its key feature; it has no Engineer, but instead it has a Project Manager who fulfils a stronger management function than an Engineer under a traditional contract.

*Advantages of management contracts*   Concerns over traditional contracts are frequently expressed. It is assumed that consulting engineers lack the political adroitness, commercial vision and day to day management skills to manage a costly and politically sensitive project. While this premise is debatable in its generality, and cannot be right in every case, it must be

---

49   *Copthorne Hotel (Newcastle) Ltd v. Arup Associates* (1997) 85 BLR 50 (CA) per Judge L.J. at 53.
50   In *Bernhard's Rugby Landscapes Ltd v. Stockley Park Consortium Ltd* (1997) 82 BLR 39, an ICE 5th Edition was used. The Engineer became the Project Manager and other incidental changes were made.

accepted that the term 'engineer' does throw emphasis on the technical component of the work, when in fact the project will also have other objectives. These objectives can be more effectively pursued if the person charged with that task has power to control programmes and to coordinate activities on site. This may result in a higher price, but cost is rarely the employer's only objective and, where it is, the Project Manager can take his decision accordingly.

*Problems with management contracts*   The employer seeks to ensure that his project manager can direct the works programme and coordinate on-site activities; this intervention (which, is viewed by contractors as interference) means that additional payments are inevitable. The employer also seeks to ensure that he can trust a project manager absolutely to act in his, the employer's, best interests. In the case of management contract, this is done by limiting the project manager's liability for the performance of sub-contractors; but this can create its own risks when claiming for defective work, and can lead to highly complex litigation where the works contractors or designers join the manager in the litigation looking for a contribution.[51] There is no recognised list of standard duties for construction managers and from time to time question may arise as to the extent of the project manager's role.[52]

## 5. Term contracts

*Nature of a term contract*   A term contract is an agreement between the parties that work of a certain type[53] will be undertaken by the contractor at agreed rates for a stated period.

*Payment*   Payment is normally made in accordance with a schedule of agreed rates. Each class of work may have several associated rates, with the applicable rate depending on the quantity of work in that class to be undertaken in each location. In term contracts for a period exceeding two years, it is common for there to be a price adjustment formula. Where there is no express price adjustment, none will be implied.

---

51   *Copthorne Hotel (Newcastle) Ltd v. Arup Associates* (1997) 85 BLR 50 (CA).
52   See *Pozzolanic Lytag Ltd v. Bryan Hobson Associates* [1999] BLR 267, where the Project Manager's responsibility for advising on insurances was in issue.
53   Term contracts are widely used for civil engineering maintenance contracts for local authorities. They are sometimes referred to as maintenance contracts or standing orders.

*The amount of work* Where a contractor maintains resources on standby, he will wish to ensure that there is a good and steady flow of orders. Unless it is expressly agreed, however, it will not readily be implied either from historical figures or the resources which the contractor is obliged to have ready, that any minimum amount of work will be ordered.[54] Thus, in one case,[55] a Council awarded a local highway maintenance term contract, to carry out works formerly carried out by the Council itself. The contractor agreed to take on former council employees.[56] It was estimated that about £400 000 worth of orders per month was required to ensure employment for all the employees. The start date was 1 June 1996. By that date only £15 000 worth of orders had been placed for that first month. The contractor refused to start the work or take on the workforce. Dyson J. said:

> [The contractor] was in breach of contract in failing to accept the transfer of the workforce and start work on 1 June. There was some work to do, albeit not as much as the contractor would have wished. If (which I doubt) the volume of work available was such as to put the Council in breach of contract, that might have given rise to a claim for damages by the contractor. It did not entitle the contractor to refuse to perform the contract.

*Exclusive entitlement to the work* The parties may agree that all work of the specified type will be awarded to the contractor[57] but the degree of exclusivity is a matter of interpreting the contract as a whole.

*After the term has expired* Where the parties continue to operate the arrangement after the term has expired, it will normally be inferred that both parties agree to continue on the contract basis, unless some other intention is clear. The contractor will be entitled to the contract rates rather than a *quantum meruit*. Either party may discontinue the arrangement upon reasonable notice.[58]

## 6. Concession (BOT) contracts

### The nature of concession contracts

*General* A concession (or Build–Operate–Transfer) contract is one where the concession-grantor (frequently, but not necessarily, a

---

54 *R. v. Demers* [1900] AC 103 (PC).
55 *Bedfordshire County Council v. Fitzpatrick Contractors Ltd*, TCC, 16 October 1998.
56 This is a common requirement, deriving from the TUPE—Transfer of Undertakings (Protection of Employment)— Regulations 1981.
57 *Kelly Pipelines v. British Gas* (1989) 48 BLR 126. But for a term contract which was held not to be exclusive, see *Bonnell's Electrical Contractors v. London Underground* (1995) CILL 1110.
58 *Bonnell's Electrical Contractors v. London Underground* (1995) CILL 1110.

government body) grants to the concessionaire a concession to develop a piece of infrastructure (often called an 'asset' or the 'facility') and to hold that facility for a defined period and in a defined way so as to recoup the initial cost of investment and also to make a profit. The facility is usually constructed using a turnkey contract[59] and the concession-grantor usually takes the facility over at the end of the concession period. A concession contract is not primarily a construction contract. It is, in large part, a service contract in which the concessionaire provides to the concession-grantor (directly or indirectly) a service. In addition, it is a finance mechanism, enabling the concession-grantor to have the service (and, ultimately, the facility) without having to find the initial capital (although it may have to underwrite it, directly or indirectly).

*The service nature of concession contracts*   Concession contracts usually entitle and/or oblige the concessionaire to render a service. Under power and water supply contracts, power/water is produced for supply either to a state or private distribution company; 'offtake'[60] agreements may fix prices and quantities of power or water and may require the offtaker to guarantee the purchase of agreed quantities.[61] Road, rail and bridge contracts require the concessionaire to make these facilities available for defined classes of user; direct or shadow[62] tolls may be charged and a variable fee (depending on the quality of service provided) may also be payable. The service nature of the project becomes most obvious when the facility is a school, hospital or prison, with the concessionaire not only building the facility, but managing and providing support staff and upgrades to computing, communication and services during the concession period.

*Concession contracts as a financing mechanism*   A concession contract enables a party who desires the construction of a facility to have it done without having to find the initial capital. For example, private healthcare companies or school trusts may have facilities constructed for them, which

---

59   Although this is often achieved using familiar contract conditions as a platform; the contract for the Queen Elizabeth II Bridge, as an early example, used the ICE Conditions of Contract to provide the core construction obligations.
60   This expression usually refers to the output from a power, water or gas facility. Here, it is used more generally to mean the eventual user of any concession contract.
61   These are sometimes called 'hell or high water clauses' because the offtaker is obliged to take the service whether or not it is needed.
62   A shadow toll is one where the user does not pay directly; the state authority pays a fee for the use of the facility. The usage may be measured by sampling or more accurately by electronic tracking systems.

are paid for over the term of the concession from recurrent funds. But the most significant promoters are governments. Concessions are used both in developing countries and in developed countries. In the former, governments may find it difficult to finance important but non-urgent infrastructural projects and private investment can assist them; private finance is not a panacea here, of course, because while the money invested into the project is not sovereign debt, its repayment relies on the future availability of hard, rather than local, currency and concessionaires frequently look to the government to underwrite the availability of that money.[63] In some ways, the major success of concession contracts is their use by the governments of developed countries (e.g. under the UK Private Finance Initiative) to produce an efficient private–public partnership for the development of the infrastructure.

*History*   Concession contracts were first introduced on a major scale during the industrial revolution for bridge, canal, road and railway construction. Transport companies promoted statutes allowing them to buy continuous tracts of land and to exploit them for profit by building transport schemes. In many cases, the term of the concession was unlimited. During the late nineteenth and first three-quarters of the twentieth centuries, governments nationalised important transport assets; concession contracts then became impracticable, except for the exploitation of mineral resources (e.g. in the oil and gas industries). In the late 1970s, however, there was a resurgence of interest in the use of private capital for projects used by the public and the modern ascendancy of concession contracts dates from that time.

*Terminology: employers and contractors*   The service nature of a concession agreement, together with its focus on finance, means that the designations 'contractor' and 'employer' are often inappropriate. In the case of a government-promoted facility, the Government Authority might appear to occupy the role of employer, while the party responsible for constructing, owning and operating the facility may be thought of as the contractor. Such designations may, however, be misleading. The contractor tends to be a company set-up specifically to deal with the construction and management of this one project—the project company.[64] This company may let the work on a turnkey basis and to that extent occupies the employer's role under a

---

63   Stein SJ, Build–Operate–Transfer (BOT)—a re-evaluation [1994] ICLR 101.
64   Sometimes called a 'special project vehicle'.

conventional contract. To add to this complication, the project company is frequently a joint venture of various companies, including the construction companies who will take on the construction work and so, in some senses, part of the employer may also be the contractor. In addition, the service-user or offtaker may not be the concession grantor; in the case of a road project, the user will be the motorist, but the Government is clearly a bene-ficiary. In the case of power production and mineral concessions, the party who agrees or is assumed to want the product may not be the Government; the latter may benefit only indirectly in the form of sales taxes.

*Risk under concession contracts*[65]    Concession contracts often represent the ultimate risk assumption for a 'contractor' (concessionaire). Each con-tract involves its own unique risks because of:

(1) *The nature of the product being constructed*; for example, where a bridge is being constructed and the concessionaire is to recover his initial outlay by charging tolls, he takes the risk that there will in fact be demand at the rate of toll he proposes[66]
(2) *The nature of the construction process*; projects such as tunnels often involve great cost uncertainty because of the uncertain nature of the ground conditions
(3) *The risk division and safeguards in the contract*; for example, where the project is a light rail system, the risks may be largely associated with the take up by prospective passengers. Since the concession-grantor is likely to be an arm of government which can influence the attractive-ness of the scheme through its transportation taxes and subsidies, the concessionaire may seek comfort as to the base number of passengers.

The risks to banks (who usually take no equity stake in the project) and equity investors needs to be considered also. Banks, in particular, are care-ful to carry out full risk assessments before lending money and may require 'step in' rights, enabling them to take over the facility if there is a default by the contractor, with a view to letting the work to another contractor.[67]

---

65   Nielsen KR, Trends and evolving risks in design–build, BOT and BOOT projects [1997] ICLR 188.
66   The long term nature of a concession contract can even create a serious possibility that the scheme will become obsolete, just as telex is rendered obsolete by fax, which in turn is being rendered obsolete by electronic mail. See Knutson R, Common law development of the doctrine of performance and frustration of contracts and their use and application to long-term concession contracts [1997] ICLR 298.
67   Scriven J, A banking perspective on construction risks in BOT schemes [1994] ICLR 313.

*The contract description* A bewildering array of names and acronyms are used. These include:

(1) Design, Build, Own, Operate and Transfer (DBOOT)—this expression is frequently used for bridge and tunnel projects.
(2) Design, Build, Finance and Operate (DBFO)—this expression is frequently used for school and hospital projects.
(3) Design, Construct, Manage and Finance (DCMF)—this expression is frequently used for facilities such as prisons, where the project company not only constructs but also provides specialist management in the medium term.
(4) Build, Lease, Transfer (BLT)—here the project company retains legal ownership, but the facility is leased back to the government. The government runs the facility, usually on a full repairing and insuring basis.

None of the above are terms of art. Each seeks to encapsulate what the concessionaire is required to achieve. In practice, there are so many variables and the contracts can be used in so many different contexts that such descriptions are of little value in themselves. The project company's obligations as always are determined by an interpretation of the contract.

*Recouping the outlay* The concessionaire finances the project and must therefore recover his outlay plus profit when the works are complete. This may be achieved in a number of ways:

(1) *tolls*; for example, on a bridge project, the contractor will charge a crossing toll which will, over a number of years, pay back the initial outlay, cover maintenance and bring in a profit
(2) *shadow tolls*; for example, on a road for public use, the government may pay the contractor for the level of usage, for example, at a defined rate for each type of vehicle
(3) *rent-back or lease-back*; for example, on a hospital or school project, the contractor may rent the completed facility back to the NHS or the education authority, thus enabling the authority to make accurate provision for its future expenditure from recurrent funds rather than using its capital which can be focused elsewhere.

The income may be generated from a combination of these; for example, a light railway project may generate income from a combination of passenger usage, attainment of various performance targets and for attracting custom from environmentally less-efficient modes of transport.

*Service level agreements*   The concession-grantor requires a service. This should, as far as he is concerned, be explicitly set out in the concession contract. It should deal with:

(1)  measurable standards of service, including availability, punctuality, throughput, etc., and compensation for failure to provide
(2)  maintenance throughout the project and condition at the date of transfer of the facility to the concession-grantor
(3)  upgrading, if new technologies require it (e.g. new signalling on railway projects, new signing on roads, etc.)
(4)  sufficient incentives for the concessionaire to provide the required service up to and including termination rights—these may be achieved using 'liquidated damages' but the shortcomings of these should be noted; optional standards with significant incentives for providing a good service are normally preferable.[698]

## UK Government initiatives

The UK Government is an enthusiastic supporter of privately financed projects. One concern has been the lack of expertise in the public sector about risk managing these projects and a new private-sector led body is being established. Partnerships UK will provide advice to the public sector and act as a project manager for public–private concession deals to provide public sector organisations with back-up expertise.

## Contract forms

*Contract forms*   Each concession contract is unique. Standard forms will provide no more than the 'inner core' of a contract. In practice, existing models for turnkey contracts tend to be used. The UK Treasury Taskforce has issued (July 1999) Standard Contract Guidance.[69]

---

68   For example, where the concessionaire provides and runs a prison, shortfall in service may occur due to lack of staff recruitment and yet the liquidated damages payable to the concession-grantor (the government) will—being a genuine pre-estimate of loss—be small. This leads to the situation where it is to the concessionaire's advantage habitually to supply poor service.
69   Also, the UK Treasury has published (a) the full contract for the Agecroft Prison development at Salford: www.treasury-projects-taskforce.gov.uk/series-other/contract, (b) guidelines on procurement including the need to publish in the *European Journal*, (c) use of competitive negotiated procedures etc.

*Basic components* A successful concession contract deals with the following:

(1) the quality of the possession of the land over which the concession is granted
(2) the duration of the concession and the obligations of each party for that land (e.g. in respect of contaminants, nuisances, etc.) during that period
(3) the opportunities to acquire additional land if, for any technical reason, it becomes necessary
(4) planning consents
(5) the concession-grantor's requirements as to time, quality and environmental protection
(6) the concession-grantor's involvement during the construction and operation phases
(7) maintenance standards during the concession possession, including dilapidation surveys etc.
(8) upgrading works as new technologies come on stream (e.g. signs and lighting for road schemes, signalling and safety equipment for rail schemes)
(9) the standard of the assets at the date of transfer back to the employer and on going liability for defects which arise immediately after transfer
(10) facilities for 'consumers'
(11) concession-grantor's remedies if the contractor fails to operate the works in accordance with the concession agreement
(12) risk sharing in the event that the works are more difficult or costly to construct than envisaged due e.g. to lengthy enquiries, interference by protesters, price fluctuations due partly to government action, increased technical requirements as a result of new legislation, etc.
(13) risk sharing in the event that the works do not achieve the levels of use envisaged, e.g. because of changes in governmental tax and subsidy regimes, competing schemes promoted by the government, new technologies, etc.
(14) 'force majeure'
(15) early termination provisions should the project prove non-viable or if the concessionaire is not providing the service required
(16) rights of the concession-grantor to change the use or standards (this is particularly important e.g. in a school or prison project)
(17) compensation if the concession-grantor re-privatises the scheme or takes it over on an emergency basis

(18) information transfer between the parties during construction, during operation and at the end of the concession period

(19) transfer of staff, computer systems, databases and know-how from the contractor to the concession-grantor at the end of the concession and/ or phased transfer

(20) permitted sources of raising capital; for example if financed by debt what security will lenders be able to acquire

(21) assignability of the concessionaire's interest.

### Financing concession contracts

*Finance and bankability*    The money to pay for the initial construction can be raised in four main ways: (*a*) government grant, (*b*) development agency loan, (*c*) equity participation or (*d*) 'debt' (i.e. an agreed bank lend). Government grants may be appropriate where a government wishes to promote a project which is not inherently 'bankable'; that is, a risk and reward analysis indicates that money lent to build it would not be recovered sufficiently certainly and sufficiently quickly for a bank to find the prospect attractive. Where the project is key to the development of a country, development agency (e.g. World Bank) loans may be available. Ordinarily, the finance is raised by a mixture of equity participation (e.g. a share issue, in which investors buy a stake in the project) or debt (e.g. bank loans). Equity participation may succeed by force of investor confidence in the project; in some cases, where investors are not convinced, governments may underwrite a bond issue. Banks tend to shy away from becoming major equity participators and so take on a risk of default by the project company without the corresponding opportunity to make a large profit if the project is successful; this means that banks tend to require stringent tests of viability to be met, often backed up by a range of devices, such as sponsor or government guarantees[70] and 'step-in' rights (i.e. the right to take over and administer the project if the project company defaults).

*The concession contract and bankability*    The 'bankability' of a project depends on its full engineering, market, finance and political context. The banks will examine the concession agreement with an eye to the following:

---

70  Parent company or government guarantees may be limited; there is always a tension because the owners of the project company will wish it to be a 'limited recourse' vehicle, shielding them from excessive project risks. In addition, government guarantees may be unenforceable because of the principle of sovereign immunity; the nature of the government is thus often more important than the wording of the guarantee.

(1) *The extent and terminability of the concession* The concession should run for a significant period, it should be extensible if there are any *force majeure* conditions during construction or operation and should not be terminable except upon terms which enable the project concessionaire to pay off outstanding debt. Thus, even where termination is brought about by the default of the concessionaire through his construction contractor, the banks prefer either that the part-completed work be paid for (with the bank having first call on that payment) or the concession should run with the project rather than the concessionaire project company (with the banks having 'step-in' rights, so that they can sell the project on).

(2) *Insulation from change of law risks should be significant* New environmental and safety regulations often required time as well as money to implement. Direct tax regimes should be dealt with explicitly so that arbitrary tax rises cannot be imposed without compensation. Indirect tax changes (e.g. where government policy, though taxation, changes transportation demand) may also be considered. Where the concession-grantor will not agree to stand these risks, banks prefer that they be passed on to suppliers and offtakers.

*The civil engineering/construction contracts and bankability* The bankability of a project also depends on arrangements in the construction contracts. A bank considering lending against a project will inspect the agreements for the following:

(1) Full risk allocation, with complete coverage across and down. The construction contract should deal comprehensively with all risks. Banks prefer that all risks that can sensibly be transferred to the construction contractor should be, usually by a fixed price turnkey arrangement. 'Coverage across' means that no risks fall between the cracks at the interface between contracts; 'coverage down' means the proper defraying and management of risks down to sub-contractors and construction suppliers, using back-to-back contracts, sub-contractor direct warranties, guarantees, etc.

(2) Time conditions should be clear, and monitored using milestones (which may also be linked to payment). 'Liquidated damages' (or, more commonly these days, optional completion type arrangements[71]) should ensure that the contractor has the incentive to keep to

---

71  See Chapter 5, Section 6, for the problems associated with liquidated damages.

programme and an obligation to compensate adequately if he fails to to do so

(3)  Quality and throughput criteria (e.g. in the case of a power facility: maximum and normal working power output, fuel efficiency, etc.) should be clearly stated and the contractor's obligations should be tested with damages payable as for breach of time obligations.

(4)  Limited grounds for claim and a restricted definition of *force majeure*.

(5)  Provision for contractor (and sub-contractor) guarantees.

*Bankability and supply and offtake contracts*   Supply contracts mean here the supply of raw materials for the operation of the plant. In the case of a fuelled power or other process plant, this means the fuel or raw material. In the case of hydro-electric power plant it means access to water head and volume. In the case of a road, it means access to units of vehicles at the proposed toll levels. In each case there are risks and the banks will wish to see provisions which deal explicitly with any increase in supply costs or shortage. Risks may be passed on to offtakers (e.g. shortage of vehicles may—if this is economically viable—result in higher tolls) or to the concession-grantor who underwrites the supply. This is the bank's preferred solution, but, in practice, the concessionaires tend to retain many of these risks, except for those which are directly within the control of the concession-grantor. As for the offtaker contracts, the banks prefer to see 'hell or high water' provisions so that the offtaker is required to accept and pay for the service offered whether or not it is required.[72] Since off-takers tend to be major companies, often in a close relationship with the concession-grantor, banks are uneasy where the project concessionaire ends up carrying a major risk here.

---

72   For example, the cold-fusion 'scare' worried many owners of power facilities that they were in possession of soon-to-be obsolete plant. 'Hell or high water' provisions defeat claims that *force majeure* applies or that frustration has rendered the contract void.

# 13

# The ICE Conditions of Contract— Measurement Version (7th Edition, 1999)

The ICE Conditions of Contract, 7th Edition, were published in September 1999. The contract is drawn up by the Conditions of Contract Standing Joint Committee (the CCSJC); this is a committee set up by and with the approval and sponsorship of the Institution of Civil Engineers, the Civil Engineering Contractors Association and the Association of Consulting Engineers.

The ICE Conditions of Contract have always been based on the measure and value principle—that is, the value of the work was computed by measuring the quantities actually required to complete the work. The express sub-heading 'Measurement Version' indicates the CCSJC's intention to bring out further versions to allow for different payment schemes.

## 1. Comparison with the 6th Edition

The ICE Conditions of Contract, 6th Edition, were published in 1991 and are in widespread use. A number of amendments have been published since then. The 7th Edition contains a number of novel features, but in many ways it is a consolidating edition, sweeping up all these amendments. The new features—when compared with the 6th Edition as originally released in 1991—include:

(1) Wording changes—a number of changes have been introduced to bring the language up to date. For example 'execute' (as in 'execute the works') has been changed to 'carry out'.

(2) Clause 11 of the 6th Edition caused confusion. It required the Employer to supply to the Contractor all relevant site investigation documents. This created a number of difficulties. First, there was a practical difficulty: local authorities had to scour through their records to make sure they didn't miss anything. Second, the clause was unclear as to the effect of non-compliance. Some suggested that this non-compliance was a breach of contract; this book took the view that it set up an evidential presumption that the Employer considered the withheld documents irrelevant at the time of making the Contract, which could assist the Contractor in a claim based on unforeseeability. A third difficulty was introduced by dispute resolution developments; adjudication is a quick method of dispute resolution and the evidence of non-compliance with Clause 11 was unlikely to be to hand, so that the adjudicator might be forced to make his decision on a different basis to that of the arbitrator or judge who heard the matter subsequently. The 7th Edition solves this problem simply by removing the requirement for the Employer to supply all relevant documents; the Contractor tenders on the basis of the documents with which he is furnished.

(3) Clause 52 of the 6th Edition contained provisions relating to: (*a*) valuation of variations and (*b*) additional payments, including notices. This clause has now been split with the former as Clause 52 and the latter as Clause 53. The old Clauses 53 and 54 which dealt with similar matters have been consolidated as Clause 54.

(4) Clauses 64 (Frustration) and 65 (War) of the 6th Edition have been consolidated into a newly numbered Clause 63. Clause 63 of the 6th Edition (Forfeiture) has been expanded and a new symmetrical provision dealing with defaults by the Employer has been created: these new clauses are numbered 64 and 65 respectively.

(5) Changes have been made to a number of clauses, most notably Clause 60 (Payment) and 66 (Disputes) to comply with the provisions of the Housing Grants, Construction and Regeneration Act 1996

(6) Clause 71 has been introduced to account for the Construction (Design and Management) Regulations 1994

## 2. The scheme of the ICE Conditions of Contract

The scheme of the contract is broadly as follows.

*Personnel and administration*    The parties to the contract are the Employer and the Contractor. Other persons described by the contract include:

(1) the Engineer
(2) the Engineer's Representative
(3) the Engineer's Representative's assistants
(4) the Contractor's agent.

The contract is administered by the Engineer. The Employer's role is limited to matters such as nominating the Engineer, consenting to assignments, making payment upon certificates and giving notice to determine the Contractor's employment.

*The Contract*   The Contract is defined as comprising:

(1) Conditions of Contract
(2) Specification
(3) Drawings
(4) Bill of Quantities
(5) The Tender and the written acceptance thereof
(6) Contract Agreement (if completed).

All parts of the Contract carry equal weight and any inconsistencies are to be explained and adjusted by the Engineer.

*Time*   The time scheme of the Contract is:

(1) Works Commencement Date: the date on which the works physically commence or should commence.
(2) The Time for Completion is the period stated in the Contract as being the time in which the Contractor has substantially to complete the Works. This time may be extended if circumstance arise entitling the Contractor to an extension of time.
(3) Substantial completion: the Contractor must achieve substantial completion by the end of the Time for Completion (or extended time if appropriate). Upon its attainment the Engineer issues a Certificate of Substantial Completion. There is no definition of substantial completion in the Contract.
(4) Defects Correction Period: upon the issue of the Certificate of Substantial Completion, the 'Defects Correction Period' begins to run. This is a period specified in the Contract during which the Contractor is obliged and entitled to return to correct any problems which become apparent in the Works. Upon the expiry of the Defects Correction Period and the making good of any work which is to be made good, the Engineer issues a Defects Correction Certificate.

*Payment*  The ICE Conditions of Contract have the following important features:

(1)  It creates a measure and value contract. The quantities in the Bill of Quantities are estimated. The prices for elements of work are given as rates per unit of work. The Engineer determines the value by 'admeasurement'. The units which are measured are specified in a Standard Method of Measurement (see Clause 57) and the unit rates are set out in the Bill of Quantities.

(2)  The Contractor is paid monthly, approximately two months in arrears. The amount paid is the value of work done to date less a proportion which is retained by the Employer; the proportion is stated in the Appendix. The first half of this retained sum is released upon the issue of the Certificate of Substantial Completion; the second half is released upon the issue of the Defects Correction Certificate. Where any sum is due from the Contractor to the Employer (e.g. as liquidated damages), the Employer may deduct it by bona fide set-off against sums certified.

(3)  The mechanism of payment is as follows. (1) The Contractor submits an account setting out the value (in his opinion) of the total work which has been performed to date. The units of work in each class defined in the Bill of Quantities are multiplied by the unit rates for that work and the total is given. From this is deducted the retention money and the amounts paid by the Employer to date. The remainder is the sum for which the Contractor makes application. (2) The Engineer considers this application and modifies any quantities, rates or items which to him appear inaccurate or inapplicable. He prepares an account based on his approved items and rates and certifies an amount which in his opinion is due from the Employer to the Contractor. (3) Upon issue of the certificate, the Employer has 28 days from the date of the application in which to pay the certified amount. (4) In the event the sum certified is less that should have been certified, or where the Employer fails to pay it in full, the Contractor is entitled to recover interest at the rate specified in Clause 60(7).

*Planning of operations*  The Contractor is obliged to supply a method statement and a programme. He must advise the Engineer of any proposed changes to the method and update the programme where the Engineer requires this. He is entitled to have possession of the site so as to execute the Works in accordance with the programme. There are provisions whereby the Contractor may give notice to the Engineer of the need for further information.

*Unforeseen conditions*    Where unforeseen ground or other physical conditions (except weather conditions) or other obstructions are encountered, the Contractor is entitled to recover the additional cost of dealing with these, plus reasonable profit. Where other employees of the Employer hinder the Contractor, the Contractor may recover for any consequent disruption.


# 3.  Commentary on the Conditions of Contract

## *Definitions and interpretation*

### Definitions    1

(1)  In the Contract (as hereinafter defined) the following words and expressions shall have the meanings hereby assigned to them except where the context otherwise requires.

    (a)  "Employer" means the person or persons firm company or other body named in the Appendix to the Form of Tender and includes the Employer's personal representatives successors and permitted assignees.

    (b)  "Contractor" means the person or persons firm or company to whom the Contract has been awarded by the Employer and includes the Contractor's personal representatives successors and permitted assignees.

    (c)  "Engineer" means the person firm or company appointed by the Employer to act as Engineer for the purposes of the Contract and named in the Appendix to the Form of Tender or any other person firm or company so appointed from time to time by the Employer and notified in writing as such to the Contractor.

    (d)  "Engineer's Representative" means a person notified as such from time to time by the Engineer under Clause 2(3)(a).

    (e)  "Contract" means the Conditions of Contract Specification Drawings Bill of Quantities the Form of Tender the written acceptance thereof and the Form of Agreement (if completed).

    (f)  "Specification" means the specification referred to in the Form of Tender and any modification thereof or addition thereto as may from time to time be furnished or approved in writing by the Engineer.

(g) "Drawings" means the drawings referred to in the Specification and any modification of such drawings approved in writing by the Engineer and such other drawings as may from time to time be furnished by or approved in writing by the Engineer.

(h) "Bill of Quantities" means the priced and completed Bill of Quantities.

(i) "Tender Total" means the total of the Bill of Quantities at the date of award of the Contract or in the absence of a Bill of Quantities the agreed estimated total value of the Works at that date.

(j) "Contract Price" means the sum to be ascertained and paid in accordance with the provisions hereinafter contained for the construction and completion of the Works in accordance with the Contract.

(k) "Prime Cost (PC) Item" means an item in the Contract which contains (either wholly or in part) a sum referred to as Prime Cost (PC) which will be used for the carrying out of work or the supply of goods materials or services for the Works.

(l) "Provisional Sum" means a sum included and so designated in the Contract as a specific contingency for the carrying out of work or the supply of goods materials or services which may be used in whole or in part or not at all at the direction and discretion of the Engineer.

(m) "Nominated Sub-contractor" means any merchant tradesman specialist or other person firm or company nominated in accordance with the Contract to be employed by the Contractor for the carrying out of work or supply of goods materials or services for which a Prime Cost or a Provisional Sum has been included in the Contract.

(n) "Permanent Works" means the permanent works to be constructed and completed in accordance with the Contract.

(o) "Temporary Works" means all temporary works of every kind required in or about the construction and completion of the Works.

(p) "Works" means the Permanent Works together with the Temporary Works.

(q) "Works Commencement Date"—as defined in Clause 41 (1).

(r) "Certificate of Substantial Completion" means a certificate issued under Clause 48.

(s) "Defects Correction Period" means that period stated in the Appendix to the Form of Tender calculated from the date on which the Contractor becomes entitled to a Certificate of Substantial Completion for the Works or any Section or part thereof.

(t) "Defects Correction Certificate"—as defined in Clause 61 (1).

(u) "Section" means a part of the Works separately identified in the Appendix to the Form of Tender.

(v) "Site" means the lands and other places on under in or through which the Works are to be constructed and any other lands or places provided by the Employer for the purposes of the Contract together with such other places as may be designated in the Contract or subsequently agreed by the Engineer as forming part of the Site.

(w) "Contractor's Equipment" means all appliances or things of whatsoever nature required in or about the construction and completion of the Works but does not include materials or other things intended to form or forming part of the Permanent Works.

## Singular and plural

(2) Words importing the singular also include the plural and vice-versa where the context requires.

## Headings and marginal notes

(3) The headings and marginal notes in the Conditions of Contract shall not be deemed to be part thereof or be taken into consideration in the interpretation or construction thereof or of the Contract.

## Clause references

(4) All references herein to clauses are references to clauses numbered in the Conditions of Contract and not to those in any other document forming part of the Contract.

## Cost

(5) The word "cost" when used in the Conditions of Contract means all expenditure properly incurred or to be incurred whether on or off the Site including overhead finance and other charges properly allocatable thereto but does not include any allowance for profit.

## Communications in writing

(6) Communications which under the Contract are required to be "in writing" may be hand-written typewritten or printed and sent by hand post telex cable facsimile or other means resulting in a permanent record.

*General*   Clause 1(1) sets out the defined terms used, which are indicated by upper case initial letters. However, Clause 1(5) defines the term 'cost' which appears with a lower case initial letter. Clauses 1(2), (3) and (4) deal with matters of interpretation.

*Clauses 1(1)(e), (f), (g), (h)—the Contract*   The Contract is defined at Clause 1(1)(e). The subsidiary definitions of Specification, Drawings and Bill of Quantities are given in Clauses 1(1)(f), (g) and (h). Read together with Clause 5: 'The several documents forming the Contract are to be taken as mutually explanatory of one another …', these clauses produce a scheme in which all documents carry equal weight and importance. Where additional documents are to be incorporated into the Contract, this should be done expressly and clearly so as to override the presumption created by Clause 1(1)(e) that other documents have no direct contractual effect.

*Clauses 1(1)(n), (o)—the Works*   These clauses define the Permanent and Temporary Works respectively. These definitions may not be mutually exclusive since elements of structure used as Temporary Works may eventually be incorporated into the Permanent Works. Clause 8(2) provides, *inter alia*, that 'the Contractor shall not be responsible for the design or specification of the Permanent Works … or of any Temporary Works design supplied by the Engineer'. Thus where Permanent Works are used in a temporary mode, and the Contractor is required or specifically entitled to use them in this mode, any damage they sustain in this mode will be at the Employer's risk. This is notwithstanding Clause 8(3) ('The Contractor shall take full responsibility for the adequacy stability and safety of all site operations and methods of construction'), Clause 14(9) ('consent of the Engineer to the Contractor's proposed methods of construction … shall not relieve the Contractor of any of his duties or responsibilities under the Contract') or Clause 20 ('(1) The Contractor shall … take full responsibility for the care of the Works … [save for] … (2)(b) any fault defect error or omission in the design of the Works').

*Clauses 1(1)(s), (t)—Defects Correction*   The term 'Defects Correction' was introduced by the 6th Edition. Previous editions used the term 'maintenance'. The new expression is to be preferred as it describes more accurately the function of the period and avoids confusion where maintenance and servicing of the works is a contract obligation.

*Clause 1(1)(v)—the Site*   The term 'Site' appears in Clauses 1(5), 11(2), 19, 22(2)(a), 32, 42, 54, 60(1) and 65. The extent of the Site may be uncertain, for instance where land outside the immediate area of construction is used for purposes such as storage, materials processing, etc. Where land is designated at part of the the Site, the Engineer may require the Contractor to provide at the Contractor's expense lights, guards, fencing, warning signs and watching: Clause 19(1).[1] Clause 54(1) provides that materials, equipment, etc., brought onto the Site shall not be removed with the written consent of the Engineer. And Clause 60(1)(b) envisages that goods or materials 'delivered to the Site' form a special category for payment. It should be noted that where the Engineer requires the Contractor to remove any person or sub-contractor employed on the Works, that person is not just to be removed from the Site, but from the Works—Clause 16.

*Clause 1(5)—cost*   The terms 'cost' or 'costs' appears in relation to claims in Clauses 7(4), 12(3), 13(3), 14(6), 17, 27(6), 31(2), 36(2), 36(3), 38(2), 40(1), 42(1) and 50. For any item of cost to become recoverable it must be 'properly incurred or to be incurred'. It means not just direct costs but also indirect costs such as overhead charges, finance charges and any other such charges. It does not include any element of profit; however, in several claims clauses profit is expressly allowed in addition to the cost as defined: see, for example, Clauses 12(6) and 13(3).

## Engineer and Engineer's representative

### Duties and authority of Engineer   2

(1)  (a)  The Engineer shall carry out the duties specified in or necessarily to be implied from the Contract.

(b)  The Engineer may exercise the authority specified in or necessarily to be implied from the Contract. If the Engineer is required under the terms of his appointment by the Employer to obtain the specific approval of the Employer before exercising any such authority particulars of such requirements shall be those set out in the Appendix to the Form of Tender. Any requisite approval shall be deemed to have been given by the Employer for any such authority exercised by the Engineer.

---

1    See also the Occupiers' Liability Acts 1957 and 1984. These place obligations upon occupiers in respect of invitees and trespassers. The definition of Site may be important in defining the Contractor's responsibility. Generally the question turns on considerations of actual control rather than the definition of terms such as 'Site'.

(c)   Except as expressly stated in the Contract the Engineer shall have no authority to amend the Contract nor to relieve the Contractor of any of his obligations under the Contract.

(d)   The giving of any consent or approval by or on behalf of the Engineer shall not in any way relieve the Contractor of any of his obligations under the Contract or of his duty to ensure the correctness or accuracy of the matter or thing which is the subject of the consent or approval.

## Named individual

(2)   (a)   Where the Engineer as defined in Clause 1 (1)(c) is not a single named Chartered Engineer the Engineer shall within 7 days of the award of the Contract and in any event before the Works Commencement Date notify to the Contractor in writing the name of the Chartered Engineer who will act on his behalf and assume the full responsibilities of the Engineer under the Contract.

(b)   The Engineer shall thereafter in like manner notify the Contractor of any replacement of the named Chartered Engineer.

## Engineer's Representative

(3)   (a)   The Engineer's Representative shall be responsible to the Engineer who shall notify his appointment to the Contractor in writing.

(b)   The Engineer's Representative shall watch and supervise the construction and completion of the Works. He shall have no authority

(i)   to relieve the Contractor of any of his duties or obligations under the Contract

nor except as expressly provided for in sub-clause (4) of this Clause

(ii)   to order any work involving delay or any extra payment by the Employer or

(iii)  to make any variation of or in the Works.

## Delegation by Engineer

(4)   The Engineer may from time to time delegate to the Engineer's Representative or any other person responsible to the Engineer any of the duties and authorities vested in the Engineer and he may at any time revoke such delegation. Any such delegation

(a)   shall be in writing and shall not take effect until such time as a copy thereof has been delivered to the Contractor or his agent appointed under Clause 15(2)

(b)   shall continue in force until such time as the Engineer shall notify the Contractor in writing that the same has been revoked

(c) shall not be given in respect of any decision to be taken or certificate to be issued under Clauses 12(6) 44 46(3) 48 60(4) 61 65 or 66.

## Assistants

(5) (a) The Engineer or the Engineer's Representative may appoint any number of persons to assist the Engineer's Representative in the carrying out of his duties under sub-clause (3)(b) or (4) of this Clause. He shall notify to the Contractor the names duties and scope of authority of such persons.

(b) Such assistants shall have no authority to issue any instructions to the Contractor save insofar as such instructions may be necessary to enable them to carry out their duties and to secure the acceptance of materials and workmanship as being in accordance with the Contract. Any instructions given by an assistant for these purposes shall where appropriate be in writing and be deemed to have been given by the Engineer's Representative.

(c) If the Contractor is dissatisfied by reason of any instruction of any assistant of the Engineer's Representative appointed under sub-clause (5)(a) of this Clause he shall be entitled to refer the matter to the Engineer's Representative who shall thereupon confirm reverse or vary such instruction.

## Instructions

(6) (a) Instructions given by the Engineer or by any person exercising delegated duties and authorities under sub-clause (4) of this Clause shall be in writing. Provided that if for any reason it is considered necessary to give any such instruction orally the Contractor shall comply therewith.

(b) Any such oral instruction shall be confirmed in writing as soon as is possible under the circumstances. Provided that if the Contractor confirms in writing any such oral instruction which confirmation is not contradicted in writing by the Engineer or the Engineer's Representative forthwith it shall be deemed to be an instruction in writing by the Engineer.

(c) Upon the written request of the Contractor the Engineer or the person exercising delegated duties or authorities under sub-clause (4) of this Clause shall specify in writing under which of his duties and authorities the instruction is given.

## Impartiality

(7) The Engineer shall act impartially within the terms of the Contract having regard to all the circumstances. In like manner the Engineer's Representative and any person exercising delegated duties and authorities shall also act impartially.

*Clauses 2(1)(a), (7)—the function of the Engineer*   The Engineer is required to act 'impartially'. He has two distinct types of function under the Contract:[2]

(1)   He is the Employer's agent. In this context it is thought that 'impartiality' means an honest approach. It does not mean that the interests of the Employer must be compromised. It is thought that the Engineer acts in this capacity when he orders action under Clause 12(4), issues instructions under Clause 13(3), issues design criteria under Clause 14(5), instructs the uncovering of work under Clause 38(2), issues orders for the suspension of the Works under Clause 40(1), specifies the Commencement Date under Clause 41(1), requests accelerated completion under Clause 46(3), issues instructions to undertake searches or trials under Clause 50, issues variations under Clause 51, orders Provisional and/or Prime Cost work under Clause 58 or issues instructions regarding nominated sub-contractors under Clause 59. In his capacity as the Employer's authorised agent, the Employer is responsible for the Engineer's acts or omissions. Thus, for instance, where no commencement date is agreed, a failure by the Engineer to specify a Commencement Date in accordance with Clause 41(1)(b) is a breach of contract by the Employer.

(2)   In the exercise of many of his powers, however, the Engineer acts as an impartial decision-maker. Here the concept of 'impartiality' means judicial impartiality. He must take decisions based only on the evidence and his interpretation of the Contract. The exercise of these powers includes wherever he values any work, variation or instruction; or where he issues any certificate or exercises any power of opinion or takes any decision under Clause 66. It is thought that the effect of Clause 2(7) is that the Employer warrants that the Engineer will make his decision impartially. A failure by the Engineer to act impartially will, therefore, be a breach of contract and, if loss flows, will entitle the Contractor to damages. Subject to this, the Engineer is not an agent of the Employer when he acts as an impartial decision-maker and the Employer is not responsible to the Contractor for his decisions.[3]

---

2   See *Brodie* v. *Cardiff Corporation* [1919] AC 337 (HL); *London Borough of Merton* v. *Hugh Stanley Leach* (1985) 32 BLR 51.

3   Nevertheless, the Engineer may be responsible to the Employer for any loss due to negligent decision making or certification: *Sutcliffe* v. *Thackrah* [1974] AC 727 (HL).

*Clause 2(1)(b), (c)—restrictions on the Engineer's authority*   Clause 2(1)(b) enables the Employer to state explicitly what limitations he has placed upon the Engineer's authority under the terms of the Engineer's appointment. Without more, this would deprive the Engineer of authority in respect of those matters.[4] However, the final sentence deems the Engineer to have received the requisite approval; in the absence of further express notice of lack of authority, therefore, the Contractor may assume that the Engineer is fully authorised notwithstanding any limitation stated in the Appendix. Clause 2(1)(c) states the general rule that the Engineer is neither authorised to vary the terms of the Contract nor to relieve the Contractor of any of his obligations under it.[5]

*Clause 2(2)—Engineer to be chartered engineer*   It seems that where the Engineer or the person designated to act on his behalf is not a chartered engineer, this will be a breach of contract. No damage will ordinarily flow directly from this. However, the consequential effects may be: (*a*) that the Contractor is not obliged to comply with the instructions, variations and the like of the person who purports to act as Engineer; and (*b*) that decisions, certificates, etc., issued are invalid.[6] Where the Contractor waives his objection to the Employer's continuing breach, he will be unable to claim that decisions etc. made during the relevant period were invalid.

*Clauses 2(3), (4), (5)—delegation*   These clauses establish a scheme for the delegation of powers by the Engineer or his delegates. Clause 2(3) establishes a defined appointment of an Engineer's Representative (who is the person normally known as the resident engineer). Despite the apparently mandatory wording of Clause 2(3) (the Engineer 'shall notify his appointment to the Contractor in writing') the appointment of an Engineer's Representative seems to be discretionary; where no such appointment is made it is thought that the Engineer carries out the Engineer's Representative's functions, such as to 'watch and supervise the construction and completion of the Works': Clause 2(3)(b). Where appointed, the Engineer's Representative is granted a series of powers by virtue of his status, such as the authority to require safety measures in regard to the Site and the

---

4   Where any agent has no actual authority, the Contractor will have to rely on his ostensible or apparent authority. Clearly, he cannot do this where he has notice of the limitation.

5   *Toepfer* v. *Warinco* [1978] 2 Lloyd's Rep. 569.

6   Any certificate etc. which is made by an unauthorised person is invalid: see *Croudace* v. *London Borough of Lambeth* (1986) 33 BLR 20 (CA).

Works: Clause 19(1). Where the Engineer's Representative is to take any further authority, an express written delegation must be made under Clause 2(4), with notice to the Contractor. However, such powers as may be delegated may also be delegated to 'any other person responsible to the Engineer'. Certain powers may be exercised only by the Engineer: these are Clause 12(6)—the time and financial consequences of unforeseen physical conditions; Clause 44—extensions of time; Clause 46(3)—accelerated completion; Clause 48—Certificate of Substantial Completion; Clause 60(4)—certificate of sums due under the final account; Clause 61— Defects Correction Certificate; Clause 65—certificate prior to the determination of the Contractor's employment; Clause 66—Engineer's decision on a matter of dissatisfaction. In broad terms these are the duties of the Engineer as an impartial decision-maker which are of greatest significance. Nevertheless, a great number of discretionary powers remain delegable, including issuing certificates for payment, except on the final account. Furthermore, all the powers as Employer's agent are delegable, including issuing instructions (Clause 13) and variation orders (Clause 51).

Clause 2(5) provides that either the Engineer or Engineer's Representative may appoint assistants. The person delegating shall 'notify to the Contractor the names duties and scope of authority of such persons'; Clause 68 requires that any such notification must be in writing at the Contractor's principal place of business.

*Clause 2(5)(c)—reference*   Clause 2(5)(c) provides for a reference from an assistant's instruction to the Engineer's Representative. No procedure is set; it is thought that the proper time is a reasonable time in all the circumstances. Note that there is no provision for the Employer to seek a reference under these provisions.

*Clause 2(6)—instructions in writing*   The proviso to Clause 2(6)(a) suggests that, in some circumstances, an oral instruction shall be complied with. The relevant circumstances are 'if for any reason it is considered necessary'. The person who must consider it necessary is presumably the person authorised to issue the instruction. This may mean that any oral instruction issued to the Contractor, in circumstances where it is clearly intended that it should be acted upon, is valid. The procedures in Clauses 2(6)(b) and (c) seem to be directed at attaining regularity, clarity and a degree of evidential certainty. When an oral instruction is issued, the Contractor is entitled to a written confirmation immediately.

## Assignment and subcontracting

### Assignment    3

(1)   Neither the Employer nor the Contractor shall assign the Contract or any part thereof or any benefit or interest therein or thereunder without the prior written consent of the other party which consent shall not unreasonably be withheld.

(2)   Nothing in this Contract confers or purports to confer on any third party any benefit or any right to enforce any term of the Contract.

*Prohibition on assignment*   The prohibition is on 'the Contract or any part thereof or any benefit or interest therein or thereunder …'. This is comprehensive and extends to any monies due under the Contract, including, for instance, retention monies.[7] It should be noted that the Employer and Contractor are defined in Clauses 1(1)(a),(b) as including their 'personal representatives [and] successors'. It is thought that where the name, status or identity of the Employer changes by operation of law (e.g. by an enactment transferring authority to a new body) the new body will automatically assume the role of Employer. Where, however, the Employer or Contractor wishes to reorganise his corporate structure, an automatic assignment between corporate persons will not take effect.

*Consent not to be unreasonably withheld*   This provision clearly envisages that some assignments are possible. The use of the term 'assignment' presumably precludes the assignments of burdens, including the Employer's obligation of cooperation and the Contractor's obligation of construction. No guidance is given for the term 'unreasonably'. It may mean, for example, that consent must be given save where the other party is likely to be materially disadvantaged; or it may mean that consent may be withheld if there is any conceivable disadvantage to the party whose consent is required— which amounts to a veto. Where a dispute arises as to whether or not any consent is reasonably withheld, it seems it must be referred to the Engineer under Clause 66 for his decision;[8] the parties are then required to give effect to his decision, unless and until it might be revised by an arbitrator.

---

7    *Helstan Securities Ltd* v. *Hertfordshire County Council* [1978] 3 All ER 262; *Linden Gardens Trust Ltd* v. *Linesta Sludge Disposals* [1994] 1 AC 85 (HL).

8    On such a technical issue, it may be that a court will entertain an application for a declaration without staying the matter to arbitration: see dicta in *Bristol Corporation* v. *Aird* [1913] AC 241 (HL); *Lakers Mechanical Services* v. *Boskalis Westminster* (1989) 5 Constr. L.J. 139.

**Sub-contracting   4**

(1)   The Contractor shall not sub-contract the whole of the Works without the prior written consent of the Employer.

(2)   Except where otherwise provided in the Appendix to the Form of Tender the Contractor may sub-contract any part of the Works or their design. The extent of the work to be sub-contracted and the name and address of the sub-contractor must be notified in writing to the Engineer as soon as practicable and in any event not later than 14 days prior to the sub-contractor's entry on to the Site or in the case of design on appointment.

Provided that if not later than 7 days after receipt of such notification the Engineer for good reason objects to the Contractor in writing the sub-contractor so notified shall not be employed on or in connection with the Works. Such objection must be accompanied by reasons in writing.

(3)   The employment of labour-only sub-contractors does not require notification to the Engineer under sub-clause (2) of this Clause.

(4)   The Contractor shall be and remain liable under the Contract for all work sub-contracted by him and for acts defaults or neglects of any sub-contractor his agents servants or workpeople.

(5)   The Engineer shall be at liberty after due warning in writing to require the Contractor to remove from the Works or their design any sub-contractor who mis-conducts himself or is incompetent or negligent in the performance of his duties or fails to conform with any particular provisions with regard to safety which may be set out in the Contract or persists in any conduct which is prejudicial to safety or health and such sub-contractor shall not be again employed upon the Works without the permission of the Engineer.

This provision sets out a scheme for specifying which work may not be sub-contracted and processes for notification. Otherwise, all work, including design work may be sub-contracted; this provision defeats the general presumption that design work may not be vicariously performed.[9]

*Clause 4(4)—Contractor to remain liable for neglects*   A contractor is generally liable for the breaches of contract caused by his sub-contractors; he is not, however, generally liable for their torts,[10] unless those torts are

---

9    *Moresk Cleaners v. Hicks* [1966] 2 Lloyd's Rep. 338.

10   *Honeywell and Stein Ltd v. Larkin (London's Commercial Photographers) Ltd* [1933] All ER 77 (CA)
     *D & F Estates v. Church Commissioners* [1989] AC 177 (HL).

authorised by him[11] or the work is of a particularly hazardous nature.[12] The term 'neglect' in Clause 4, however, seems to suggest the tort of negligence and it may be that this provision renders the Contractor liable to the Employer for the torts of all his sub-contractors, even where such torts are not breaches of contract.

*Clause 4(5)—removing sub-contractors*   This provision entitles the Engineer, after 'due warning' in writing to require that sub-contractors be removed for a variety of reasons. Contractors should ensure that all sub-contracts provide that a Contractor shall not be in default of the sub-contract if the Contractor is required to remove a sub-contractor from the Works. The clause envisages that the removal will proceed in two stages:

(1)   *The warning.* The warning must be given to the Contractor rather than the sub-contractor. It is submitted that this 'warning' is not a notice within the ambit of Clause 68, so can be handed informally to the Contractor's senior site staff. Due warning means, it is submitted, proper warning in the sense that it must: (*a*) specify what the problem is and hence explicitly or implicitly set out what can be done to rectify it and (*b*) must be given in a reasonable time in all the circumstances.

(2)   *The removal.* A strict reading of the clause suggests that the Engineer can require a sub-contractor's removal by oral instruction provided that a written warning has been given. It is thought, however, that this amounts to an instruction which must be in writing—or at least followed up in writing under Clause 2(6).

## Contract documents

### Documents mutually explanatory   5

The several documents forming the Contract are to be taken as mutually explanatory of one another and in case of ambiguities or discrepancies the same shall be explained and adjusted by the Engineer who shall thereupon issue to the Contractor appropriate instructions in writing which shall be regarded as instructions issued in accordance with Clause 13.

---

11   *Ellis v. Sheffield Gas Consumers Co.* (1853) 2 E&B 767.
12   *Salisbury v. Woodland* [1970] 1 QB 324 (CA) *Honeywell and Stein Ltd v. Larkin (London's Commercial Photographers) Ltd* [1933] All ER 77 (CA) *Dalton v. Angus* (1881) 6 App Cas 740 (HL).

*Documents to be mutually explanatory*   This provision appears to override the canon of construction that specially prepared documents are to be given precedence.[13]

*Engineer to explain and adjust ambiguities or discrepancies*   This appears to require the Engineer to generate certainty in the case where the technical requirements are unclear or where two or more reasonably clear requirements conflict. Where the Engineer is of the opinion that an ambiguity or discrepancy exists, he is obliged to issue a written instruction. Any explanation or adjustment may have financial consequences, in particular where the content of the written instruction is 'beyond that reasonably to have been foreseen by an experienced contractor at the time of tender'—Clause 13(3). Where appropriate, the Contractor is entitled to be paid the extra cost, profit ('in respect of any additional permanent or temporary work') and to be granted an extension of time.

**Supply of documents   6**

(1)   Upon award of the Contract the following shall be furnished to the Contractor free of charge

    (a)   four copies of the Conditions of Contract Specification and (unpriced) bill of quantities and

    (b)   the number and type of copies as entered in the Appendix to the Form of Tender of all Drawings listed in the Specification.

(2)   Upon approval by the Engineer in accordance with Clause 7(6) the Contractor shall supply to the Engineer four copies of all Drawings Specifications and other documents submitted by the Contractor. In addition the Contractor shall supply at the Employer's expense such further copies of such Drawings Specifications and other documents as the Engineer may request in writing for his use.

(3)   Copyright of all Drawings Specifications and the Bill of Quantities (except the pricing thereof) supplied by the Employer or the Engineer shall not pass to the Contractor but the Contractor may obtain or make at his own expense any further copies required by him for the purposes of the Contract. Similarly copyright in all documents supplied by the Contractor under Clause 7(6) shall remain with the Contractor but the Employer and the Engineer shall have full power to reproduce and use the same for the purpose of completing operating maintaining and adjusting the Works.

---

13   *Glynn v. Margetson* [1893] AC 351 (HL); *English Industrial Estates v. Wimpey* [1973] 1 Lloyd's Rep. 118 (CA).

*Copyright*   Copyright does not pass, merely a licence to use documents and designs for the purpose of the Works.

### Further Drawings Specifications and instructions   7

(1)   The Engineer shall from time to time during the progress of the Works supply to the Contractor such modified or further Drawings Specifications and instructions as shall in the Engineer's opinion be necessary for the purpose of the proper and adequate construction and completion of the Works and the Contractor shall carry out and be bound by the same.

If such Drawings Specifications or instructions require any variation to any part of the Works the same shall be deemed to have been issued pursuant to Clause 51.

### Contractor to provide further documents

(2)   Where sub-clause (6) of this Clause applies the Engineer may require the Contractor to supply such further documents as shall in the Engineer's opinion be necessary for the purpose of the proper and adequate construction completion and maintenance of the Works and when accepted by the Engineer the Contractor shall be bound by the same.

### Notice by Contractor

(3)   The Contractor shall give adequate notice in writing to the Engineer of any further Drawing or Specification that the Contractor may require for the construction and completion of the Works or otherwise under the Contract.

### Delay in issue

(4)   (a)   If by reason of any failure or inability of the Engineer to issue at a time reasonable in all the circumstances Drawings Specifications or instructions requested by the Contractor and considered necessary by the Engineer in accordance with sub-clause (1) of this Clause the Contractor suffers delay or incurs additional cost then the Engineer shall take such delay into account in determining any extension of time to which the Contractor is entitled under Clause 44 and the Contractor shall subject to Clause 53 be paid in accordance with Clause 60 the amount of such cost as may be reasonable.

(b)   If the failure of the Engineer to issue any Drawing Specification or instruction is caused in whole or in part by the failure of the Contractor after due notice in writing to submit drawings specifications or other documents which he is required to submit under the Contract the Engineer shall take into account such failure by the Contractor in taking any action under sub-clause (4)(a) of this Clause.

### One copy of documents to be kept on Site

(5) One copy of the Drawings and Specification furnished to the Contractor as aforesaid and of all Drawings Specifications and other documents required to be provided by the Contractor under sub-clause (6) of this Clause shall at all reasonable times be available on the Site for inspection and use by the Engineer and the Engineer's Representative and by any other person authorized by the Engineer in writing.

### Permanent Works designed by Contractor

(6) Where the Contract expressly provides that part of the Permanent Works shall be designed by the Contractor he shall submit to the Engineer for acceptance

   (a) such drawings specifications calculations and other information as shall be necessary to satisfy the Engineer that the Contractor's design generally complies with the requirements of the Contract and

   (b) operation and maintenance manuals together with as completed drawings of that part of the Permanent Works in sufficient detail to enable the Employer to operate maintain dismantle reassemble and adjust the Permanent Works incorporating that design. No certificate under Clause 48 covering any part of the Permanent Works designed by the Contractor shall be issued until manuals and drawings in such detail have been submitted to and accepted by the Engineer.

### Responsibility unaffected by approval

(7) Acceptance by the Engineer in accordance with sub-clause (6) of this Clause shall not relieve the Contractor of any of his responsibilities under the Contract. The Engineer shall be responsible for the integration and co-ordination of the Contractor's design with the rest of the Works.

*Supply of additional documents*   It is common in civil engineering projects for many detailed drawings and some specifications to be provided or re-issued after the Contract is made. Not only is this convenient because of the interaction of structures with the detailed geology etc. but re-issues are frequently necessitated by variations to the Works. Where documents are to be issued, they must be issued in a reasonable time given all the circumstances, including the constraints under which the Engineer and his designers are working[14] and the agreed Time for Completion.[15] Any

---

14   See *Neodox v. Swinton & Pendlebury Borough Council* (1958) 5 BLR 34; *London Borough of Merton v. Hugh Stanley Leach* (1985) 32 BLR 51.
15   *Glenlion Construction v. Guinness Trust* (1988) 39 BLR 89.

approved programme (see Clause 14) will suggest the time at which documents need to be supplied. Where the approved Clause 14 programme provides for completion in a shorter time than the Contract Time for Completion, it is thought that the Contractor is entitled to receive information in line with this shortened programme until the Engineer gives reasonable notice that information will only be released to meet completion in a longer period (not exceeding the full Time for Completion).

*Delay in issue*   The Engineer is not obliged to issue Drawings and/or Specifications save as is necessary for the 'proper and adequate construction and completion of the Works'—Clause 7(1). Clause 7(4) provides for addition costs incurred as a result of late supply to be paid.

*Notice by Contractor*   By Clause 7(3) the Contractor is to give adequate notice in writing of his need for these. It is unclear whether an approved Clause 14 programme is of itself sufficient, where it is reasonably clear from this that information must be provided for the works to be progressed. It is always preferable for the Contractor to give specific notification.

*Clauses 7(2), (6), (7)—Contractor's design*   These clauses apply where the Contractor is required by express terms to design any element of the Permanent Works. The Engineer may require the Contractor, under Clause 7(2) to supply working drawings etc. In addition, Clause 7(6) provides that two classes of document be submitted to the Engineer for approval: (*a*) Drawings etc. as shall be necessary to satisfy the Engineer that the design complies with the Contract and (*b*) manuals. The level of detail required is a matter of professional judgment. Excessive demands by the Engineer must, it is thought, be complied with: Clause 13(1) 'the Contractor ... shall comply with and adhere strictly to the Engineer's instructions ...'. However such demands will amount to a instruction entitling the Contractor to the additional cost incurred: Clauses 13(3) and 51. Where the Engineer is not satisfied, he may call for additional documents under Clause 7(2). Approval by the Engineer will not relieve the Contractor of any of his obligations under the Contract. Nevertheless, written comments which amount in effect to an instruction in regard to designs may rank as an instruction pursuant to Clause 13. No time is stipulated for the approval of designs (Compare this with Clause 14, approval of programmes) and it seems, therefore, that a reasonable time is allowed. The reasonable time will be determined by factors such as the importance, scale and complexity of the design as well as the urgency for compliance with the programme.

Failure by the Engineer to act within a reasonable time will be a breach of contract by the Employer. Before any Certificate of Substantial Completion is issued, the Contractor must supply documents required by Clause 7(6)(b).

## General obligations

### Contractor's general responsibilities   8

(1)   The Contractor shall subject to the provisions of the Contract

    (a)   construct and complete the Works and

    (b)   provide all labour materials Contractor's Equipment Temporary Works transport to and from and in or about the Site and everything whether of a temporary or permanent nature required in and for such construction and completion so far as the necessity for providing the same is specified in or reasonably to be inferred from the Contract.

### Design responsibility

(2)   The Contractor shall not be responsible for the design or specification of the Permanent Works or any part thereof (except as may be expressly provided in the Contract) or of any Temporary Works design supplied by the Engineer. The Contractor shall exercise all reasonable skill care and diligence in designing any part of the Permanent Works for which he is responsible.

### Contractor responsible for safety of site operations

(3)   The Contractor shall take full responsibility for the adequacy stability and safety of all site operations and methods of construction.

*General responsibilities*   Clause 8(1) requires the Contractor to construct all the Works in accordance with the Contract and do everything needed to achieve that aim. It provides a clear restatement of the general principle, that a contractor is required to do all things reasonably to be inferred, whether mentioned in the Contract or not.[16] Where the Contract is operated unamended (i.e. as a measure and value contract with a Bill of Quantities), any items of work which are not mentioned in the Bill of Quantities are omissions and the Engineer must add them in, which entitles the Contractor to be paid for them—Clause 55. Where, however, the contract is amended to a lump sum contract without a Bill of Quantities,

---

16   *Williams v. Fitzmaurice* (1858) 3 H&N 844.

the effect of Clause 8(1) is to require the Contractor to complete the Works for the agreed price.

*Contractor's design and site operation*   Clause 8(2) reflects the growing practice of passing design responsibility onto the Contractor. Together with Clause 8(3) it states the general position under the Contract—the Employer is responsible for the Permanent Works and the Contractor is responsible for the Temporary Works. However, both of these assumptions may be expressly overridden. The Contract9r may take on design responsibility where expressly stated in the Contract; thus a voluntary performance of some design will not apparently cause the Contractor to be liable for it.[17] The Engineer may however design elements of the Temporary Works, in which case the Employer is responsible for them.

*Standard of design care*   The expression 'all reasonable skill and care and diligence ...' should be contrasted with the words 'reasonable skill and care' generally used to characterise a professional's standard of care.[18] It is thought that the use of 'all' and 'diligence' does not create any higher standard of care. The word 'diligence' may simply oblige the Contractor to ensure that his designers have all the relevant and up to date information. This closes down the slight (but possible) argument that the Contractor is not responsible since he delegated the design to a competent designer and he is not liable under Clause 4(4) because his designer is not in 'default' or 'neglect'.

*Exceptions to the Contractor's responsibility for site operations*   Clause 8(3) purports to place the full responsibility for site operations and methods of construction on the Contractor. There are, however, a number of exceptions. Where the interaction between the method of construction and reasonably unforeseeable physical conditions (as defined in Clause 12(1)) means that the method of construction is more costly that could have been foreseen, the Contractor may be entitled to recover pursuant to Clause 12.[19] Also, where specified methods of construction become legally or physically impossible, the Contractor is entitled to an instruction under Clause 13 or a variation order or damages for breach of contract.[20] Furthermore,

---

17   In contract at any rate. The Contractor may owe a duty in the tort of negligence to third parties and possibly even to the Employer.
18   See Chapter 9.
19   *Humber Oil Terminals Trustee* v. *Harbour and General* (1991) 59 BLR 1 (CA).
20   See Clause 13. *Yorkshire Water Authority* v. *Sir Alfred McAlpine* (1985) 32 BLR 114; *Holland Dredging* v. *Dredging and Construction Co.* (1987) 37 BLR 1 (CA).

the responsibility of the Employer under, *inter alia*, the Health and Safety at Work etc. Act 1974, the Construction (Design and Management) Regulations and the Occupiers Liability Acts 1957 and 1984 will not be excluded.

### Form of Agreement 9

The Contractor shall if called upon so to do enter into and execute an agreement to be prepared at the cost of the Employer in the form annexed to these Conditions.

*Form annexed to these Conditions* This form includes a provision for the agreement to be made under seal. The execution of a sealed contract is frequently in the Employer's interest as it extends the limitation period from 6 years to 12 years.

### Performance security 10

(1) If the Contract requires the Contractor to provide security for the proper performance of the Contract he shall obtain and provide to the Employer such security in a sum not exceeding 10% of the Tender Total within 28 days of the award of the Contract. The security shall be provided by a body approved by the Employer and be in the Form of Bond annexed to these Conditions. The Contractor shall pay the cost of the security unless the Contract provides otherwise.

### Dispute resolution upon security

(2) For the purposes of the dispute resolution provisions in such security

   (a) the Employer shall be deemed to be a party to the security for the purpose of doing everything necessary to give effect to such provisions and

   (b) any agreement decision award or other determination touching or concerning the relevant date for the discharge of the security shall be wholly without prejudice to the resolution or determination of any dispute between the Employer and the Contractor under Clause 66.

*Effect of non-compliance* Where the Contractor fails to provide the security within 28 days, it is thought that this does not amount immediately to a breach which is sufficiently serious to enable the Employer to terminate the Contract. Obligations as to time are generally treated as 'not of the essence'; in other words, their breach does not entitle the Employer to terminate. Where, however, the Employer serves notice on the Contractor that he will insist, and the Contractor fails to comply within a reasonable period, the Employer will be entitled to terminate the Contract and claim damages.[21]

*The Form of Bond*   This bond has been redrafted for the 7th Edition. The former document attracted a lot of criticism for its archaic language and unclear intent.[22] The bond is now clearly a performance bond; any call by the Employer requires proof of breach and damage. The bond contains its own internal arbitration provisions, hence the reference in Clause 10(2).

### Provision and interpretation of information   11

(1)   As between the Employer and the Contractor and without prejudice to sub-clause (2) of this Clause information on

   (a)   the nature of the ground and subsoil and hydrological conditions and

   (b)   pipes and cables in on or over the ground

   obtained by or on behalf of the Employer from investigations under-taken relevant to the Works shall only be taken into account to the extent that it was made available to the Contractor before the submission of his tender.

   The Contractor shall be responsible for the interpretation of all such information for the purposes of constructing the Works and for any design which is the Contractor's responsibility under the Contract.

### Inspection of Site

(2)   The Contractor shall be deemed to have inspected and examined the Site and its surroundings and information available in connection there-with and to have satisfied himself so far as is practicable and reasonable before submitting his tender as to

   (a)   the form and nature thereof including the ground and sub-soil and hydrological conditions and

   (b)   the extent and nature of work and materials necessary for con-structing and completing the Works and

   (c)   the means of communication with and access to the Site and the accommodation he may require

   and in general to have obtained for himself all necessary information as to risks contingencies and all other circumstances which may influence or affect his tender.

---

21   As to service of notice making time of the essence and its effects see *United Scientific Holdings* v. *Burnley Council* [1978] AC 904.

22   See *Trafalgar House Construction* v. *General Surety & Guarantee Co.* [1996] 1 AC 199.

**Basis and sufficiency of tender**

(3)  The Contractor shall be deemed to have

    (a)  based his tender on his own inspection and examination as afore-said and on all information whether obtainable by him or made available by the Employer and

    (b)  satisfied himself before submitting his tender as to the correctness and sufficiency of the rates and prices stated by him in the Bill of Quantities which shall (unless otherwise provided in the Con-tract) cover all his obligations under the Contract.

*Clause 11(1)—information which is withheld by the Employer*    This provision does not oblige the Employer to provide to the Contractor all the information he has obtained on site conditions. But, where the Employer withholds any information that information may not be 'taken into account'. Thus, where the Employer withholds a site investigation or a drawing showing locations of services, then such information shall not be taken into account when considering whether ground or services encountered were foreseeable.

*Clause 11(1)—responsibility and interpretation of information supplied*    By Clause 11(1) the Employer does not warrant that the information is suffi-cient to design any of the Works. Interpretation for the purpose of design is to be undertaken with all reasonable skill and care and diligence (see Clause 8(2)) and with proper caution.[23]

*Clause 11(2)—inspection of the site and its surroundings*    The extent to which it is 'practicable and reasonable' to inspect and examine the site etc. is a question of fact. It is submitted that, in the ordinary case, the location of the Contractor's main office has no bearing on the matter; otherwise local contractors would take on a higher obligation. Where the Works are costly, additional effort can reasonably be expected. Likewise, where an experi-enced contractor would reasonably foresee that practical difficulties may be encountered, there is a greater obligation of inspection. In practice, how-ever, there is often little opportunity to do more than examine the ground surface, open water courses, vegetation and areas surrounding the Site which are accessible to the public.

---

23    *Moneypenny v. Hartland* (1824) 2 C&P 378; *Sealand of the Pacific v. Robert C McHaffie Ltd* [1974] 51 DLR (3d) 702 (British Colombia Court of Appeal).

*Clause 11(3)(b)—sufficiency of the tender*   This appears to add little to the Contractor's obligations. Where, however, the tender rates are weighted for any reason (e.g. where 'front-end-loaded' in order to promote early positive cash flow) the Contractor will be held to these prices even where they turn out to work to his disadvantage.

### Adverse physical conditions and artificial obstructions    12

(1)  If during the carrying out of the Works the Contractor encounters physical conditions (other than weather conditions or conditions due to weather conditions) or artificial obstructions which conditions or obstructions could not in his opinion reasonably have been foreseen by an experienced contractor the Contractor shall as early as practicable give written notice thereof to the Engineer.

### Intention to claim

(2)  If in addition the Contractor intends to make any claim for additional payment or extension of time arising from any such condition or obstruction he shall at the same time or as soon thereafter as may be reasonable inform the Engineer in writing pursuant to Clause 53 and/or Clause 44(1) as may be appropriate specifying the condition or obstruction to which the claim relates.

### Measures being taken

(3)  When giving notification or information in accordance with sub-clauses (1) and/or (2) of this Clause or as soon as practicable thereafter the Contractor shall give details of any anticipated effects of the condition or obstruction the measures he has taken is taking or is proposing to take their estimated cost and the extent of the anticipated delay in or interference with the carrying out of the Works.

### Action by Engineer

(4)  Following receipt of any notification under sub-clauses (1) or (2) or receipt of details in accordance with sub-clause (3) of this Clause the Engineer may if he thinks fit among other things

   (a)  require the Contractor to investigate and report upon the practicality cost and timing of alternative measures which may be available

   (b)  give written consent to measures notified under sub-clause (3) of this Clause with or without modification

   (c)  give written instructions as to how the physical conditions or artificial obstructions are to be dealt with

   (d)  order a suspension under Clause 40 or a variation under Clause 51.

### Conditions reasonably foreseeable

(5)   If the Engineer shall decide that the physical conditions or artificial obstructions could in whole or in part have been reasonably foreseen by an experienced contractor he shall so inform the Contractor in writing as soon as he shall have reached that decision but the value of any variation previously ordered by him pursuant to sub-clause (4)(d) of this Clause shall be ascertained in accordance with Clause 52 and included in the Contract Price.

### Delay and extra cost

(6)   Where an extension of time or additional payment is claimed pursuant to sub-clause (2) of this Clause the Engineer shall if in his opinion such conditions or obstructions could not reasonably have been foreseen by an experienced contractor

(a)   determine any delay which the Contractor has suffered and

(b)   determine the amount of any costs which may reasonably have been incurred by the Contractor (together with a reasonable percentage addition thereto in respect of profit)

by reason of such conditions or obstructions and shall notify the Contractor accordingly with a copy to the Employer.

Any delay so determined shall forthwith be considered under Clause 44(3) for an appropriate extension of time and the Contractor shall subject to Clause 53 be paid in accordance with Clause 60 the amount so determined.

*General*   Clauses 11 and 12 should be read together. Clause 11 sets out the position in regard to information about the Site and it surroundings. Clause 12 divides the risk of unforeseen conditions between the Contractor and Employer and establishes a right for the Contractor to claim should unforeseeable conditions occur. The right to recover under Clause 12 is in addition to any claims for (*a*) misrepresentation,[24] (*b*) implied warranties in borehole or other survey data,[25] or (*c*) breach of contract where survey information which is incorporated into the Contract is materially wrong.

*Clause 12(1)—general*   The test applied in Clause 12 is based on the reasoning that it is efficient for the Employer to receive tenders which do not include speculative contingencies for conditions which are unlikely to

---

24   See Chapter 8.
25   *Bacal v. Northampton Development Corporation* (1975) 8 BLR 88 (CA).

arise. Accordingly, the clause places the risk of adverse conditions substantially on the Employer.

*Avoidance of adverse physical conditions*   During the works, the Contractor will often become aware of adverse conditions which may affect works yet to be commenced. For example, during the works, a survey for a haul road may reveal that a section of the haul road will—unless diverted—run over a marsh. The Contractor may divert the road to avoid expensive piling. The question arises whether Clause 12 allows the costs of diversion to be recovered since the conditions from which the costs flow have not actually been 'encountered' because they have been avoided. It is suggested that a strictly literal interpretation is not appropriate and that a Contractor who properly mitigates the costs should not be penalised. A real problem exists, however, where avoidance decisions are taken in good faith only later to be proved wrong. For example, evidence of large water flows into the line of a proposed inclined tunnel may be uncovered. As a result, the direction of an inclined tunnel drive may be reversed to excavate uphill (this is safer for operatives); if, in the event, the water source is found to be of very limited volume so that the expected problem does not arise, what is the position? It is submitted that each case must be treated in the light of its own circumstances. In particular, when decisions are taken to safeguard the safety of operatives, it is submitted that a real likelihood of dangerous physical conditions may be treated as equivalent to an encounter with such conditions.

*'Physical conditions', 'artificial obstructions'*   'Physical conditions' are not limited to pre-existing conditions and may include an absence of the anticipated physical conditions,[26] a transient condition or a combination of physical conditions and applied loading.[27] The term 'artificial obstructions' clearly includes such things as buried services etc. It is not clear whether or not intangible obstructions brought about artificially (i.e. by human agency) such as local authority or pressure-group interventions fall within the scope of Clause 12.

*'Other than weather conditions or conditions due to weather conditions'*   It is not clear whether conditions which are contributed to by weather, such as

---

26   *Atlantic Civil Pty Ltd* v. *Water Administration Ministerial Corporation* (1992) 83 BLR 116 (High Court of NSW).
27   *Humber Oil Terminals Trustee* v. *Harbour and General* (1991) 59 BLR 1 (CA).

rising water tables and floods caused by breached river banks are included. It is submitted that the test is whether or not the weather was the dominant cause of the condition.

*'Could not ... reasonably have been foreseen by an experienced contractor'*   This test is at the heart of Clause 12. It is given in objective terms. It is submitted that the test is to be viewed from the perspective of a contractor with reasonable skill and average experience of the work envisaged at the time of tender; the skill and experience of the particular contractor is irrelevant. Commentators frequently attempt to recast the test into different words which they feel are easier to apply. It is submitted that this is a mistake. The test has been broadly formulated to allow it to be used in a wide range of circumstances and the courts have rightly been wary of embellishing the test.

*Illustrations of the test*   Where a contractor undertook to excavate under an old sewer, the existence of which was known at the time of tender, the court found that the poor condition of the sewer was reasonably foreseeable.[28] Where, however, the conditions encountered or their impact on the work is surprising, this is indicative that those conditions were not reasonably foreseeable in the sense of Clause 12.[29] In a South African case[30] the wording was 'sub-surface conditions which in the opinion of the engineer could not reasonably have been foreseen'. The contract was for a railway tunnel. Rock was classified into six classes, ranging from Very Good (I) to Very Poor (VI). A report was provided at the time of tender which suggested that the rock condition was reasonably good. During construction the average rock encountered was 1½ classes worse than suggested by the report. Corbett C.J. said:[31]

> The differences between the geomechanical classes of rock predicted in the report and those alleged to have been encountered are so substantial that if the report predictions represent approximately what was reasonably foreseeable, the actual conditions were clearly not reasonably foreseeable. ... It may be that in evaluating the report, Comiat should have made some allowance for predictions being overly optimistic and thus built a safety margin into its

28   *CJ Pearce & Co. Ltd* v. *Hereford Corporation* (1968) 66 LGR 647.
29   Award No. 5 (1989) CLY 1994, 98 (Mr Uff QC, arbitrator). On appeal to the Court of Appeal: *Humber Oil Terminals Trustee* v. *Harbour and General* (1991) 59 BLR 1 (CA).
30   *Companie Interafricaine de Travaux (Comiat)* v. *South African Transport Services* (1991) CLY 1994, 149 (South African Supreme Court).
31   At p. 169. The entire court agreed with this analysis.

tender (I make no finding in this regard), but it seems to me to be unlikely that any such allowance would have come anywhere near to bridging the gap between the report predictions and actuality.

In an Australian case,[32] Clause 12 of the contract was closely modelled on Clause 12 of the ICE contract. The contractor was required to obtain fill materials from a borrow pit; when he began to excavate, the material was unsuitable. This was held to fall squarely within Clause 12.

*Partly foreseeable conditions*   Conditions encountered may be reasonably foreseeable up to a point and unforeseeable thereafter, so that only the additional costs etc. are recoverable under Clause 12(6). Thus, where water inflow into an excavation is greater than could reasonably have been foreseen, this does not entitle the Contractor to all the costs of pumping it dry, merely the costs in excess of what was reasonably foreseeable.

*Notifications*   Clause 12(1) provides that the Contractor shall give notice of any condition etc. which is in his opinion not reasonably foreseeable. Clause 12(2) provides for a notice of intention to claim. These notices are not necessary conditions to recovery, but failure to serve them in accordance with the Contract may reduce the amount recoverable. Clause 53(5) provides that the Contractor shall be entitled to recover 'only to the extent that the Engineer has not been prevented from or substantially prejudiced by such failure in investigating the said claim'.

*Measures and action*   Clause 12(3) provides for the Contractor to supply details of anticipated effects or measures being taken; there is no express requirement for him to supply written details. Clause 12(4) provides for the Engineer to take action to control the situation. Engineers are reluctant to take proactive decisions under Clause 12(4); such action, they believe, may be perceived as evidence that the conditions are not reasonably foreseeable, or may amount to a variation entitling the Contractor to recover without having to demonstrate that conditions are unforeseeable.[33] These risks may be overstated; in any event Clause 12(5) expressly contemplates the Engineer altering his view after issuing a variation order.

---

32   *Atlantic Civil Pty Ltd* v. *Water Administration Ministerial Corporation* (1992) 83 BLR 116 (High Court of NSW).

33   See *Simplex Piling* v. *St Pancras Borough Council* (1958) 14 BLR 80: here the Contractor was entitled to be paid for the variation despite requesting it for his own benefit; the court held that since the contractual machinery was operated, the right to payment automatically followed.

*Clause 12(6)—recovery by the contractor*   Where the Engineer decides that the conditions etc. could not reasonably have been foreseen, he is to determine the

> costs which may reasonably have been incurred by the Contractor (together with a reasonable percentage addition thereto in respect of profit) ... by reason of such conditions or obstructions.

It is submitted that this means:

(1)   To the extent that any conditions are not reasonably foreseeable the Contractor is entitled to reasonable cost plus profit.

(2)   This cost plus profit may be in addition to or, in substitution for (in part or in whole) the billed rates. To the extent that the original billed work is still required, despite the unforeseen conditions, cost plus profit will be paid only for the new/additional work; the original work will be paid at billed rates.

(3)   The Contractor must demonstrate that the costs were actually incurred by showing proper cost breakdowns for items of work. The fact that an item costs £x does not mean an automatic entitlement to £x; the Contractor must show that all costs were reasonable and neither extravagant nor wasteful and that he took all reasonable steps to mitigate the cost.

(4)   Any costs claimed and the need for them, are to viewed at the time they were incurred and not in hindsight.

Thus, upon encountering adverse conditions, the Contractor may reasonably decide to employ a construction technique X, which may prove inappropriate as circumstances unfold, requiring a revised technique Y. Here the cost of X will be recoverable in addition to the cost of Y. This is not unreasonable as the Engineer has the option of controlling the measures being taken by instruction or variation order in accordance with Clause 12(4).

### Work to be to satisfaction of Engineer    13

(1)   Save insofar as it is legally or physically impossible the Contractor shall construct and complete the Works in strict accordance with the Contract to the satisfaction of the Engineer and shall comply with and adhere strictly to the Engineer's instructions on any matter connected therewith (whether mentioned in the Contract or not). The Contractor shall take instructions only from the Engineer or subject to Clause 2(4) from his duly appointed delegate.

### Mode and manner of construction

(2)  The whole of the materials Contractor's Equipment and labour to be provided by the Contractor under Clause 8 and the mode manner and speed of construction of the Works are to be of a kind and conducted in a manner acceptable to the Engineer.

### Delay and extra cost

(3)  If in pursuance of Clause 5 or sub-clause (1) of this Clause the Engineer shall issue instructions which involve the Contractor in delay or disrupt his arrangements or methods of construction so as to cause him to incur cost beyond that reasonably to have been foreseen by an experienced contractor at the time of tender then the Engineer shall take such delay into account in determining any extension of time to which the Contractor is entitled under Clause 44 and the Contractor shall subject to Clause 53 be paid in accordance with Clause 60 the amount of such cost as may be reasonable except to the extent that such delay and extra cost result from the Contractor's default. Profit shall be added thereto in respect of any additional permanent or temporary work. If such instructions require any variation to any part of the Works the same shall be deemed to have been given pursuant to Clause 51.

*General*   This clause contains a number of distinct powers and provisions, including a repetition of the Contractor's primary obligation to construct and complete the Works.

*Legal or physical impossibility*[34]   A legal impossibility arises where any specified method of construction or necessary structure contravenes the law (e.g. the Health and Safety at Work etc. Act 1974 or the Construction (Design and Management) Regulations 1994). A physical impossibility arises where any specified method of construction cannot physically be accomplished because of any restriction on access etc.[35] Where an impossibility arises, the Contractor is not obliged to perform the relevant work as originally specified. In such a case, it seems that Clause 13(1) requires the Engineer to issue an instruction as to how progress is to be achieved, although this is not expressly stated. Where such an instruction is not forthcoming, this may well amount to a breach of contract; the quantum of

---

34   The expression 'impossible' must not be interpreted too rigidly; it does not mean absolutely impossible: *Turriff Ltd v. Welsh National Water Development Authority* (1979) CLY 1994, 122.

35   As where the Works cannot be constructed in accordance with a programme or method statement which forms part of the contract terms: *Award No. 2* (1985), CLY 1994, 58 (Mr Hawker, arbitrator); on appeal to the Court of Appeal: *Yorkshire Water Authority v. Sir Alfred McAlpine* (1985) 32 BLR 114; *Holland Dredging v. Dredging and Construction Co.* (1987) 37 BLR 1 (CA).

damages will be the value of the instruction or variation order to which the contractor is entitled, valued in accordance with Clause 52.

*Contractor to comply with the Engineer's instructions*   The Contractor is required to comply with any Engineer's instruction on any matter 'whether mentioned in the Contract or not'—Clause 13(1). The instruction is to be in writing save where 'for any reason it is considered necessary to give any such instruction orally'—Clause 2(6)(a). Where the instruction requires a variation to the Works it shall be deemed to have been given as a variation order under Clause 51—Clause 13(3).

*'To the satisfaction of the Engineer', 'of a kind and conducted in a manner acceptable to the Engineer'*   These references are not thought to add anything of substance to the Contractor's general obligation. They make it clear, however, that during the currency of the project, any instructions by the Engineer are to be given effect, whether or not justified by an objective reading of the specification. Where the Contractor complies with the objective meaning of the Contract but the Engineer instructs him to perform to a more onerous specification, this will be an instruction amounting a variation and the Contractor will be entitled to claim for it.

### Programme to be furnished   14

(1) (a)   Within 21 days after the award of the Contract the Contractor shall submit to the Engineer for his acceptance a programme showing the order in which he proposes to carry out the Works having regard to the provisions of Clause 42(1).

(b)   At the same time the Contractor shall also provide in writing for the information of the Engineer a general description of the arrangements and methods of construction which the Contractor proposes to adopt for the carrying out of the Works.

(c)   Should the Engineer reject any programme under sub-clause (2)(b) of this Clause the Contractor shall within 21 days of such rejection submit a revised programme.

### Action by Engineer

(2)   The Engineer shall within 21 days after receipt of the Contractor's programme

(a)   accept the programme in writing or

(b)   reject the programme in writing with reasons or

(c)   request the Contractor to supply further information to clarify or substantiate the programme or to satisfy the Engineer as to its reasonableness having regard to the Contractor's obligations under the Contract.

Provided that if none of the above actions is taken within the said period of 21 days the Engineer shall be deemed to have accepted the programme as submitted.

**Provision of further information**

(3)   The Contractor shall within 21 days after receiving from the Engineer any request under sub-clause (2)(c) of this Clause or within such further period as the Engineer may allow provide the further information requested failing which the relevant programme shall be deemed to be rejected.

Upon receipt of such further information the Engineer shall within a further 21 days accept or reject the programme in accordance with sub-clauses (2)(a) or (2)(b) of this Clause.

**Revision of programme**

(4)   Should it appear to the Engineer at any time that the actual progress of the work does not conform with the accepted programme referred to in sub-clause (1) of this Clause the Engineer shall be entitled to require the Contractor to produce a revised programme showing such modifications to the original programme as may be necessary to ensure completion of the Works or any Section within the time for completion as defined in Clause 43 or extended time granted pursuant to Clause 44. In such event the Contractor shall submit his revised programme within 21 days or within such further period as the Engineer may allow. Thereafter the provisions of sub-clauses (2) and (3) of this Clause shall apply.

**Design criteria**

(5)   The Engineer shall provide to the Contractor such design criteria relevant to the Permanent Works or any Temporary Works design supplied by the Engineer as may be necessary to enable the Contractor to comply with sub-clauses (6) and (7) of this Clause.

**Methods of construction**

(6)   If requested by the Engineer the Contractor shall submit at such times and in such further detail as the Engineer may reasonably require information pertaining to the methods of construction (including Temporary Works and the use of Contractor's Equipment) which the Contractor proposes to adopt or use and calculations of stresses strains and deflections that will arise in the Permanent Works or any parts thereof during construction so as to enable the Engineer to decide whether if these methods are adhered to the Works can be constructed

and completed in accordance with the Contract and without detriment to the Permanent Works when completed.

### Engineer's consent

(7)  The Engineer shall inform the Contractor in writing within 21 days after receipt of the information submitted in accordance with sub-clauses (1)(b) and (6) of this Clause either

   (a)  that the Contractor's proposed methods have the consent of the Engineer or

   (b)  in what respects in the opinion of the Engineer they fail to meet the requirements of the Contract or will be detrimental to the Permanent Works.

In the latter event the Contractor shall take such steps or make such changes in the said methods as may be necessary to meet the Engineer's requirements and to obtain his consent. The Contractor shall not change the methods which have received the Engineer's consent without the further consent in writing of the Engineer which shall not be unreasonably withheld.

### Delay and extra cost

(8)  If the Contractor unavoidably incurs delay or extra cost because

   (a)  the Engineer's consent to the proposed methods of construction is unreasonably delayed or

   (b)  the Engineer's requirements pursuant to sub-clause (7) of this Clause or any limitations imposed by any of the design criteria supplied by the Engineer pursuant to sub-clause (5) of this Clause could not reasonably have been foreseen by an experienced contractor at the time of tender

then the Engineer shall take such delay into account in determining any extension of time to which the Contractor is entitled under Clause 44 and the Contractor shall subject to Clause 53 be paid in accordance with Clause 60 the amount of such cost as may be reasonable except to the extent that such delay and extra cost result from the Contractor's default. Profit shall be added thereto in respect of any additional permanent or temporary work.

### Responsibility unaffected by acceptance or consent

(9)  Acceptance (or deemed acceptance) by the Engineer of the Contractor's programme in accordance with sub-clauses (2)(3) or (4) of this Clause and the consent of the Engineer to the Contractor's proposed methods of construction in accordance with sub-clause (7) of this Clause shall not relieve the Contractor of any of his duties or responsibilities under the Contract.

*What the Contractor must provide*   Clause 14 calls for two distinct documents to be provided:

(1)  A *programme*—the format of the programme is not specified, save that it must show the order in which the Contractor proposes to carry out the Works.[36] Many contractors supply detailed programmes including bar charts and networks; it seems, however, that a simple list of activities which shows the proposed order may suffice. Upon receipt of the programme, however, the Engineer may request the Contractor to supply further information to 'clarify or substantiate the programme or to satisfy the Engineer as to its reasonableness'—Clause 14(2)(c).

(2)  A *'method statement'*—the expression 'method statement' does not appear in Clause 14; it is used here as convenient shorthand. What is required is a 'general description of the arrangements and methods of construction which the Contractor proposes to adopt for the carrying out of the Works'. The expression 'general description' is thought to indicate that the statement needs to set out the broad proposals, rather than a detailed statement of them. Note, however, that Clause 14(6) entitles the Engineer to call for further reasonable information as to methods of construction. While Clause 14(6) addresses principally the impact of temporary working on the Permanent Works, its effect appears wider than that.

*The importance of the programme etc. to the Engineer*   The programme and method statement assist the Engineer in safeguarding the Employer's interest in having the work done expeditiously, safely and efficiently. They also enables him to plan the flow of design information and possession of the Site to suit the programme.

*The Construction (Design and Management) Regulations 1994*   These regulations provide for a Health and Safety Plan which will contain details in common with the Clause 14 programme. The plan is prepared in the first instance by the Planning Supervisor (who may be the Engineer—see Clause 71). The principal contractor (who may be the Contractor—see Clause 71) then takes over responsibility for maintaining its accuracy 'until the end of the construction phase'—Reg. 15(4).

---

36   Having regard to Clause 42 which deals with possession, access, etc.

*The nature of the Clause 14 programme*    The Clause 14 programme and method statement are not terms of the Contract. They do not create rights and obligations for the parties in the ordinary sense. Once accepted, however, the Contractor is entitled to assume that the Employer/Engineer will provide such reasonable cooperation as will enable him to complete in accordance with the programme. There is no breach of the Contract if the Contractor decides later to reschedule his work;[37] nor is it a breach of the Contract if he fails to meet any of the dates shown on the programme or to execute the work in a manner other than as shown in the original programme[38] unless the parties have agreed to sectional completion—see Appendix Part 2. Likewise, the Employer will not be in breach if it is not possible to execute the Works in accordance with the programme or method statement.[39]

*Revision of the programme*    Once the programme has been accepted by the Engineer, it seems that it may only be formally revised where it appears to the Engineer 'that the actual progress of the work does not conform with the accepted programme'—Clause 14(4). In this event, the 'Engineer shall be entitled to require the Contractor to produce a revised programme' showing how the works may be completed within the Contract time for completion.[40] Where the Engineer or arbitrator later grants an extension of time, the Contractor may be entitled to claim a re-rate for acceleration under Clause 52. The Engineer may, it is submitted, unilaterally vary the programme under Clause 51 provided that this is for the benefit of the construction and/or functioning of the Works;[41] in this event he must value the variation. Clause 14(7) provides:

---

37    The general principle is that a Contractor may, subject to express terms, conduct his operations as he wishes: e.g. *Greater London Council* v. *Cleveland Bridge Engineering Co. Ltd* (1984) 34 BLR 50.

38    The Contractor does however have an obligation under 41(2) to 'proceed with the Works with due expedition and without delay in accordance with the Contract'. Furthermore, failure to proceed with 'due diligence' is one of the grounds upon which the Contractor's employment may be terminated under Clause 63(1).

39    The Contractor may, however, be entitled under the Contract, e.g. under Clause 51 if the inability to perform the work amounts to a variation or under Clause 42(2) where the Employer fails to provide the Site in accordance with the accepted programme.

40    Where the Contractor is so late that he cannot possibly complete within the Contract time, it is thought that he must show a completion date as early as is practicable.

41    See Clause 51 and the commentary there. Variations of the programme for other reasons (e.g. to match the Employer's cash flow with the expenditure plan or to accommodate other contractors) are not permissible.

The Contractor shall not change the methods which have received the Engineer's consent without further consent in writing of the Engineer which shall not be unreasonably withheld.

It is not clear whether the 'methods' include the programme, although a method probably includes a sequence of work and hence entails the programme.

*Timing of programmes and information*    The timing in respect of the provision and approval of the programme is as follows:

(1)    Within 21 of the award of the Contract, the Contractor shall furnish a programme—Clause 14(1)(a).
(2)    Within 21 days of the receipt of the programme, the Engineer shall accept or reject it or request further information. If none of these actions is taken, the Engineer is deemed to have accepted the programme—Clause 14(2).
(3)    If the Engineer rejects the programme, the Contractor shall submit a revised programme within 21 days of the rejection—Clause 14(1)(c).
(4)    If the Engineer perceives that the actual progress does not accord with the accepted programme he is entitled to require the Contractor to provide a revised programme within 21 days—Clause 14(4).

The timing in respect of the method statement is slightly different:

(1)    The method statement is to be provided 'at the same time' as the programme—Clause 14(1)(b).
(2)    The Engineer shall inform the Contractor within 21 days whether the method statement has the consent of the Engineer or in what respects it fails to meet the Contract. In the latter case, the Contractor shall make appropriate changes, but no time is set for this—Clause 14(7).
(3)    Further details called for under Clause 14(6) are to be submitted 'at such times … as the Engineer may reasonably require'.

*Consent and responsibility*    The Engineer is obliged to indicate his view on the programme; otherwise he is deemed to have accepted it—proviso to Clause 14(2). Likewise, he must indicate his view on the method statement, though there is no equivalent provision by which the method statement become automatically accepted—Clause 14(7). It is not clear why the terminology in each case differs: a programme is to be 'accepted', while a method statement receives 'consent'. In either event, acceptance or consent do not relieve the Contractor of his responsibility under the Contract—Clause 14(9).

*Design obligations*   Where the Contractor has express design obligations, Clause 14(5) provides for the supply of relevant information to and by the Contractor.

### Contractor's superintendence   15

(1)   The Contractor shall provide all necessary superintendence during the construction and completion of the Works and for as long thereafter as the Engineer may reasonably consider necessary.

Such superintendence shall be given by sufficient persons having adequate knowledge of the operations to be carried out (including the methods and techniques required the hazards likely to be encountered and methods of preventing accidents) for the satisfactory and safe construction of the Works.

### Contractor's agent

(2)   The Contractor or a competent and authorized agent or representative approved of in writing by the Engineer (which approval may at any time be withdrawn) is to be constantly on the Works and shall give his whole time to the superintendence of the same. Such authorized agent or representative shall be in full charge of the Works and shall receive on behalf of the Contractor directions and instructions from the Engineer or (subject to the limitations of Clause 2) the Engineer's Representative. The Contractor or such authorized agent or representative shall be responsible for the safety of all operations.

*Where the Contractor fails to provide proper superintendence*   It is clearly in the Employer's interest that the Contractor should provide proper supervision, and that there be a person on the site who is authorised to act on the Contractor's behalf. Where the Contractor fails to maintain an agent on Site in accordance with Clause 15, this will be a breach of contract. No damages will flow automatically from this breach. However, the Engineer will be able to suspend the Work until proper supervision is provided—Clause 40(1). Where the Contractor persistently fails to provide a proper agent, he will be liable to have the Contract determined pursuant to Clause 65(1)(j).

### Removal of Contractor's employees   16

The Contractor shall employ or cause to be employed in and about the construction and completion of the Works and in the superintendence thereof only persons who are careful skilled and experienced in their several trades and callings.

The Engineer shall be at liberty to object to and require the Contractor to remove or cause to be removed from the Works any person employed thereon who in the opinion of the Engineer mis-conducts himself or is incompetent or negligent in the performance of his duties or fails to conform with any particular provisions with regard to safety which may be set out in the Contract or persists in any conduct which is prejudicial to safety or health and such persons shall not be again employed upon the Works without the permission of the Engineer.

It is a precondition of removal that the person must 'in the opinion of the Engineer' (*a*) misconduct himself or (*b*) be incompetent or (*c*) fail to conform with safety procedures or (*d*) persist in conduct prejudicial to safety. Hence, where the Engineer cites any other reason or indicates that he is acting at the behest of the Employer without having formed the requisite opinion, any removal will be a breach of Contract. The Engineer must act impartially—Clause 2(7); where he acts otherwise than impartially, this will be a breach of contract by the Employer. Thus, where a person is improperly removed from the Works and this causes a loss to the Contractor, damages will be recoverable.

**Setting out    17**

(1)  The Contractor shall be responsible for the true and proper setting-out of the Works and for the correctness of the position levels dimensions and alignment of all parts of the Works and for the provision of all necessary instruments appliances and labour in connection therewith.

(2)  If at any time during the progress of the Works any error shall appear or arise in the position levels dimensions or alignment of any part of the Works the Contractor on being required so to do by the Engineer shall at his own cost rectify such error to the satisfaction of the Engineer unless such error is based on incorrect data supplied in writing by the Engineer or the Engineer's Representative in which case the cost of rectifying the same shall be borne by the Employer.

(3)  The checking of any setting-out or of any line or level by the Engineer or the Engineer's Representative shall not in any way relieve the Contractor of his responsibility for the correctness thereof and the Contractor shall carefully protect and preserve all bench-marks sight rails pegs and other things used in setting out the Works.

'Setting out' is the transfer of design dimensions etc. from plans and drawings to the actual work in progress by way of marks, lines, levels, etc. Incorrect setting out may lead to the Works being incorrectly located. This clause indicates clearly that the Contractor is to be responsible for his own setting out except where the error is introduced by 'incorrect data supplied in writing by the Engineer or the Engineer's Representative'. Note that, in this instance, the Engineer's Representative needs no specific authorisation under Clause 2(4) to fix the Employer with liability.

### Boreholes and exploratory excavation 18

If at any time during the construction of the Works the Engineer shall require the Contractor to make boreholes or to carry out exploratory excavation such requirement shall be ordered in writing and shall be deemed to be a variation under Clause 51 unless a Provisional Sum or Prime Cost Item in respect of such anticipated work shall have been included in the Bill of Quantities.

Where difficult ground is encountered or suspected, it is frequently useful for the Engineer to order boreholes to be drilled or other exploratory excavations to be made. It is thought that the Engineer is not entitled to require the Contractor to provide an interpretive report on the boreholes.

### Safety and security 19

(1) The Contractor shall throughout the progress of the Works have full regard for the safety of all persons entitled to be upon the Site and shall keep the Site (so far as the same is under his control) and the Works (so far as the same are not completed or occupied by the Employer) in an orderly state appropriate to the avoidance of danger to such persons and shall among other things in connection with the Works provide and maintain at his own cost all lights guards fencing warning signs and watching when and where necessary or required by the Engineer or the Engineer's Representative or by any competent statutory or other authority for the protection of the Works or for the safety and convenience of the public or others.

### Employer's responsibilities

(2) If the Employer carries out work on the Site with his own workpeople he shall in respect of such work

(a) have full regard for the safety of all persons entitled to be upon the Site and

(b) keep the Site in an orderly state appropriate to the avoidance of danger to such persons.

If the Employer employs other contractors on the Site he shall require them to have the same regard for safety and avoidance of danger.

This clause sets out the general principle that the Contractor is responsible for safety on the Site, but that persons employed by the Employer shall keep their work safe and take proper care for the safety of others. This provision must be read together with the requirements of the health and safety legislation[42] and the legislation dealing with occupation of land.[43] The fact

---

42 Especially the Health and Safety at Work etc. Act 1974 and the Construction 'Design and Management' Regulations 1994; see also Clause 71.
43 Occupiers' Liability Acts 1957 and 1984.

that the provisions of Clause 19 extend only to 'all persons entitled to be on the Site' will not prevent the Contractor (or indeed the Employer) from being liable in appropriate cases to trespassers under the Occupier's Liability Act 1984.[44]

### Care of the Works    20

(1) (a) The Contractor shall save as in paragraph (b) hereof and subject to sub-clause (2) of this Clause take full responsibility for the care of the Works and materials plant and equipment for incorporation therein from the Works Commencement Date until the date of issue of a Certificate of Substantial Completion for the whole of the Works when the responsibility for the said care shall pass to the Employer.

(b) If the Engineer issues a Certificate of Substantial Completion for any Section or part of the Permanent Works the Contractor shall cease to be responsible for the care of that Section or part from the date of issue of that Certificate of Substantial Completion when the responsibility for the care of that Section or part shall pass to the Employer.

(c) The Contractor shall take full responsibility for the care of any work and materials plant and equipment for incorporation therein which he undertakes during the Defects Correction Period until such work has been completed.

### Excepted Risks

(2) The Excepted Risks for which the Contractor is not liable are loss or damage to the extent that it is due to

(a) the use or occupation by the Employer his agents servants or other contractors (not being employed by the Contractor) of any part of the Permanent Works

(b) any fault defect error or omission in the design of the Works (other than a design provided by the Contractor pursuant to his obligations under the Contract)

(c) riot war invasion act of foreign enemies or hostilities (whether war be declared or not)

---

44   See also Reg. 16(1)(c) of the Construction (Design and Management) Regulations 1994. This provides that

The principal contractor [the Contractor may be the principal contractor—see Clause 71] appointed for any project shall— ... (c) take reasonable steps to ensure that only authorised persons are allowed into any premises or part of premises where construction work is being carried out.

It is possible that this creates a civil liability—see Reg. 21.

(d)  civil war rebellion revolution insurrection or military or usurped power

(e)  ionizing radiations or contamination by radioactivity from any nuclear fuel or from any nuclear waste from the combustion of nuclear fuel radioactive toxic explosive or other hazardous properties of any explosive nuclear assembly or nuclear component thereof and

(f)  pressure waves caused by aircraft or other aerial devices travelling at sonic or supersonic speeds.

**Rectification of loss or damage**

(3)  (a)  In the event of any loss or damage to

(i)   the Works or any Section or part thereof or

(ii)  materials plant or equipment for incorporation therein

while the Contractor is responsible for the care thereof (except as provided in sub-clause (2) of this Clause) the Contractor shall at his own cost rectify such loss or damage so that the Permanent Works conform in every respect with the provisions of the Contract and the Engineer's instructions. The Contractor shall also be liable for any loss or damage to the Works occasioned by him in the course of any operations carried out by him for the purpose of complying with his obligations under Clauses 49 and 50.

(b)  Should any such loss or damage arise from any of the Excepted Risks defined in sub-clause (2) of this Clause the Contractor shall if and to the extent required by the Engineer rectify the loss or damage at the expense of the Employer.

(c)  In the event of loss or damage arising from an Excepted Risk and a risk for which the Contractor is responsible under sub-clause (1) (a) of this Clause then the Engineer shall when determining the expense to be borne by the Employer under the Contract apportion the cost of rectification into that part caused by the Excepted Risk and that part which is the responsibility of the Contractor.

**Insurance of Works etc.    21**

(1)  The Contractor shall without limiting his or the Employer's obligations and responsibilities under Clause 20 insure in the joint names of the Contractor and the Employer the Works together with materials plant and equipment for incorporation therein to the full replacement cost plus an additional 10% to cover any additional costs that may arise incidental to the rectification of any loss or damage including professional fees cost of demolition and removal of debris.

### Extent of cover

(2) (a) The insurance required under sub-clause (1) of this Clause shall cover the Employer and the Contractor against all loss or damage from whatsoever cause arising other than the Excepted Risks defined in Clause 20(2) from the Works Commencement Date until the date of issue of the relevant Certificate of Substantial Completion.

(b) The insurance shall extend to cover any loss or damage arising during the Defects Correction Period from a cause occurring prior to the issue of any Certificate of Substantial Completion and any loss or damage occasioned by the Contractor in the course of any operation carried out by him for the purpose of complying with his obligations under Clauses 49, 50 and 51.

(c) Nothing in this Clause shall render the Contractor liable to insure against the necessity for the repair or reconstruction of any work constructed with materials or workmanship not in accordance with the requirements of the Contract unless the Bill of Quantities provides a special item for this insurance.

(d) Any amounts not insured or not recovered from insurers whether as excesses carried under the policy or otherwise shall be borne by the Contractor or the Employer in accordance with their respective responsibilities under Clause 20.

### Damage to persons and property 22

(1) The Contractor shall except if and so far as the Contract provides otherwise and subject to the exceptions set out in sub-clause (2) of this Clause indemnify and keep indemnified the Employer against all losses and claims in respect of

(a) death of or injury to any person or

(b) loss of or damage to any property (other than the Works)

which may arise out of or in consequence of the construction of the Works and the remedying of any defects therein and against all claims demands proceedings damages costs charges and expenses whatsoever in respect thereof or in relation thereto.

### Exceptions

(2) The exceptions referred to in sub-clause (1) of this Clause which are the responsibility of the Employer are

(a) damage to crops being on the Site (save in so far as possession has not been given to the Contractor)

(b) the use or occupation of land provided by the Employer for the purposes of the Contract (including consequent losses of crops) or interference whether temporary or permanent with any right of way light air or water or other easement or quasi-easement which are the unavoidable result of the construction of the Works in accordance with the Contract

(c) the right of the Employer to construct the Works or any part thereof on over under in or through any land

(d) damage which is the unavoidable result of the construction of the Works in accordance with the Contract and

(e) death of or injury to persons or loss of or damage to property resulting from any act neglect or breach of statutory duty done or committed by the Employer his agents servants or other contractors (not being employed by the Contractor) or for or in respect of any claims demands proceedings damages costs charges and expenses in respect thereof or in relation thereto.

**Indemnity by Employer**

(3) The Employer shall subject to sub-clause (4) of this Clause indemnify the Contractor against all claims demands proceedings damages costs charges and expenses in respect of the matters referred to in the exceptions defined in sub-clause (2) of this Clause.

**Shared responsibility**

(4) (a) The Contractor's liability to indemnify the Employer under sub-clause (1) of this Clause shall be reduced in proportion to the extent that the act or neglect of the Employer his agents servants or other contractors (not being employed by the Contractor) may have contributed to the said death injury loss or damage.

(b) The Employer's liability to indemnify the Contractor under sub-clause (3) of this Clause in respect of matters referred to in sub-clause (2)(e) of this Clause shall be reduced in proportion to the extent that the act or neglect of the Contractor or his sub-contractors servants or agents may have contributed to the said death injury loss or damage.

**Third party insurance    22**

(1) The Contractor shall without limiting his or the Employer's obligations and responsibilities under Clause 22 insure in the joint names of the Contractor and the Employer against liabilities for death of or injury to any person (other than any operative or other person in the employment of the Contractor or any of his sub-contractors) or loss of or damage to any property (other than the Works) arising out of the performance of the Contract other than those liabilities arising out of the exceptions defined in Clause 22(2)(a) (b) (c) and (d).

### Cross liability clause

(2)  The insurance policy shall include a cross liability clause such that the insurance shall apply to the Contractor and to the Employer as separate insured.

### Amount of insurance

(3)  Such insurance shall be for at least the amount stated in the Appendix to the Form of Tender.

### Accident or injury to operatives etc.   24

The Employer shall not be liable for or in respect of any damages or compensation payable at law in respect or in consequence of any accident or injury to any operative or other person in the employment of the Contractor or any of his sub-contractors save and except to the extent that such accident or injury results from or is contributed to by any act or default of the Employer his agents or servants and the Contractor shall indemnify and keep indemnified the Employer against all such damages and compensation (save and except as aforesaid) and against all claims demands proceedings costs charges and expenses whatsoever in respect thereof or in relation thereto.

### Evidence and terms of insurance   25

(1)  The Contractor shall provide satisfactory evidence to the Employer prior to the Works Commencement Date that the insurances required under the Contract have been effected and shall if so required produce the insurance policies for inspection. The terms of all such insurances shall be subject to the approval of the Employer (which approval shall not unreasonably be withheld). The Contractor shall upon request produce to the Employer receipts for the payment of current insurance premiums.

### Excesses

(2)  Any excesses on the policies of insurance effected under Clauses 21 and 23 shall be no greater than those stated in the Appendix to the Form of Tender.

### Remedy on Contractor's failure to insure

(3)  If the Contractor shall fail upon request to produce to the Employer satisfactory evidence that there is in force any of the insurances required under the Contract then the Employer may effect and keep in force any such insurance and pay such premium or premiums as may be necessary for that purpose and from time to time deduct the amount so paid from any monies due or which may become due to the Contractor or recover the same as a debt due from the Contractor.

### Compliance with policy conditions

(4)  Both the Employer and the Contractor shall comply with all conditions
laid down in the insurance policies. Should the Contractor or the
Employer fail to comply with any condition imposed by the insurance
policies effected pursuant to the Contract each shall indemnify the
other against all losses and claims arising from such failure.

*The care of the Works and insurance*    Clauses 20 to 25 cover all aspects of
the care and insurance of:

(1)  the Works
(2)  plant, materials, etc.
(3)  third parties and their property
(4)  workpeople.

The clauses allocate risk between the Contractor and Employer.

*Synopsis of risk allocation*    Clauses 20, 22 and 24 provide the general prin-
ciples of risk, namely that the Contractor is responsible for damage to the
Works (Clause 20), persons and property (Clause 22) and workpeople
(Clause 24). There are exceptions given in Clauses 20(2) and 22(2).

*Synopsis of insurance provisions*    Clauses 21, 23 and 25 deal with insurance. The basic principle is that the the Contractor shall insure against
damage to the Works (Clause 21), persons and property (Clause 23). The
insurance shall be in the joint names of the Contractor and the Employer.
The terms shall be subject to the approval of the Employer (Clause 25(1)).
Evidence of insurance shall be provided to the Employer and if this is not
done, the Employer may effect the insurance and countercharge the
Contractor.

### Giving of notices and payment of fees    26

(1)  Except where otherwise provided in the Contract the Contractor shall
give all notices and pay all fees required to be given or paid by any Act of
Parliament or any Regulation or Bye-law of any local or other statutory
authority in relation to the construction and completion of the Works
and by the rules and regulations of all public bodies and companies
whose property or rights are or may be affected in any way by the Works.

### Repayment by Employer

(2)  The Employer shall repay or allow to the Contractor all such sums as
the Engineer shall certify to have been properly payable and paid by the
Contractor in respect of such fees and also all rates and taxes paid by the

Contractor in respect of the Site or any part thereof or anything constructed or erected thereon or on any part thereof or any temporary structures situated elsewhere but used exclusively for the purposes of the Works or any structures used temporarily and exclusively for the purposes of the Works.

## Contractor to conform with Statutes etc.

(3) The Contractor shall ascertain and conform in all respects with the provisions of any general or local Act of Parliament and the Regulations and Bye-laws of any local or other statutory authority which may be applicable to the Works and with such rules and regulations of public bodies and companies as aforesaid and shall keep the Employer indemnified against all penalties and liability of every kind for breach of any such Act Regulation or Bye-law. Provided always that

(a) the Contractor shall not be required to indemnify the Employer against the consequences of any such breach which is the unavoidable result of complying with the Contract or instructions of the Engineer

(b) if the Contract or instructions of the Engineer shall at any time be found not to be in conformity with any such Act Regulation or Bye-law the Engineer shall issue such instructions including the ordering of a variation under Clause 51 as may be necessary to ensure conformity with such Act Regulation or Bye-law and

(c) the Contractor shall not be responsible for obtaining any planning permission which may be necessary in respect of the Permanent Works in their final position or of any Temporary Works designed by the Engineer in their designated position on Site. The Employer hereby warrants that all such permissions have been or will in due time be obtained.

(4) If the Contractor incurs delay or extra cost arising from matters referred to in sub-clause (3)(b) or failure of the Employer to comply with sub-clause (3)(c) of this Clause the Engineer shall take such delay into account in determining any extension of time to which the Contractor may be entitled under Clause 44 and the Contractor shall subject to Clause 53 be paid in accordance with Clause 60 the amount of such extra cost as may be reasonable except to the extent that such delay or extra cost result from the Contractor's default.

*Statutory obligations* Many statutory and similar provisions apply to civil engineering work.[45] Clause 26(3) requires that the Contractor shall 'ascertain and conform ... with' all statutes, regulations, etc. The Contractor

---

45   See Chapter 1, Section 4, for a general review.

must indemnify the Employer against any consequences of any breach of statutory requirements, save for the exceptions given in Clause 26(3). Clauses 26(1) and 26(2) deal with two common matters which arise in relation to such controls, namely notices (e.g. notices under the Building Regulations, the New Roads and Street Works Act 1991, etc.) and payment of fees, taxes, etc.

### New Roads and Street Works Act 1991—Definitions  27

(1) (a) In this Clause "the Act" shall mean the New Roads and Street Works Act 1991 and any statutory modification or re-enactment thereof for the time being in force.

(b) For the purpose of obtaining any licence under the Act required for the Permanent Works the undertaker shall be the Employer who for the purposes of the Act will be the licensee.

Provided that where the license contains a prohibition against assignment which is notified to the Contractor then the Contractor shall give the Employer all notices required to be given by the undertaker and shall indemnify the Employer from and against all costs and charges which may arise from any failure by him so to do.

(c) All other expressions common to the Act and to this Clause shall have the same meaning as those assigned to them by the Act.

### Licences

(2) (a) The Employer shall obtain any street works licence and any other consent licence or permission that may be required for the carrying out of the Permanent Works and shall supply the Contractor with copies thereof including details of any conditions or limitations imposed.

(b) Any condition or limitation in any licence obtained after the award of the Contract shall be deemed to be an instruction under Clause 13.

### Notices

(3) The Contractor shall be responsible for giving to any relevant authority any required notice (or advance notice where prescribed) of his proposal to commence any work. A copy of each such notice shall be given to the Employer.

### Delays attributable to variations

(4) If any instruction pursuant to sub-clause (2)(b) of this Clause results in delay to the construction and completion of the Works because the Contractor needs to comply with sub-clause (3) of this Clause the Engineer shall in addition to valuing the variation under Clause 52 take

such delay into account in determining any extension of time to which the Contractor is entitled under Clause 44 and the Contractor shall subject to Clause 53 be paid in accordance with Clause 60 such additional cost as the Engineer shall consider to have been reasonably attributable to such delay.

The New Roads and Street Works Act 1991 provides that an 'undertaker' shall obtain a 'streetworks licence' from the 'street authority' before carrying out any 'streetworks'. The Act requires the undertaker to obtain a licence before placing, retaining and inspecting 'apparatus' in the street. Apparatus includes a sewer, drain or tunnel. The Clause provides that both the Employer and the Contractor are to act as undertaker in different cases. The Employer is to obtain the licence and the Contractor shall be responsible for giving notices.[46]

### Patent rights    28

(1)   The Contractor shall indemnify and keep indemnified the Employer from and against all claims and proceedings for or on account of infringement of any patent right design trademark or name or other protected right in respect of any

(a)   Contractor's Equipment used for or in connection with the Works

(b)   materials plant and equipment for incorporation in the Works

and from and against all claims demands proceedings damages costs charges and expenses whatsoever in respect thereof or in relation thereto except where such infringement results from compliance with the design or Specification provided other than by the Contractor. In the latter event the Employer shall indemnify the Contractor from and against all claims and proceedings for or on account of infringement of any patent right design trademark or name or other protected right aforesaid.

### Royalties

(2)   Except where otherwise stated the Contractor shall pay all tonnage and other royalties rent and other payments or compensation (if any) for getting stone sand gravel clay or other materials required for the Works.

*Intellectual property*    This clause deals with the protection of property in ideas and designs.

---

46   See Sauvain SJ, *Highway Law*, 2nd edn, Sweet & Maxwell, 1997. The Streetworks Bill 1999 is currently (August 1999) before Parliament. When enacted it will require an undertaker of streetworks to supply specific information to the public.

### Interference with traffic and adjoining properties   29

(1)  All operations necessary for the construction and completion of the Works shall so far as compliance with the requirements of the Contract permits be carried on so as not to interfere unnecessarily or improperly with

   (a)  the convenience of the public or

   (b)  the access to public or private roads footpaths or properties whether in the possession of the Employer or of any other person and with the use or occupation thereof.

The Contractor shall indemnify and keep indemnified the Employer in respect of all claims demands proceedings damages costs charges and expenses whatsoever arising out of or in relation to any such matters.

### Noise disturbance and pollution

(2)  All work shall be carried out without unreasonable noise disturbance or other pollution.

### Indemnity by Contractor

(3)  To the extent that noise disturbance or other pollution is not the unavoidable consequence of constructing and completing the Works or performing the Contract the Contractor shall indemnify the Employer from and against any liability for damages on that account and against all claims demands proceedings damages costs charges and expenses whatsoever in regard or in relation to such liability.

### Indemnity by Employer

(4)  The Employer shall indemnify the Contractor from and against any liability for damages on account of noise disturbance or other pollution which is the unavoidable consequence of carrying out the Works and from and against all claims demands proceedings damages costs charges and expenses whatsoever in regard or in relation to such liability.

### Avoidance of damage to highways etc.   30

(1)  The Contractor shall use every reasonable means to prevent any of the highways or bridges communicating with or on the routes to the Site from being subjected to extraordinary traffic within the meaning of the Highways Act 1980 or in Scotland the Roads (Scotland) Act 1984 or any statutory modification or re-enactment thereof by any traffic of the Contractor or any of his sub-contractors and in particular shall select routes and use vehicles and restrict and distribute loads so that any such extraordinary traffic as will inevitably arise from the moving of Contractor's Equipment and materials or manufactured or fabricated articles from and to the Site shall be limited as far as reasonably possible and so that no unnecessary damage or injury may be occasioned to such highways and bridges.

### Transport of Contractor's Equipment

(2)  Save insofar as the Contract otherwise provides the Contractor shall be responsible for and shall pay the cost of strengthening any bridges or altering or improving any highway communicating with the Site to facilitate the movement of Contractor's Equipment or Temporary Works required in the carrying out of the Works and the Contractor shall indemnify and keep indemnified the Employer against all claims for damage to any highway or bridge communicating with the Site caused by such movement including such claims as may be made by any competent authority directly against the Employer pursuant to any Act of Parliament or other Statutory Instrument and shall negotiate and pay all claims arising solely out of such damage.

### Transport of materials

(3)  If notwithstanding sub-clause (1) of this Clause any damage occurs to any bridge or highway communicating with the Site arising from the transport of materials or manufactured or fabricated articles being or intended to form part of the Permanent Works or any Temporary Works designed by the Engineer the Contractor shall notify the Engineer as soon as he becomes aware of such damage or as soon as he receives any claim from the authority entitled to make such claim.

Where under any Act of Parliament or other Statutory Instrument the haulier of such materials or manufactured or fabricated articles is required to indemnify the highway authority against damage the Employer shall not be liable for any costs charges or expenses in respect thereof or in relation thereto.

In other cases the Employer shall negotiate the settlement of and pay all sums due in respect of such claim and shall indemnify the Contractor in respect thereof and in respect of all claims demands proceedings damages costs charges and expenses in relation thereto. Provided always that if and so far as any such claim or part thereof is in the opinion of the Engineer due to any failure on the part of the Contractor to observe and perform his obligations under sub-clause (1) of this Clause then the amount certified by the Engineer to be due to such failure shall be paid by the Contractor to the Employer or deducted from any sum due or which may become due to the Contractor.

These clauses deal with the impact of the Contractor's operations on traffic and adjoining properties and highway structures.

*Nuisance etc. claims*  Claims in respect of traffic and adjoining properties are dealt with in Clause 29. A system of cross-indemnities is established. The Contractor indemnifies the Employer against any claim, save where the claim is the 'unavoidable consequence of carrying out the Works'. In the latter case, it is the Employer who indemnifies the Contractor.

*Highways*   A system of cross-indemnities is also established by Clause 30. Clause 30(1) requires the Contractor to 'use every reasonable means' to prevent overloading and 'unnecessary damage'. Where damage occurs despite the Contractor complying with these provisions, the Employer negotiates and settles and pays any claims—Clause 30(3). Where, however, the damage occurs as a result of a breach of Clause 30(1), the Contractor indemnifies the Employer against claims—Clause 30(2).

### Facilities for other contractors   31

(1)   The Contractor shall in accordance with the requirements of the Engineer or Engineer's Representative afford all reasonable facilities for any other contractors employed by the Employer and their workmen and for the workmen of the Employer and of any other properly authorised authorities or statutory bodies who may be employed in the carrying out on or near the Site of any work not in the Contract or of any contract which the Employer may enter into in connection with or ancillary to the Works.

### Delay and extra cost

(2)   If compliance with sub-clause (1) of this Clause involves the Contractor in delay or cost beyond that reasonably to have been foreseen by an experienced contractor at the time of tender then the Engineer shall take such delay into account in determining any extension of time to which the Contractor is entitled under Clause 44 and the Contractor shall subject to Clause 53 be paid in accordance with Clause 60 the amount of such cost as may be reasonable. Profit shall be added thereto in respect of any additional permanent or temporary work.

*The extent of the Contractor's licence*   This clause saves for the Employer an important right, namely to allow other contractors onto the Site. Accordingly, the Contractor is not to be regarded as having an exclusive licence. It should be noted, however, that Clause 42(1) envisages that the Contract will prescribe the extent of portions of the Site of which the Contractor is to be given possession from time to time. Clause 42(2)(a),(b) requires the Employer to give the Contractor possession of the whole of the Site subject to any restrictions specified and access 'as is necessary to enable the Contractor to proceed with the construction of the Works with due despatch'.

*Clause 31(2) notice*   Where the Contractor intends to claim additional cost, Clause 53 requires him to give notice in writing. Where he fails to do so, he will be entitled only to the extent that the Engineer has been substantially prejudiced in investigating the claim.

**Fossils etc.   32**

All fossils coins articles of value or antiquity and structures or other remains or things of geological or archaeological interest discovered on the Site shall as between the Employer and the Contractor be deemed to be the absolute property of the Employer and the Contractor shall take reasonable precautions to prevent his workmen or any other persons from removing or damaging any such article or thing and shall immediately upon discovery thereof and before removal acquaint the Engineer of such discovery and carry out at the expense of the Employer the Engineer's orders as to the disposal of the same.

A find of fossils etc. which requires a delay or extra costs may found a claim under Clause 12. Where the Engineer issues an instruction (the clause speaks of the Engineer's 'orders') then this is to be in writing and may amount to a variation entitling the Contractor to additional payment.

**Clearance of Site on completion   33**

On completion of the Works the Contractor shall clear away and remove from the Site all Contractor's Equipment surplus material rubbish and Temporary Works of every kind and leave the whole of the Site and Permanent Works clean and in a workmanlike condition to the satisfaction of the Engineer.

This clause states what the Contract would otherwise imply.

**34**   (Not used)

This clause formerly dealt with the fair wages resolution, which is not longer in effect.

**Returns of labour and Contractor's Equipment   35**

The Contractor shall if required by the Engineer deliver to the Engineer or the Engineer's Representative a return in such form and at such intervals as the Engineer may prescribe showing in detail the numbers of the several classes of labour from time to time employed by the Contractor on the Site and such information respecting Contractor's Equipment as the Engineer may require. The Contractor shall require his sub-contractors to observe the provisions of this Clause.

The Engineer's right to call upon the Contractor to give returns of labour and equipment is a valuable one. It enables the Employer to keeps such records as may be required, e.g. for assessing and valuing claims.

## *Workmanship and materials*

### Quality of materials and workmanship and tests    36

(1)  All materials and workmanship shall be of the respective kinds
     described in the Contract and in accordance with the Engineer's
     instructions and shall be subjected from time to time to such tests as the
     Engineer may direct at the place of manufacture or fabrication or on the
     Site or such other place or places as may be specified in the Contract.
     The Contractor shall provide such assistance instruments machines
     labour and materials as are normally required for examining measuring
     and testing any work and the quality weight or quantity of any materials
     used and shall supply samples of materials before incorporation in the
     Works for testing as may be selected and required by the Engineer.

### Cost of samples

(2)  All samples shall be supplied by the Contractor at his own cost if the
     supply thereof is clearly intended by or provided for in the Contract but
     if not then at the cost of the Employer.

### Cost of tests

(3)  The cost of making any test shall be borne by the Contractor if such test
     is clearly intended by or provided for in the Contract and (in the cases
     only of a test under load or of a test to ascertain whether the design of
     any finished or partially finished work is appropriate for the purposes
     which it was intended to fulfil) is particularized in the Contract in suffi-
     cient detail to enable the Contractor to have priced or allowed for the
     same in his tender.

(4)  In all other cases the cost of making any test shall be borne by the
     Employer unless the need for such test results from the Contractor's
     default or from failure on the part of the Contractor to observe and per-
     form his obligations under the Contract.

### Access to site    37

The Engineer and any person authorized by him shall at all times have
access to the Works and to the Site and to all workshops and places
where work is being prepared or whence materials manufactured articles and machinery are being obtained for the Works and the
Contractor shall afford every facility for and every assistance in obtaining such access or the right to such access.

### Examination of work before covering up    38

(1)  No work shall be covered up or put out of view without the consent of
     the Engineer and the Contractor shall afford full opportunity for the
     Engineer to examine and measure any work which is about to be cov-
     ered up or put out of view and to examine foundations before

permanent work is placed thereon. The Contractor shall give due notice to the Engineer whenever any such work or foundations is or are ready or about to be ready for examination and the Engineer shall without unreasonable delay unless he considers it unnecessary and advises the Contractor accordingly attend for the purpose of examining and measuring such work or of examining such foundations.

### Uncovering and making openings

(2)   The Contractor shall uncover any part or parts of the Works or make openings in or through the same as the Engineer may from time to time direct and shall reinstate and make good such part or parts to the satisfaction of the Engineer. If any such part or parts have been covered up or put out of view after compliance with the requirements of sub-clause (1) of this Clause and are found to have been carried out in accordance with the Contract the cost of uncovering making openings in or through reinstating and making good the same shall be borne by the Employer but in any other case all such cost shall be borne by the Contractor.

### Removal of unsatisfactory work and materials    39

(1)   The Engineer shall during the progress of the Works have power to instruct in writing the

(a)   removal from the Site within such time or times specified in the instruction of any materials which in the opinion of the Engineer are not in accordance with the Contract

(b)   substitution with materials in accordance with the Contract and

(c)   removal and proper replacement (notwithstanding any previous test thereof or interim payment therefor) of any work which in respect of

(i)   material or workmanship or

(ii)   design by the Contractor or for which he is responsible

is not in the opinion of the Engineer in accordance with the Contract.

### Default of Contractor in compliance

(2)   In case of default on the part of the Contractor in carrying out such instruction the Employer shall be entitled to employ and pay other persons to carry out the same and all costs consequent thereon or incidental thereto as determined by the Engineer shall be recoverable from the Contractor by the Employer and may be deducted by the Employer from any monies due or to become due to him and the Engineer shall notify the Contractor accordingly with a copy to the Employer.

### Failure to disapprove

(3) Failure of the Engineer or any person acting under him pursuant to
Clause 2 to disapprove any work or materials shall not prejudice the
power of the Engineer or any such person subsequently to take action
under this Clause.

*Synopsis*   Clauses 36 to 39 deal with workmanship and materials generally. Clause 36(1) states the general principle that materials and
workmanship shall be as described in the Contract. Clause 36 also deals
with testing and sampling and states who is to pay for tests and samples.
Clause 37 requires the Contractor to allow the Engineer access for the purpose of inspecting work. Clause 38 requires that the Contractor shall allow
the Engineer to examine work before it is covered up, and Clause 38(2) provides for uncovering work which has been covered. Clause 39 requires that
the Contractor shall remove unsatisfactory work and materials.

*Clause 36—quality*   The Contractor's obligation is to provide work and
materials in accordance with the Contract. Where the Contract is not specific, the obligation implied is that materials will be sound and fit for their
intended purpose and the work will be done in a good and workmanlike
manner.[47] The obligation to supply and perform the work 'in accordance
with the Engineer's instructions' does not, it is thought, alter the Contractor's obligations in any way. It merely emphasises that the Contractor is
obliged to give effect to the instructions of the Engineer. Thus, if the Engineer requires materials to be removed, the Contractor must remove them
notwithstanding that they comply with the Contract. The Contractor will
be entitled to recover his loss, for example, as a variation.[48]

*Testing*   The Works may be subject to sampling or tests as required by the
Engineer. The Contractor shall provide any equipment required to carry
out the tests. Where the Contract 'clearly' intends or provides for such tests
or samples, the Contractor shall bear the cost. Otherwise, the cost is to be
borne by the Employer. The one exception is where the test is directed at
determining whether the materials or workmanship correspond to the Contract—where the tests show that the materials or workmanship do not
correspond, the cost is to be borne by the Contractor.

---

47   *Young & Marten Ltd* v. *McManus Childs Ltd* [1969] 1 AC 454 (HL).
48   See Clauses 13(3) and 51.

*Examination*   The Contractor is required to give the Employer a 'full opportunity' to 'examine and measure' any work which is about to be covered up.

### Suspension of work   40

(1)   The Contractor shall on the written order of the Engineer suspend the progress of the Works or any part thereof for such time or times and in such manner as the Engineer may consider necessary and shall during such suspension properly protect and secure the work so far as is necessary in the opinion of the Engineer. Except to the extent that such suspension is

   (a)   otherwise provided for in the Contract or

   (b)   necessary by reason of weather conditions or by some default on the part of the Contractor or

   (c)   necessary for the proper construction and completion or for the safety of the Works or any part thereof in as much as such necessity does not arise from any act or default of the Engineer or the Employer or from any of the Excepted Risks defined in Clause 20(2)

then if compliance with the Engineer's instructions under this clause involves the Contractor in delay or extra cost the Engineer shall take such delay into account in determining any extension of time to which the Contractor is entitled under Clause 44 and the Contractor shall subject to Clause 53 be paid in accordance with Clause 60 the amount of such extra cost as may be reasonable. Profit shall be added thereto in respect of any additional permanent or temporary work.

### Suspension lasting more than three months

(2)   If the progress of the Works or any part thereof is suspended on the written order of the Engineer and if permission to resume work is not given by the Engineer within a period of 3 months from the date of suspension then the Contractor may unless such suspension is otherwise provided for in the Contract or continues to be necessary by reason of some default on the part of the Contractor serve a written notice on the Engineer requiring permission within 28 days from the receipt of such notice to proceed with the Works or that part thereof in regard to which progress is suspended. If within the said 28 days the Engineer does not grant such permission the Contractor by a further written notice so served may (but is not bound to) elect to treat the suspension where it affects part only of the Works as an omission of such part under Clause 51 or where it affects the whole Works as an abandonment of the Contract by the Employer.

*Suspension*   Without this provision, suspension may amount to a breach, and may indeed amount to a fundamental breach entitling the Contractor to terminate the contract and to sue for damages. Here, the Contractor is entitled under Clause 40(1) to costs and profit plus an extension of time except to the extent that the suspension is due to one or more of the factors in Clause 40(1)(a), (b), (c). Where the work has been suspended for three months, the Contractor may (unless the suspension is due to the Contractor's default or is otherwise provided for) serve a 28 day notice upon the Engineer—Clause 40(2). Where the Engineer does not grant permission for the Contractor to proceed with the Works, the Contractor may by serving a further notice elect to treat the relevant Works as abandoned or omitted. It is thought that the Employer will not be in breach where this happens; however, where part of the Works are omitted this will amount to a variation which is to valued in accordance with Clause 52. Where the entire works are to be treated as an 'abandonment', it is thought that there will be an entitlement under the Contract to such sums as would be payable by way of damages for repudiation even though technically this will not amount to a breach.

## Commencement time and delays

### Works Commencement Date   41

(1)   The Works Commencement Date shall be

    (a)   the date specified in the Appendix to the Form of Tender or if no date is specified

    (b)   a date between 14 and 28 days of the award of the Contract to be notified to the Contractor by the Engineer in writing or

    (c)   such other date as may be agreed between the parties.

### Start of Works

(2)   The Contractor shall start the Works on or as soon as is reasonably practicable after the Works Commencement Date. Thereafter the Contractor shall proceed with the Works with due expedition and without delay in accordance with the Contract.

*The obligation to start on time and to 'proceed ... with due expedition and without delay'*   It is in the Employer's interests that the Contractor should control the Site until the project is complete and should complete expeditiously. Clause 41 requires the Contractor to start as soon as is reasonably practicable after the Works Commencement Date and thereafter to proceed with the Works with due expedition. Where the Contractor fails to start or to maintain progress, the Employer may determine the Contract under Clause 65. Clause 41 should generally be read in conjunction with Clause 46(1) which deals with the related issue of rate of progress which is too slow to ensure substantial completion within the time for completion.

**Possession of Site and access    42**

(1)   The Contract may prescribe

    (a)   the extent of portions of the Site of which the Contractor is to be given possession from time to time

    (b)   the order in which such portions of the Site shall be made available to the Contractor

    (c)   the availability and the nature of the access which is to be provided by the Employer

    (d)   the order in which the Works shall be constructed.

(2)   (a)   Subject to sub-clause (1) of this Clause the Employer shall give to the Contractor on the Works Commencement Date possession of the whole of the Site together with such access thereto as may be necessary to enable the Contractor to commence and proceed with the construction of the Works.

    (b)   Thereafter the Employer shall during the course of the Works give to the Contractor such further access in accordance with the Contract as is necessary to enable the Contractor to proceed with the construction of the Works with due despatch.

**Failure to give possession**

(3)   If the Contractor suffers delay and/or incurs extra cost from failure on the part of the Employer to give possession or access in accordance with the terms of this Clause the Engineer shall take such delay into account in determining any extension of time to which the Contractor is entitled under Clause 44 and the Contractor shall subject to Clause 53 be paid in accordance with Clause 60 the amount of any extra cost to which he may be entitled. Profit shall be added thereto in respect of any additional permanent or temporary work.

**Access and facilities provided by the Contractor**

(4)   The Contractor shall bear all costs and charges for any access required by him additional to those provided by the Employer. The Contractor shall also provide at his own cost any additional facilities outside the Site required by him for the purposes of the Works.

*Possession of the Site*    The Contractor requires access to and possession of such part of the Site as are necessary for him to undertake and complete the Works. The Contract may specify the extent of possession to be given. Where the Contract is silent, Clause 42(2)(a) requires that 'the whole of the Site ...' be given. Clause 42(2)(b) also requires the Employer to give such 'access' as is necessary to enable the Contractor to proceed with due despatch. Where such possession is not given, the Contractor is entitled to cost, profit and an extension of time—Clause 42(3).

### Time for completion    43

The whole of the Works and any Section required to be completed within a particular time as stated in the Appendix to the Form of Tender shall be substantially completed within the time so stated (or such extended time as may be allowed under Clause 44 or revised time agreed under Clause 46(3)) calculated from the Works Commencement Date.

The Contractor is obliged to complete by the time given in the Appendix to the Form of Tender, as extended or agreed under Clause 46(3). The Employer is entitled[49] to deduct liquidated damages at the rate specified in the Appendix where the Contractor completes late. Where no time for completion is included in the Appendix, a reasonable time for completion will be implied and unliquidated damages may be claimed.

### Extension of time for completion    44

(1)   Should the Contractor consider that

  (a)   any variation ordered under Clause 51(1) or

  (b)   increased quantities referred to in Clause 51(4) or

  (c)   any cause of delay referred to in these Conditions or

  (d)   exceptional adverse weather conditions or

  (e)   any delay impediment prevention or default by the Employer or

  (f)   other special circumstances of any kind whatsoever which may occur

  be such as to entitle him to an extension of time for the substantial completion of the Works or any Section thereof he shall within 28 days after the cause of any delay has arisen or as soon thereafter as is reasonable

---

49    It seems that the Engineer's role is limited to granting extensions of time and issuing the Certificate of Substantial Completion. He does not deduct liquidated damages in the Clause 60 certificate. It is for the Employer to decide whether or not he will deduct the damages.

deliver to the Engineer full and detailed particulars in justification of the period of extension claimed in order that the claim may be investigated at the time.

## Assessment of delay

(2) (a) The Engineer shall upon receipt of such particulars consider all the circumstances known to him at that time and make an assessment of the delay (if any) that has been suffered by the Contractor as a result of the alleged cause and shall so notify the Contractor in writing.

  (b) The Engineer may in the absence of any claim make an assessment of the delay that he considers has been suffered by the Contractor as a result of any of the circumstances listed in sub-clause (1) of this Clause and shall so notify the Contractor in writing.

## Interim grant of extension of time

(3) Should the Engineer consider that the delay suffered fairly entitles the Contractor to an extension of the time for the substantial completion of the Works or any Section thereof such interim extension shall be granted forthwith and be notified to the Contractor in writing with a copy to the Employer. In the event that the Contractor has made a claim for an extension of time but the Engineer does not consider the Contractor entitled to an extension of time he shall so inform the Contractor without delay.

## Assessment at due date for completion

(4) The Engineer shall not later than 14 days after the due date or extended date for completion of the Works or any Section thereof (and whether or not the Contractor shall have made any claim for an extension of time) consider all the circumstances known to him at that time and take action similar to that provided for in sub-clause (3) of this Clause. Should the Engineer consider that the Contractor is not entitled to an extension of time he shall so notify the Employer and the Contractor.

## Final determination of extension

(5) The Engineer shall within 28 days of the issue of the Certificate of Substantial Completion for the Works or for any Section thereof review all the circumstances of the kind referred to in sub-clause (1) of this Clause and shall finally determine and certify to the Contractor with a copy to the Employer the overall extension of time (if any) to which he considers the Contractor entitled in respect of the Works or the relevant Section. No such final review of the circumstances shall result in a decrease in any extension of time already granted by the Engineer pursuant to sub-clauses (3) or (4) of this Clause.

*The mechanism for granting extensions of time*   Where the Contractor claims an extension of time he should, in the first instance, deliver full and detailed particulars to the Engineer within 28 days 'or as soon thereafter as is reasonable'—Clause 44(1). The scheme of the clause suggests that this is not a condition precedent; for instance Clause 44(2)(b) entitles and Clause 44(4) requires the Engineer to make an assessment of extensions whether or not a claim for an extension has been received.

*The grounds for granting an extension*   The Engineer may extend the time for completion in a number of given situations. Clause 44(1)(c) refers to a number of clauses: 7(4), 12(2), 13(3), 14(8), 31(2), 40(1), 42(3) and 59(4)(f). The provision in Clause 44(1)(f) as to 'special circumstance of any kind whatsoever' is designed to protect the clause against any matter arising for which the Employer may be liable and which would cause the liquidated damages provisions to become ineffective.[50] There is a view that a 'sweeping up' provision such as this is ineffective.[51] It is submitted that this view is incorrect; it derives partly from the assumption that the liquidated damages provisions of a contract are for the Employer's benefit and hence are to be construed strictly against the Employer.[52] In fact liquidated damages provisions are valuable to the Contractor; they provide a valuable limit to his exposure for damages.[53] It is submitted, therefore, that the provision should be given a commercial meaning and that all events, including breaches by the Employer, entitle and require the Engineer to grant an extension, where appropriate.

*Assessment*   Clauses 44(2) and (3) imply a two stage approach:

(1)   An assessment by the Engineer of the delay suffered by the Contractor—Clause 44(2).
(2)   A consideration whether any delay suffered fairly entitled the Contractor to an extension of the time for substantial completion—Clause 44(3). The mere fact that the Contractor has been delayed in some

---

50   *Peak Construction (Liverpool) Ltd* v. *McKinney Foundations Ltd* (1971) 1 BLR 111 (CA).
51   See *Fernbrook Trading Co. Ltd* v. *Taggart* [1979] 1 NZLR 556 (New Zealand Supreme Court); *Perini Pacific Ltd* v. *Greater Vancouver Sewerage and Drainage District* (1966) 57 DLR (2d) 307 (British Columbia Court of Appeal). In these cases the 'sweeping up' provisions were similar to that found in Clause 44. In each case the court found that they had to be construed narrowly and in each case breaches of the employer were held to fall outside their scope; time was put at large.
52   *Peak Construction (Liverpool) Ltd* v. *McKinney Foundations Ltd* (1970) 1 BLR 111 (CA).
53   *Temloc* v. *Errill Properties Ltd* (1988) 39 BLR 30 (CA).

elements of the Works does not mean that the overall completion date is delayed.

*Interim, due date and final determinations*  The Contract entitles and/or requires the Engineer to make an assessment of the extension of time due in three different circumstances:

(1)  The Engineer shall assess any claim made by the Contractor during the currency of the Works; in addition he may of his own volition make such an assessment—Clause 44(2). As a result of this determination he may grant an 'interim extension'—Clause 44(3).
(2)  At the date which is the currently due date for completion of the Works, the Engineer is obliged to consider whether an extension of time is due. This obligation arises whether or not the Contractor has claimed an extension of time—Clause 44(4).
(3)  Within 28 days of the issue of the Certificate of Substantial Completion the Engineer certifies his final determination of the overall extension of time to which he considers the Contractor entitled—Clause 44(5). He may not reduce any extension already granted.

### Night and Sunday work   45

Subject to any provision to the contrary contained in the Contract none of the Works shall be carried out during the night or on Sundays without the permission in writing of the Engineer save when the work is unavoidable or absolutely necessary for the saving of life or property or for the safety of the Works in which case the Contractor shall immediately advise the Engineer or the Engineer's Representative. Provided always that this Clause shall not be applicable in the case of any work which it is customary to carry out outside normal working hours or by rotary or double shifts.

Generally speaking, the Engineer may withhold his permission for night and/or Sunday work. Where, however, the conditions in Clause 46(1) are met, the permission shall not be unreasonably withheld—see Clause 46(2).

### Rate of progress   46

(1)  If for any reason which does not entitle the Contractor to an extension of time the rate of progress of the Works or any Section is at any time in the opinion of the Engineer too slow to ensure substantial completion by the time or extended time for completion prescribed by Clause 43 and 44 as appropriate or the revised time for completion agreed under sub-clause (3) of this Clause the Engineer shall notify the Contractor in writing and the Contractor shall thereupon take such steps as are necessary and to which the Engineer may consent to expedite the progress so

as substantially to complete the Works or such Section by that prescribed time or extended time. The Contractor shall not be entitled to
any additional payment for taking such steps.

**Permission to work at night or on Sundays**

(2)   If as a result of any notice given by the Engineer under sub-clause (1) of
      this Clause the Contractor seeks the Engineer's permission to do any
      work on Site at night or on Sundays such permission shall not be unrea-
      sonably refused.

**Provision for accelerated completion**

(3)   If the Contractor is requested by the Employer or the Engineer to com-
      plete the Works or any Section within a revised time being less than the
      time or extended time for completion prescribed by Clauses 43 and 44
      as appropriate and the Contractor agrees so to do then any special terms
      and conditions of payment shall be agreed between the Contractor and
      the Employer before any such action is taken.

*Catching up notice under Clause 46(1)*   Where the Engineer gives notice
under Clause 46(1), the Contractor may incur costs associated with catching up with the programme. Where an extension of time is later granted by
the Engineer or arbitrator, so that the Clause 46(1) notice was wrongly
given, the Contractor may claim that he has incurred cost as a result of
complying with the Engineer's instruction and hence is entitled to recover
under Clause 13(3).

*Acceleration agreement—Clause 46(3)*   This clause acknowledges that the
parties may make an agreement to accelerate the work. It is thought that in
the absence of agreement, the Engineer may in any event order a variation
which has the effect of reducing the time for completion under Clause 51.
Such a variation must, however, be 'desirable for the completion of and/or
improved functioning of the Works'.

## Liquidated damages for delay

**Liquidated damages for delay in substantial completion of the whole of
the Works    47**

(1)   (a)   Where the whole of the Works is not divided into Sections the
            Appendix to the Form of Tender shall include a sum which repre-
            sents the Employer's genuine pre-estimate (expressed per week or
            per day as the case may be) of the damages likely to be suffered by
            him if the whole of the Works is not substantially completed
            within the time prescribed by Clause 43 or by any extension

thereof granted under Clause 44 or by any revision thereof agreed under Clause 46(3) as the case may be.

(b) If the Contractor fails to achieve substantial completion of the whole of the Works within the time so prescribed he shall pay to the Employer the said sum for every week or day (as the case may be) which shall elapse between the date on which the prescribed time expired and the date the whole of the Works is substantially completed.

Provided that if any part of the Works is certified as substantially complete pursuant to Clause 48 before the completion of the whole of the Works the said sum shall be reduced by the proportion which the value of the part so completed bears to the value of the whole of the Works.

## Liquidated damages for delay in substantial completion where the whole of the Works is divided into Sections

(2) (a) Where the Works is divided into Sections (together comprising the whole of the Works) which are required to be completed within particular times as stated in the Appendix to the Form of Tender sub-clause (1) of this Clause shall not apply and the said Appendix shall include a sum in respect of each Section which represents the Employer's genuine pre-estimate (expressed per week or per day as the case may be) of the damages likely to be suffered by him if that Section is not substantially completed within the time prescribed by Clause 43 or by any extension thereof granted under Clause 44 or by any revision thereof agreed under Clause 46(3) as the case may be.

(b) If the Contractor fails to achieve substantial completion of any Section within the time so prescribed he shall pay to the Employer the appropriate stated sum for every week or day (as the case may be) which shall elapse between the date on which the prescribed time expired and the date of substantial completion of that Section.

Provided that if any part of that Section is certified as substantially complete pursuant to Clause 48 before the completion of the whole thereof the appropriate stated sum shall be reduced by the proportion which the value of the part so completed bears to the value of the whole of that Section.

(c) Liquidated damages in respect of two or more Sections may where circumstances so dictate run concurrently.

## Damages not a penalty

(3) All sums payable by the Contractor to the Employer pursuant to this Clause shall be paid as liquidated damages for delay and not as a penalty.

### Limitation of liquidated damages

(4)  (a)  The total amount of liquidated damages in respect of the whole of the Works or any Section thereof shall be limited to the appropriate sum stated in the Appendix to the Form of Tender. If no such limit is stated therein then liquidated damages without limit shall apply.

(b)  Should there be omitted from the Appendix to the Form of Tender any sum required to be inserted therein either by sub-clause (1)(a) or by sub-clause (2)(a) of this Clause as the case may be or if any such sum is stated to be "nil" then to that extent damages shall not be payable.

### Recovery and reimbursement of liquidated damages

(5)  The Employer may

(a)  deduct and retain the amount of any liquidated damages becoming due under the provision of this Clause from any sums due or which become due to the Contractor or

(b)  require the Contractor to pay such amount to the Employer forthwith.

If upon a subsequent or final review of the circumstances causing delay the Engineer grants a relevant extension or further extension of time the Employer shall no longer be entitled to liquidated damages in respect of the period of such extension.

Any sum in respect of such period which may already have been recovered under this Clause shall be reimbursed forthwith to the Contractor together with interest compounded monthly at the rate provided for in Clause 60(7) from the date on which such sums were recovered from the Contractor.

### Intervention of variations etc.

(6)  If after liquidated damages have become payable in respect of any part of the Works the Engineer orders a variation under Clause 51 or adverse physical conditions or artificial obstructions within the meaning of Clause 12 are encountered or any other situation outside the Contractor's control arises any of which in the Engineer's opinion results in further delay to that part of the Works

(a)  the Engineer shall so notify the Contractor and the Employer in writing and

(b)  the Employer's further entitlement to liquidated damages in respect of that part of the Works shall be suspended until the Engineer notifies the Contractor and the Employer in writing that the further delay has come to an end.

Such suspension shall not invalidate any entitlement to liquidated damages which accrued before the period of further delay started to run and subject to any subsequent or final review of the circumstances causing delay any monies already deducted or paid as liquidated damages under the provision of this Clause may be retained by the Employer.

*Synopsis*   This clause sets out detailed provisions as to liquidated damages for delay. There are two basic schemes. In the standard case where there is a single completion date for all the Works, Clause 47(1) applies but Clause 47(2) does not. Where there are sectional completion provisions, Clause 47(2) applies but Clause 47(1) does not.

*The general case*   Clause 47(1) applies unless the Works are divided into Sections in the Appendix. In this case the Contractor 'shall pay to the Employer' the agreed rate of liquidated damages when he has failed to complete in the time for completion. The proviso states that where any part of the Works has been certified as substantially complete,[54] the liquidated damages shall be reduced proportionately.

*Sectional completion*   Where the Works are divided into section, this is to be stated in the Appendix. In this case, Clause 47(1) does not apply. Here each section is to be treated independently and liquidated damages levied in respect to each section separately.

*Liquidated damages not a penalty—Clause 47(3)*   Where the sum is excessive so that it could not be a genuine pre-estimate of the likely damage, this clause will not save it.[55]

*Limitation*   Clause 47(4)(a) allows the parties to agree upon a ceiling to liquidated damages payable.

*Where no sum is inserted in the Appendix*   Clause 47(4)(b) provides that where the appropriate entry is left blank or where it is given as 'nil', no damages shall be payable.[56] This is an important provision, since many civil engineering contracts are drawn up informally by reference to the ICE Conditions of Contract without agreeing liquidated damages. Where liquidated damages are not agreed, the effect which ordinarily results is that the

---

54   Clause 48(4).
55   As to penalties generally see Chapter 7.
56   See *Temloc v. Errill Properties Ltd* (1988) 39 BLR 30 (CA).

Employer may recover unliquidated damages. The effect of Clause 47(4)(b), however, seems to be that no damages at all, liquidated or unliquidated, may be recovered. It is submitted, however, that very clear wording is required before an entitlement to damages is excluded and the efficacy of this purported exclusion must therefore be in doubt.

*Delays at the Employer's risk occurring during any period in which liquidated damages are to apply*    Clause 47(6) deals with the common situation where a variation is ordered (or other event at the Employer's risk occurs) during a period of the Contractor's culpable delay. The Employer's entitlement to liquidated damages are suspended during the period in which the variation is to be done or other delay is caused without invalidating the Employer's right to liquidated damages incurred before and after the event in question.[57]

## Certificate of substantial completion

### Notification of substantial completion    48

(1)   When the Contractor considers that

    (a)   the whole of the Works or

    (b)   any Section in respect of which a separate time for completion is provided in the Appendix to the Form of Tender

has been substantially completed and has satisfactorily passed any final test that may be prescribed by the Contract he may give notice in writing to that effect to the Engineer or to the Engineer's Representative. Such notice shall be accompanied by an undertaking to finish any outstanding work in accordance with the provisions of Clause 49(1).

### Certification of substantial completion

(2)   The Engineer shall within 21 days of the date of delivery of such notice either

    (a)   issue to the Contractor (with a copy to the Employer) a Certificate of Substantial Completion stating the date on which in his opinion the Works were or the Section was substantially completed in accordance with the Contract or

---

57    This may well be the general law: see *McAlpine Humberoak v. McDermott International* (1992) 58 BLR 1 (CA); *Balfour Beatty Building Limited v. Chestermount Properties Ltd* (1993) 9 Constr. L.J. 117.

(b)  give instructions in writing to the Contractor specifying all the work which in the Engineer's opinion requires to be done by the Contractor before the issue of such certificate.

If the Engineer gives such instructions the Contractor shall be entitled to receive a Certificate of Substantial Completion within 21 days of completion to the satisfaction of the Engineer of the work specified in the said instructions.

**Premature use by Employer**

(3)  If any substantial part of the Works has been occupied or used by the Employer other than as provided in the Contract the Contractor may request in writing and the Engineer shall issue a Certificate of Substantial Completion in respect thereof. Such certificate shall take effect from the date of delivery of the Contractor's request and upon the issue of such certificate the Contractor shall be deemed to have undertaken to complete any outstanding work in that part of the Works during the Defects Correction Period.

**Substantial completion of other parts of the Works**

(4)  If the Engineer considers that any part of the Works has been substantially completed and has passed any final test that may be prescribed by the Contract he may issue a Certificate of Substantial Completion in respect of that part of the Works before completion of the whole of the Works and upon the issue of such certificate the Contractor shall be deemed to have undertaken to complete any outstanding work in that part of the Works during the Defects Correction Period.

**Reinstatement of ground**

(5)  A Certificate of Substantial Completion given in respect of any Section or part of the Works before completion of the whole shall not be deemed to certify completion of any ground or surfaces requiring reinstatement unless such certificate shall expressly so state.

*Clause 48(1) '… substantially completed'*  The terms 'substantial completion' and 'substantially completed' are used in various clauses. The term is not defined.[58] It is submitted that it means a state in which all work is complete, save that which is latently defective or that which is unimportant for the reasonable functioning of the Works.[59] The use of the term 'substantial' clearly indicates that absolute completion is not required.

---

58   The expression 'substantial performance' is used in the general law of contracts to mean a state which entitled a contracting party to be paid under an entire contract: however, it is submitted that the two expressions have quite different meanings.

59   See *Westminster Corporation* v. *Jarvis* [1970] 1 WLR 637 (HL) which points to this conclusion even though the expression 'substantial completion' was not in issue.

*The certificate*   The giving of a certificate of substantial completion is important in a number of situations. Most importantly, it defines (*a*) the time at which liquidated damages cease to be incurred and (*b*) the date of commencement of the Defects Correction Period. Strictly speaking, of course, it is not the date that the Engineer gives the certificate but the date when he should have given the certificate which is important.

*Certificates issued in respect of sections or parts of the Works*   The standard case is where the whole Works are to be completed by a single completion date and a single Certificate of Substantial Completion is issued. Clause 48 provides for a number of other scenarios. Thus, where separate Sections are specified in the Appendix, the Contractor is entitled to a certificate in respect to each—Clause 48(1)(b). Furthermore, where any substantial part of the Works is used by the Employer before the issue of a Certificate of Substantial Completion, the Engineer shall issue a certificate upon the Contractor's request—Clause 48(3). In addition, where any 'part of the Works' (which does not amount to a Section) is substantially completed the Engineer may in his discretion issue a certificate in respect of that part of the Works—Clause 48(4).

## Outstanding work and defects

**Work outstanding   49**

(1)   The undertaking to be given under Clause 48(1) may after agreement between the Engineer and the Contractor specify a time or times within which the outstanding work shall be completed. If no such times are specified any outstanding work shall be completed as soon as practicable during the Defects Correction Period.

**Carrying out of work of repair etc.**

(2)   The Contractor shall deliver up to the Employer the Works and each Section and part thereof at or as soon as practicable after the end of the relevant Defects Correction Period in the condition required by the Contract (fair wear and tear excepted) to the satisfaction of the Engineer. To this end the Contractor shall as soon as practicable carry out all work of repair amendment reconstruction rectification and making good of defects of whatever nature as may be required of him in writing by the Engineer during the relevant Defects Correction Period or within 14 days after its expiry as a result of an inspection made by or on behalf of the Engineer prior to its expiry.

**Cost of work of repair etc.**

(3) All work required under sub-clause (2) of this Clause shall be carried out by the Contractor at his own expense if in the Engineer's opinion it is necessary due to the use of materials or workmanship not in accordance with the Contract or to neglect or failure by the Contractor to comply with any of his obligations under the Contract. In any other event the value of such work shall be ascertained and paid for as if it were additional work.

**Remedy on Contractor's failure to carry out work required**

(4) If the Contractor fails to do any such work as aforesaid the Employer shall be entitled to carry out that work by his own workpeople or by other contractors and if it is work which the Contractor should have carried out at his own expense the Employer shall be entitled to recover the cost thereof from the Contractor and may deduct the same from any monies that are or may become due to the Contractor.

Work which is not in accordance with the Contract will render the Contractor in breach of Contract. Minor latent defects in the Works may quickly come to light after substantial completion. The Contractor will not normally welcome the Employer engaging another contractor to remedy these defects, the costs of which will be deducted from the retention fund which the Employer holds. Such an arrangement gives the Contractor little control and little opportunity to mitigate the loss. Clause 49 provides for a period—the Defects Correction Period—following substantial completion during which the Contractor is entitled and obliged to remedy defects. The general position is that the Contractor pays for defects which result from his breach of contract. The Employer pays for defects which result from poor specification or design, as though such work were additional work—Clause 49(3).

**Contractor to search   50**

The Contractor shall if required by the Engineer in writing carry out such searches tests or trials as may be necessary to determine the cause of any defect imperfection or fault under the directions of the Engineer. Unless the defect imperfection or fault is one for which the Contractor is liable under the Contract the cost of the work carried out by the Contractor as aforesaid shall be borne by the Employer. If the defect imperfection or fault is one for which the Contractor is liable the cost of the work carried out as aforesaid shall be borne by the Contractor and he shall in such case repair rectify and make good such defect imperfection or fault at his own expense in accordance with Clause 49.

This provision clearly deals with matters analogous to those in Clauses 36(3) and 38(2).

## Alterations, additions and omissions

### Ordered variations   51

(1)   The Engineer

   (a)   shall order any variation to any part of the Works that is in his opinion necessary for the completion of the Works and

   (b)   may order any variation that for any other reason shall in his opinion be desirable for the completion and/or improved functioning of the Works.

   Such variations may include additions omissions substitutions alterations changes in quality form character kind position dimension level or line and changes in any specified sequence method or timing of construction required by the Contract and may be ordered during the Defects Correction Period.

### Ordered variations to be in writing

(2)   All variations shall be ordered in writing but the provisions of Clause 2(6) in respect of oral instructions shall apply.

### Variation not to affect Contract

(3)   No variation ordered in accordance with sub-clauses (1) and (2) of this Clause shall in any way vitiate or invalidate the Contract but the value (if any) of all such variations shall be taken into account in ascertaining the amount of the Contract Price except to the extent that such variation is necessitated by the Contractor's default.

### Changes in quantities

(4)   No order in writing shall be required for increase or decrease in the quantity of any work where such increase or decrease is not the result of an order given under this Clause but is the result of the quantities exceeding or being less than those stated in the Bill of Quantities.

*General*   A variations clause is desirable in civil engineering contracts where there are works of any complexity. Changes to the specification and design are very frequently required, not only because they may seem desirable as construction proceeds, but sometimes because the Works cannot be completed satisfactorily without a variation.

*Necessary variations—Clause 51(1)(a)*   The Engineer is obliged to order a variation which is in his opinion necessary for the completion of the Works.

*Desirable variations—Clause 51(1)(b)*   Where a variation is merely 'desirable for the completion and/or improved functioning of the Works', the Engineer's power to order variations is discretionary.

*Permitted and non-permitted variations*   The Engineer has no power to order a variation unless it is necessary or desirable for the completion or functioning of the Works. Thus, the Engineer is not entitled to vary the Works in order to suit the Employer's financial or management arrangements. For instance, an omission which will improve the performance of the Works may be ordered: but an omission which is desirable simply to save money may not.[60] A variety of types of variations are mentioned in Clause 51(1). These may be classified as follows:

(1) *physical*: a variation to the definition of the completed structure
(2) *positional*: the line, level, orientation, etc., may be varied
(3) *management*: the specified[61] sequence method or timing may be varied.

The words in Clause 51 from which this classification is derived are clearly illustrative and do not define the expression 'variation'. However, it is frequently important to be able to decide whether an instruction or constraint is or is not a variation. Where it is a variation, the Contractor is entitled to paid in accordance with Clause 52. It is thought that an instruction amounts to a variation where it imposes any new obligation or constraint.[62]

*Variations to be in writing*   It is clearly desirable that any variation be in writing and Clause 51(2) provides for this. Frequently, however, the Contractor is required to modify his work or his work pattern for reasons which he claims amount to a variation; for example, he may have to modify the

---

60   Where such an omission is in fact ordered, the Contractor will not be entitled to construct the wrongly omitted works regardless. His remedy will be as damages for breach of contract. In particular he will claim loss of profit suffered insofar as he is not able to mitigate that loss by redeploying his resources elsewhere.

61   Where a method is specified and it proves to be impossible, the Contractor is entitled to a variation: *Havant Borough Council* v. *South Coast Shipping* (No. 1) (1996) 14 Constr. L.J. 420. The significance of the expression 'changes in specified sequence' etc. is not wholly clear. One interpretation is that a sequence etc. cannot be varied unless it is specified in the Contract in the first place. This would produce the result that where no sequence is specified, the Engineer is powerless to specify a sequence (although he may still be entitled to do so under Clause 13(1)). It is thought that this cannot be the proper interpretation. While a variation clearly needs to relate to some contractual obligation or entitlement, it is thought that where no sequence etc. is specified, it may still be varied so that it becomes more closely specified.

62   See *English Industrial Estates* v. *Kier* (1991) 56 BLR 93.

design of Works for which he bears no responsibility under the Contract.[63] Where Clause 51(1)(a) applies, the issue of a variation order is obligatory; thus where the Engineer fails to order a variation which falls within the scope of Clause 51(1)(a), the Employer may not rely on a lack of written variation order to refuse to pay such additional sums as may become due.[64]

*Variations necessitated by the Contractor's default*   The Contractor is not entitled to claim the cost of any variation to the extent that it is made necessary by his own default. In other forms of contract, the absence of such a provision has in some circumstances caused the Employer to be liable for the cost of a variation which is instructed following the Contractor's default.[65]

### Valuation of ordered variations   52

(1)   If requested by the Engineer the Contractor shall submit his quotation for any proposed variation and his estimate of any consequential delay. Wherever possible the value and delay consequences (if any) of each variation shall be agreed before the order is issued or before work starts.

(2)   Where a request is not made or agreement is not reached under sub-clause (1) the valuation of variations ordered by the Engineer in accordance with Clause 51 shall be ascertained as follows.

    (a)   As soon as possible after receipt of the variation the Contractor shall submit to the Engineer

        (i)   his quotation for any extra or substituted works necessitated by the variation having due regard to any rates or prices included in the Contract and

        (ii)   his estimate of any delay occasioned thereby and

        (iii)   his estimate of the cost of any such delay.

    (b)   Within 14 days of receiving the said submissions the Engineer shall

---

63   *Shanks & McEwan (Contractors) Ltd* v. *Strathclyde Regional Council* (1994) CILL 916 (Scottish Court of Session).

64   *Brodie* v. *Cardiff Corporation* [1919] AC 337 (HL). Here the variation clause provided that no extras would be paid for unless there was a written order. The engineer required work to be done which he said was within the original scope of works and which the contractor claimed was an extra. The work was done. Upon completion, the contractor commenced arbitration proceedings and the arbitrator accepted that the works were extra to that originally agreed. But the employer claimed not to be required to pay for such extras because of the lack of a written order. The House of Lords held that the arbitrator's award perfected the missing variation order.

65   *Simplex Piling* v. *St. Pancras Borough Council* (1958) 14 BLR 80; compare with *Howard de Walden Estates Ltd* v. *Costain Management Design Ltd* (1991) 55 BLR 124.

     (i)   accept those submissions or

     (ii)  negotiate with the Contractor thereon.

 (c)  Upon reaching agreement with the Contractor the Contract Price shall be amended accordingly.

(3)  Failing agreement between the Engineer and the Contractor under either sub-clause (1) or (2) the value of variations ordered by the Engineer in accordance with Clause 51 shall be ascertained by the Engineer in accordance with the following principles and be notified to the Contractor.

 (a)  Where work is of similar character and carried out under similar conditions to work priced in the Bill of Quantities it shall be valued at such rates and prices contained therein as may be applicable.

 (b)  Where work is not of a similar character or is not carried out under similar conditions or is ordered during the Defects Correction Period the rates and prices in the Bill of Quantities shall be used as the basis for valuation so far as may be reasonable failing which a fair valuation shall be made.

**Engineer to fix rates**

(4)  If in the opinion of the Engineer or the Contractor any rate or price contained in the Contract for any item of work (not being the subject of any variation) is by reason of any variation rendered unreasonable or inapplicable either the Engineer shall give to the Contractor or the Contractor shall give to the Engineer notice before the varied work is commenced or as soon thereafter as is reasonable in all the circumstances that such rate or price should be increased or decreased and the Engineer shall fix such rate or price as in the circumstances he shall think reasonable and proper and shall so notify the Contractor.

**Daywork**

(5)  The Engineer may if in his opinion it is necessary or desirable order in writing that any additional or substituted work shall be carried out on a daywork basis in accordance with the provisions of Clause 56(4).

*General*    This clause is headed 'Valuation of ordered variations'.

*Quotations and agreement*    The provisions in Clause 52(1), (2) are new to the 7th Edition. It makes sense to attempt to agree the cost and time implications.

*The valuation of variations*    Clause 52(3) establishes a three level approach to the valuation of variations for which the value has not been agreed:

(1)  Evaluate variations using the existing rates where this is both possible and reasonable.[66]

(2)  Where appropriate use the existing rates as 'the basis for valuation'.

(3)  Where there are no applicable or analogous rates, use a fair valuation.

*Fixing rates*   Clause 52(4) provides that where rates are rendered unreasonable or inapplicable by reason of any variation, the Engineer shall fix a varied rate.[67] This procedure is to be commenced by notice given either by the Contractor or Engineer. This notice seems at first sight to be a condition precedent to be given 'before the varied work is commenced or as soon thereafter as is reasonable in all the circumstances'. However, Clause 53(5) provides that a failure to give notice will defeat the claim 'only to the extent that the Engineer has not been prevented from or substantially prejudiced by such failure in investigating the said claim'. The means or principles by which the rate is to be fixed are not set out in the clause. Normally the Engineer will request the Contractor to supply him with a breakdown of his rates into components such as labour, plant, materials, on-site and off-site overheads and profit. Using these the Engineer will determine how each of these components is or will be affected and adjust the quantities and/or proportion of the composite rate accordingly.

*Dayworks—Clause 52(5)*   Contractors frequently submit accounts for varied work in the form of 'daywork sheets'. Dayworks are valued in accordance with a schedule of rates produced by representative contractors' bodies.[68] An hourly or daily rate is specified in the schedule. This basis of valuation assumes, in essence, that the Contractor's labour, plant, etc., are being hired to the Employer, so that payment is not made on the basis of work output, but time input. Such a valuation frequently suits contractors. However, this basis is only formally applicable where the Engineer orders dayworks in writing. In other cases, where a reasonable rate is to be paid, for

---

66   Under the ICE 6th Edition, it has been held that although the rates were far too high or far too low because of a mistake in the tender, nevertheless the parties were entitled to have variations valued in accordance with the tender prices: *Henry Boot Construction Ltd* v. *Alsthom Combined Cycles* [1999] BLR 123. The ICE 7th requires 'reasonableness' and this may lead to a different result for variations where the rates are obviously included by mistake.

67   Clause 56(2) contains analogous provisions to those in Clause 52(2) for the case where the rates are rendered unreasonable or inapplicable by virtue of changes in quantities.

68   Clause 56(4) provides that the the daywork rates are those

> contained in the Schedule of Dayworks carried out incidental to Contract Work issued by the Civil Engineering Contractors Association current at the date of carrying out of the daywork.

example in accordance with Clause 52(3)(b), or where the Contractor is entitled to his cost plus profit, daywork rates may provide a starting basis for valuation in some cases.

### Additional payments   53

(1) If the Contractor intends to claim a higher rate or price than one notified to him by the Engineer pursuant to sub-clauses (3) and (4) of Clause 52 or Clause 56(2) the Contractor shall within 28 days after such notification give notice in writing of his intention to the Engineer.

(2) If the Contractor intends to claim any additional payment pursuant to any Clause of these Conditions other than sub-clauses (3) and (4) of Clause 52 or Clause 56(2) he shall give notice in writing of his intention to the Engineer as soon as may be reasonable and in any event within 28 days after the happening of the events giving rise to the claim.

Upon the happening of such events the Contractor shall keep such contemporary records as may reasonably be necessary to support any claim he may subsequently wish to make.

(3) Without necessarily admitting the Employer's liability the Engineer may upon receipt of a notice under this Clause instruct the Contractor to keep such contemporary records or further contemporary records as the case may be as are reasonable and may be material to the claim of which notice has been given and the Contractor shall keep such records.

The Contractor shall permit the Engineer to inspect all records kept pursuant to Clause 53 and shall supply him with copies thereof as and when the Engineer shall so instruct.

(4) After the giving of a notice to the Engineer under this Clause the Contractor shall as soon as is reasonable in all the circumstances send to the Engineer a first interim account giving full and detailed particulars of the amount claimed to that date and of the grounds upon which the claim is based.

Thereafter at such intervals as the Engineer may reasonably require the Contractor shall send to the Engineer further up to date accounts giving the accumulated total of the claim and any further grounds upon which it is based.

(5) If the Contractor fails to comply with any of the provisions of this Clause in respect of any claim which he shall seek to make then the Contractor shall be entitled to payment in respect thereof only to the extent that the Engineer has not been prevented from or substantially prejudiced by such failure in investigating the said claim.

(6) The Contractor shall be entitled to have included in any interim payment certified by the Engineer pursuant to Clause 60 such amount in

respect of any claim as the Engineer may consider due to the Contractor provided that the Contractor shall have supplied sufficient particulars to enable the Engineer to determine the amount due.

If such particulars are insufficient to substantiate the whole of the claim the Contractor shall be entitled to payment in respect of such part of the claim as the particulars may substantiate to the satisfaction of the Engineer.

*Notices, records and particulars*   Although this clause is entitled 'Additional Payments' it relates primarily to notices, records and particulars in support of additional payment. Its approach is to set out the basic requirement for notices to be given; Clause 53(5), however, provides that lack of notice will only defeat a claim to the extent that the failure to notify disadvantages the Engineer. In addition to notices, the clause also deals with records and particulars of claims.

## Property in materials and contractor's equipment

### Non-removal of Materials and Contractor's Equipment   54

(1)  No Contractor's Equipment Temporary Works materials for Temporary Works or other goods or materials owned by the Contractor and brought on to the Site for the purposes of the Contract shall be removed without the written consent of the Engineer which consent shall not unreasonably be withheld.

### Liability for loss or damage to Contractor's Equipment

(2)  The Employer shall not at any time be liable save as mentioned in Clauses 20(2) and 63 for the loss of or damage to any Contractor's Equipment Temporary Works goods or materials.

### Disposal of Contractor's Equipment

(3)  If the Contractor fails to remove any of the said Contractor's Equipment Temporary Works goods or materials as required by Clause 33 within such reasonable time after completion of the Works as the Engineer may allow then the Employer may sell or otherwise dispose of such items. From the proceeds of the sale of any such items the Employer shall be entitled to retain any costs or expenses incurred in connection with their sale and disposal before paying the balance (if any) to the Contractor.

### Vesting of goods and materials not on Site

(4)  With a view to securing payment under Clause 60(1)(c) the Contractor may (and shall if the Engineer so directs) transfer to the Employer the

property in goods and materials listed in the Appendix to the Form of Tender or as subsequently agreed between the Contractor and the Employer before the same are delivered to the Site provided that the goods and materials

(a)  have been manufactured or prepared and are substantially ready for incorporation in the Works and

(b)  are the property of the Contractor or the contract for the supply of the same expressly provides that the property therein shall pass unconditionally to the Contractor upon the Contractor taking the action referred to in sub-clause (5) of this Clause and

(c)  have been marked and set aside in accordance with sub-clause (5) of this Clause.

### Action by Contractor

(5)  The intention of the Contractor to transfer the property in any goods or materials to the Employer in accordance with this Clause shall be evidenced by the Contractor taking or causing the supplier of those goods or materials to take the following actions.

(a)  Provide to the Engineer documentary evidence that the property of the said goods or materials has vested in the Contractor.

(b)  Suitably mark or otherwise plainly identify the goods and materials so as to show that their destination is the Site that they are the property of the Employer and (where they are not stored at the premises of the Contractor) to whose order they are held.

(c)  Set aside and store the said goods and materials so marked and identified to the satisfaction of the Engineer.

(d)  Send to the Engineer a schedule listing and giving the value of every item of the goods and materials so set aside and stored and inviting him to inspect them.

### Vesting in the Employer

(6)  Upon the Engineer approving in writing the transfer in ownership of any goods and materials for the purposes of this Clause they shall vest in and become the absolute property of the Employer and thereafter shall be in the possession of the Contractor for the sole purpose of delivering them to the Employer and incorporating them in the Works and shall not be within the ownership control or disposition of the Contractor. Provided always that

(a)  approval by the Engineer for the purposes of this Clause or any payment certified by him in respect of goods and materials pursuant to Clause 60 shall be without prejudice to the exercise of any power of the Engineer contained in this Contract to reject any goods or materials which are not in accordance with the provisions of the Contract and upon any such rejection the property in the rejected goods or materials shall immediately re-vest in the Contractor and

(b)   the Contractor shall be responsible for any loss or damage to such goods or materials and for the cost of storing handling and transporting the same and shall effect such additional insurance as may be necessary to cover the risk of such loss or damage from any cause.

### Lien on goods and materials

(7)   Neither the Contractor nor a sub-contractor nor any person shall have a lien on any goods or materials which have vested in the Employer under sub-clause (6) of this Clause for any sum due to the Contractor sub-contractor or other person and the Contractor shall take all steps reasonably necessary to ensure that the title of the Employer and the exclusion of any such lien are brought to the notice of sub-contractors and other persons dealing with such goods or materials.

### Delivery to the Employer of vested goods or materials

(8)   Upon cessation of the employment of the Contractor under this Contract before the completion of the Works whether as a result of the operation of Clause 63 64 or 65 or otherwise the Contractor shall deliver to the Employer any goods or materials the property in which has vested in the Employer by virtue of sub-clause (6) of this Clause and if he fails to do so the Employer may enter any premises of the Contractor or of any sub-contractor and remove such goods and materials and recover the cost of doing so from the Contractor.

### Incorporation in sub-contracts

(9)   The Contractor shall incorporate provisions equivalent to those provided in this Clause in every sub-contract in which provision is to be made for the payment in respect of goods or materials before the same have been delivered to the Site.

This clause is designed to ensure: (*a*) that equipment on site is available to perform the work should the Contractor default; and (*b*) to transfer property in goods and materials to the Employer, where the Employer is to pay some element of their cost in advance. See the body of law on retention of title clauses.[69]

---

69   *Aluminium Industrie Vaasen v. Romalpa Aluminium* [1976] 1 WLR 676 (CA); *Armour v. Thyssen Edalstahlwerke AG* [1991] 2 AC 339 (HL) (Scotland).

## *Measurement*

### Quantities 55

(1) The quantities set out in the Bill of Quantities are the estimated quantities of the work but they are not to be taken as the actual and correct quantities of the Works to be carried out by the Contractor in fulfilment of his obligations under the Contract.

### Correction of errors

(2) No error in description in the Bill of Quantities or omission therefrom shall vitiate the Contract nor release the Contractor from the carrying out of the whole or any part of the Works according to the Drawings and Specification or from any of his obligations or liabilities under the Contract. Any such error or omission shall be corrected by the Engineer and the value of the work actually carried out shall be ascertained in accordance with Clause 52(2) or (3). Provided that there shall be no rectification of any errors omissions or wrong estimates in the descriptions rates and prices inserted by the Contractor in the Bill of Quantities.

### Measurement and valuation 56

(1) The Engineer shall except as otherwise stated ascertain and determine by admeasurement the value in accordance with the Contract of the work done in accordance with the Contract.

### Increase or decrease of rate

(2) Should the actual quantities carried out in respect of any item be greater or less than those stated in the Bill of Quantities and if in the opinion of the Engineer such increase or decrease of itself shall so warrant the Engineer shall after consultation with the Contractor determine an appropriate increase or decrease of any rates or prices rendered unreasonable or inapplicable in consequence thereof and shall notify the Contractor accordingly.

### Attending for measurement

(3) The Engineer shall when he requires any part or parts of the work to be measured give reasonable notice to the Contractor who shall attend or send a qualified agent to assist the Engineer or the Engineer's Representative in making such measurement and shall furnish all particulars required by either of them. Should the Contractor not attend or neglect or omit to send such agent then the measurement made by the Engineer or approved by him shall be taken to be the correct measurement of the work.

### Daywork

(4) Where any work is carried out on a daywork basis the Contractor shall be paid for such work under the conditions and at the rates and prices

set out in the daywork schedule included in the Contract or failing the inclusion of a daywork schedule he shall be paid at the rates and prices and under the conditions contained in the "Schedules of Dayworks carried out incidental to Contract Work" issued by The Civil Engineering Contractors Association (formerly issued by The Federation of Civil Engineering Contractors) current at the date of the carrying out of the daywork.

The Contractor shall furnish to the Engineer such records receipts and other documentation as may be necessary to prove amounts paid and/or costs incurred. Such returns shall be in the form and delivered at the times the Engineer shall direct and shall be agreed within a reasonable time.

Before ordering materials the Contractor shall if so required submit to the Engineer quotations for the same for his approval.

**Method of measurement    57**

Unless otherwise provided in the Contract or unless general or detailed description of the work in the Bill of Quantities or any other statement clearly shows to the contrary the Bill of Quantities shall be deemed to have been prepared and measurements shall be made according to the procedure set out in the "Civil Engineering Standard Method of Measurement Third Edition 1991" approved by the Institution of Civil Engineers and the Federation of Civil Engineering Contractors in association with the Association of Consulting Engineers or such later or amended edition thereof as may be stated in the Appendix to the Form of Tender to have been adopted in its preparation.

Clauses 55, 56 and 57 deal with measurement and should be read together. They establish a 'measure and value' arrangement. The measurement shall be done by the Engineer (after giving notice to the Contractor) in accordance with the Method of Measurement specified in Clause 57.[70]

## Provisional and prime cost sums and nominated sub-contracts

**Use of Provisional Sums    58**

(1)   In respect of every Provisional Sum the Engineer may order either or both of the following.

(a)   Work to be carried out or goods materials or services to be supplied by the Contractor the value thereof being determined in accordance with Clause 52 and included in the Contract Price.

---

70    The Method of Measurement mentioned in Clause 57 is the CESSM 3rd Edition.

(b)  Work to be carried out or goods materials or services to be supplied by a Nominated Sub-contractor in accordance with Clause 59.

### Use of Prime Cost Items

(2)  In respect of every Prime Cost Item the Engineer may order either or both of the following.

(a)  Subject to Clause 59 that the Contractor employ a sub-contractor nominated by the Engineer for the carrying out of any work or the supply of any goods materials or services included therein.

(b)  With the consent of the Contractor that the Contractor himself carry out any such work or supply any such goods materials or services in which event the Contractor shall be paid in accordance with the terms of a quotation submitted by him and accepted by the Engineer or in the absence thereof the value shall be determined in accordance with Clause 52 and included in the Contract Price.

### Design requirements to be expressly stated

(3)  If in connection with any Provisional Sum or Prime Cost Item the services to be provided include any matter of design or specification of any part of the Permanent Works or of any equipment or plant to be incorporated therein such requirement shall be expressly stated in the Contract and shall be included in any Nominated Sub-contract. The obligation of the Contractor in respect thereof shall be only that which has been expressly stated in accordance with this sub-clause.

*Provisional Sums and Prime Cost Items—general terminology*   These terms are partially defined in Clauses 1(1)(k), (l). The term 'Prime Cost Item' derives from the prime cost, or actual cost, of a piece of work. A Prime Cost Item is, therefore, a piece of work for which the actual cost is determined in advance. It relates specifically to work which is to be done by a nominated supplier or sub-contractor and for which an advance quote is obtained. The Contractor then includes a mark-up on the item for attendance and profit. In practice, however, an advance quotation is frequently not obtained and so the price may have to be adjusted. A Provisional Sum relates to work which may be executed at the discretion of the Engineer; it need not be executed.

### Nominated Sub-contractors— objection to nomination   59

(1)  The Contractor shall not be under any obligation to enter into a sub-contract with any Nominated Sub-contractor against whom the Contractor may raise reasonable objection or who declines to enter into a sub-contract with the Contractor containing provisions

(a) that in respect of the work goods materials or services the subject of the sub-contract the Nominated Sub-contractor will undertake towards the Contractor such obligations and liabilities as will enable the Contractor to discharge his own obligations and liabilities towards the Employer under the terms of the Contract

(b) that the Nominated Sub-contractor will indemnify and keep indemnified the Contractor against all claims demands and proceedings damages costs charges and expenses whatsoever arising out of or in connection with any failure by the Nominated Sub-contractor to perform such obligations or fulfil such liabilities

(c) that the Nominated Sub-contractor will indemnify and keep indemnified the Contractor from and against any negligence by the Nominated Sub-contractor his agents workmen and servants and against any misuse by him or them of any Contractor's Equipment or Temporary Works provided by the Contractor for the purposes of the Contract and for all claims as aforesaid

(d) that the Nominated Sub-contractor will provide the Contractor with security for the proper performance of the sub-contract and

(e) equivalent to those contained in Clause 63.

**Engineer's action upon objection to nomination or upon determination of Nominated Sub-contract**

(2) If pursuant to sub-clause (1) of this Clause the Contractor declines to enter into a sub-contract with a sub-contractor nominated by the Engineer or if during the course of the Nominated Sub-contract the Contractor shall validly terminate the employment of the Nominated Sub-contractor as a result of his default the Engineer shall

(a) nominate an alternative sub-contractor in which case sub-clause (1) of this Clause shall apply or

(b) by order under Clause 51 vary the Works or the work goods materials or services in question or

(c) by order under Clause 51 omit any or any part of such works goods materials or services so that they may be provided by workmen contractors or suppliers employed by the Employer either

   (i) concurrently with the Works (in which case Clause 31 shall apply) or

   (ii) at some other date

   and in either case there shall nevertheless be included in the Contract Price such sum (if any) in respect of the Contractor's charges and profit being a percentage of the estimated value of such omission as would have been payable had there been no such omission and the value thereof had been that estimated in the Bill of Quantities or inserted in the Appendix to the Form of Tender as the case may be or

(d) instruct the Contractor to secure a sub-contractor of his own choice and to submit a quotation for the work goods materials or services in question to be so performed or provided for the Engineer's consideration and action or

(e) invite the Contractor himself to carry out or supply the work goods materials or services in question under Clause 58(1)(a) or Clause 58(2)(b) or on a daywork basis as the case may be.

### Contractor responsible for Nominated Sub-contractors

(3) Except as otherwise provided in Clause 58 (3) the Contractor shall be as responsible for the work carried out or goods materials or services supplied by a Nominated Sub-contractor employed by him as if he had himself carried out such work or supplied such goods materials or services.

### Nominated Sub-contractor's default

(4) (a) If any event arises which in the opinion of the Contractor justifies the exercise of his right under any forfeiture clause to terminate the sub-contract or to treat the sub-contract as repudiated by the Nominated Sub-contractor he shall at once notify the Engineer in writing giving his reasons.

### Termination of Sub-contract

(b) With the consent in writing of the Engineer the Contractor may give notice to the Nominated Sub-contractor expelling him from the Sub-contract works pursuant to any forfeiture clause or rescinding the Sub-contract as the case may be. If however the Engineer's consent is withheld the Contractor shall be entitled to appropriate instructions under Clause 13.

### Engineer's action upon termination

(c) In the event that the Nominated Sub-contractor is expelled from the Sub-contract works the Engineer shall at once take such action as is required under sub-clause (2) of this Clause.

### Recovery of additional expense

(d) Having with the Engineer's consent terminated the Nominated Sub-contract the Contractor shall take all necessary steps and proceedings as are available to him to recover all additional expenses that are incurred from the Sub-contractor or under the security provided pursuant to sub-clause (1)(d) of this Clause. Such expenses shall include any additional expenses incurred by the Employer as a result of the termination.

### Reimbursement of Contractor's loss

(e)  If and to the extent that the Contractor fails to recover all his reasonable expenses of completing the Sub-contract works and all his proper additional expenses arising from the termination the Employer will reimburse the Contractor his unrecovered expenses.

### Consequent delay

(f)  The Engineer shall take any delay to the completion of the Works consequent upon the Nominated Sub-contractor's default into account in determining any extension of time to which the Contractor is entitled under Clause 44.

### Provisions for payment

(5)  For all work carried out or goods materials or services supplied by Nominated Sub-contractors there shall be included in the Contract Price

(a)  the actual price paid or due to be paid by the Contractor in accordance with the terms of the sub-contract (unless and to the extent that any such payment is the result of a default of the Contractor) net of all trade and other discounts, rebates and allowances other than any discount obtainable by the Contractor for prompt payment

(b)  the sum (if any) provided in the Bill of Quantities for labours in connection therewith and

(c)  in respect of all other charges and profit a sum being a percentage of the actual price paid or due to be paid calculated (where provision has been made in the Bill of Quantities for a rate to be set against the relevant item of prime cost) at the rate inserted by the Contractor against that item or (where no such provision has been made) at the rate inserted by the Contractor in the Appendix to the Form of Tender as the percentage for adjustment of sums set against Prime Cost Items.

### Production of vouchers etc.

(6)  The Contractor shall when required by the Engineer produce all quotations invoices vouchers sub-contract documents accounts and receipts in connection with expenditure in respect of work carried out by all Nominated Sub-contractors.

### Payment to Nominated Sub-contractors

(7)  Before issuing any certificate under Clause 60 the Engineer shall be entitled to demand from the Contractor reasonable proof that all sums (less retentions provided for in the Sub-contract) included in previous certificates in respect of the work carried out or goods or materials or services supplied by Nominated Sub-contractors have been paid to the

Nominated Sub-contractors or discharged by the Contractor in default whereof unless the Contractor shall

(a)   give details to the Engineer in writing of any reasonable cause he may have for withholding or refusing to make such payment and

(b)   produce to the Engineer reasonable proof that he has so informed such Nominated Sub-contractor in writing

the Employer shall be entitled to pay to such Nominated Sub-contractor direct upon the certification of the Engineer all payments (less retentions provided for in the Sub-contract) which the Contractor has failed to make to such Nominated Sub-contractor and to deduct by way of set-off the amount so paid by the Employer from any sums due or which become due from the Employer to the Contractor. Provided always that where the Engineer has certified and the Employer has made direct payment to the Nominated Sub-contractor the Engineer shall in issuing any further certificate in favour of the Contractor deduct from the amount thereof the amount so paid but shall not withhold or delay the issue of the certificate itself when due to be issued under the terms of the Contract.

*Nomination*   Clause 59 sets out a detailed scheme for the nomination of sub-contractors.

*Grounds for objection to a nomination*   Clause 59(1) provides that the Contractor shall not be obliged to enter into a contract with a prospective nominated sub-contractor where the Contractor raises a 'reasonable objection'. There is no guidance as to what might amount to reasonable objection. It is thought that the sub-contractor's reputation, experience, financial standing as well as his safety and quality procedures are relevant factors. Clause 59(1) also provides an outline of the minimum terms which the Contractor may insist on in any sub-contract; the form or amount of security in Clause 59(1)(d), which is a most material consideration is not, however, specified.

*Payment for work done by nominated sub-contractor*   Clause 59(5) sets out the elements of payment which may be applied for by the Contractor, namely: (*a*) the actual price to be paid to the nominated sub-contractor; (*b*) any billed sum for specific labours associated with the nominated work; (*c*) the mark-up inserted by the Contractor in the Bill of Quantities. Reference should also be made to Clauses 60(1), 60(2), 60(8)(a), (b) which provide technical rules concerning the application for payment in respect of nominated sub-contractors. In particular, amounts in applications for payment for nominated sub-contractors should be shown separately; and no deductions should be made from a sum in respect of work done by nominated sub-contractors if the Contractor has already paid that sum.

*Payments to nominated sub-contractors*   Clause 59(7) provides a scheme whereby the Contractor is required to show that he has paid nominated sub-contractors the relevant proportion of sums certified. If the Contractor does not provide this evidence, the Employer may pay the Nominated Sub-Contractor directly. However, the Contractor may refuse to pay the Nominated Sub-Contractor where the latter is in breach of the Sub-Contract and the Engineer and Nominated Sub-Contractor have both received details of the sums deducted. It seems that the Engineer must, in this case, determine whether the amount deducted is proper; if the Engineer values the proper deduction at a smaller sum, the Engineer should inform the Contractor of the sum which may be deducted. Should the Contractor's deduction be vindicated at a later date, the Contractor may recover the difference from the Sub-Contractor .

## Certificates and payment

### Monthly statements    60

(1)   Unless otherwise agreed the Contractor shall submit to the Engineer at monthly intervals commencing one month after the Works Commencement Date a statement (in such form if any as may be prescribed in the Specification) showing

(a)   the estimated contract value of the Permanent Works carried out up to the end of that month

(b)   a list of any goods or materials delivered to the Site for but not yet incorporated in the Permanent Works and their value

(c)   a list of any of those goods or materials identified in the Appendix to the Form of Tender which have not yet been delivered to the Site but of which the property has vested in the Employer pursuant to Clause 54 and their value and

(d)   the estimated amounts to which the Contractor considers himself entitled in connection with all other matters for which provision is made under the Contract including any Temporary Works or Contractor's Equipment for which separate amounts are included in the Bill of Quantities unless in the opinion of the Contractor such values and amounts together will not justify the issue of an interim certificate.

Amounts payable in respect of Nominated Sub-contracts are to be listed separately.

## Monthly payments

(2)   Within 25 days of the date of delivery of the Contractor's monthly statement to the Engineer or the Engineer's Representative in accordance with sub-clause (1) of this Clause the Engineer shall certify and within 28 days of the same date the Employer shall pay to the Contractor (after deducting any previous payments on account)

(a)   the amount which in the opinion of the Engineer on the basis of the monthly statement is due to the Contractor on account of sub-clauses (1)(a) and (1)(d) of this Clause less a retention as provided in sub-clause (5) of this Clause and

(b)   such amounts (if any) as the Engineer may consider proper (but in no case exceeding the percentage of the value stated in the Appendix to the Form of Tender) in respect of sub-clauses (1)(b) and (1)(c) of this Clause.

The payments become due on certification with the final date for payment being 28 days after the date of delivery of the Contractor's monthly statement.

The amounts certified in respect of Nominated Sub-contracts shall be shown separately in the certificate.

## Minimum amount of certificate

(3)   Until the whole of the Works has been certified as substantially complete in accordance with Clause 48 the Engineer shall not be bound to issue an interim certificate for a sum less than that stated in the Appendix to the Form of Tender but thereafter he shall be bound to do so and the certification and payment of amounts due to the Contractor shall be in accordance with the time limits contained in this Clause.

## Final account

(4)   Not later than 3 months after the date of the Defects Correction Certificate the Contractor shall submit to the Engineer a statement of final account and supporting documentation showing in detail the value in accordance with the Contract of the Works carried out together with all further sums which the Contractor considers to be due to him under the Contract up to the date of the Defects Correction Certificate.

Within 3 months after receipt of this final account and of all information reasonably required for its verification the Engineer shall issue a certificate stating the amount which in his opinion is finally due under the Contract from the Employer to the Contractor or from the Contractor to the Employer as the case may be up to the date of the Defects Correction Certificate and after giving credit to the Employer for all amounts previously paid by the Employer and for all sums to which the Employer is entitled under the Contract.

Such amount shall subject to Clause 47 be paid to or by the Contractor as the case may require. The payment becomes due on certification. The final date for payment is 28 days later.

### Retention

(5)  The retention to be made pursuant to sub-clause (2)(a) of this Clause shall be the difference between

(a)  an amount calculated at the rate indicated in and up to the limit set out in the Appendix to the Form of Tender upon the amount due to the Contractor on account of sub-clauses (1)(a) and (1)(d) of this Clause and

(b)  any payment which shall have become due under sub-clause (6) of this Clause.

### Payment of retention

(6)  (a)  Upon the issue of a Certificate of Substantial Completion in respect of any Section or part of the Works there shall become due to the Contractor one half of such proportion of the amount calculated to date under sub-clause (5)(a) of this Clause as the value of the Section or part bears to the value of the whole of the Works completed to date as certified under sub-clause (2)(a) of this Clause.

The total of the payments which shall become due under this sub-clause shall in no event exceed one half of the limit of retention set out in the Appendix to the Form of Tender.

(b)  Upon issue of the Certificate of Substantial Completion in respect of the whole of the Works there shall become due to the Contractor one half of the amount calculated in accordance with sub-clause (5)(a) of this Clause less any payments which shall have become due under sub-clause (6)(a) of this Clause. Within 10 days of the date of issue of the said Certificate the Engineer shall certify the amount due.

Payment becomes due on certification of the amount due with the final date for payment being 14 days after the issue of the Certificate of Substantial Completion.

(c) At the end of the Defects Correction Period or if more than one the last of such periods there shall become due to the Contractor the remainder of the retention money. Within 10 days of the date of the end of the said period the Engineer shall certify the amount due.

Payment becomes due on certification of the amount due with the final date for payment being 14 days after the end of the said period notwithstanding that at that time there may be outstanding claims by the Contractor against the Employer.

Provided that if at that time there remains to be carried out by the Contractor any outstanding work referred to under Clause 48 or any work ordered pursuant to Clauses 49 or 50 the Engineer may withhold certification until the completion of such work of so much of the said remainder as shall in the opinion of the Engineer represent the cost of the work remaining to be carried out.

### Interest on overdue payments

(7) In the event of

(a) failure by the Engineer to certify or the Employer to make payment in accordance with sub-clauses (2) (4) or (6) of this Clause or

(b) any decision of an adjudicator or any finding of an arbitrator to such effect

the Employer shall pay to the Contractor interest compounded monthly for each day on which any payment is overdue or which should have been certified and paid at a rate equivalent to 2% per annum above the base lending rate of the bank specified in the Appendix to the Form of Tender.

If in an arbitration pursuant to Clause 66 the arbitrator holds that any sum or additional sum should have been certified by a particular date in accordance with the aforementioned sub-clauses but was not so certified this shall be regarded for the purposes of this sub-clause as a failure to certify such sum or additional sum. Such sum or additional sum shall be regarded as overdue for payment 28 days after the date by which the arbitrator holds that the Engineer should have certified the sum or if no such date is identified by the arbitrator shall be regarded as overdue for payment from the date of the Certificate of Substantial Completion for the whole of the Works.

### Correction and withholding of certificates

(8) The Engineer shall have power to omit from any certificate the value of any work done goods or materials supplied or services rendered with which he may for the time being be dissatisfied and for that purpose or for any other reason which to him may seem proper may by any

certificate delete correct or modify any sum previously certified by him. Provided that

(a)  the Engineer shall not in any interim certificate delete or reduce any sum previously certified in respect of work done goods or materials supplied or services rendered by a Nominated Sub-contractor if the Contractor shall have already paid or be bound to pay that sum to the Nominated Sub-contractor and

(b)  if the Engineer in the final certificate shall delete or reduce any sum previously certified in respect of work done goods or materials supplied or services rendered by a Nominated Sub-contractor which sum shall have been already paid by the Contractor to the Nominated Sub-contractor the Employer shall reimburse to the Contractor the amount of any sum overpaid by the Contractor to the Sub-contractor in accordance with the certificates issued under sub-clause (2) of this Clause which the Contractor shall be unable to recover from the Nominated Sub-contractor together with interest thereon at the rate stated in sub-clause (7) of this Clause from 28 days after the date of the final certificate issued under sub-clause (4) of this Clause until the date of such reimbursement.

**Certificates and payment notices**

(9)  Every certificate issued by the Engineer pursuant to this Clause shall be sent to the Employer and on the Employer's behalf to the Contractor. By this certificate the Employer shall give notice to the Contractor specifying the amount (if any) of the payment proposed to be made and the basis on which it was calculated.

**Notice of intention to withhold payment**

(10) Where a payment under Clause 60(2) (4) or (6) is to differ from that certified or the Employer is to withhold payment after the final date for payment of a sum due under the Contract the Employer shall notify the Contractor in writing not less than one day before the final date for payment specifying the amount proposed to be withheld and the ground for withholding payment or if there is more than one ground each ground and the amount attributable to it.

*The HGCRA 1996*   Where the contract is subject to the Housing Grants, Construction and Regeneration Act 1996, the payment provisions must meet certain minimum requirements or the Scheme for Construction Contracts 1998 is activated. Clause 60 was revised in the 6th Edition of the Conditions of Contract in order to comply. In particular, Clause 60(9) makes it clear that the certificates issued by the Engineer under this clause shall stand as the notice to the Contractor specifying the amount of payment proposed to be made. Clause 60(10) makes new provision for a notice of intention to withhold payment.

*Synopsis* The Contractor submits monthly statements in accordance with Clause 60(1). These statements (frequently called 'interim applications'), list all the work done to date. Within 25 days of the delivery of the statement, the Engineer certifies and the Employer pays the amount which, in the opinion of the Engineer, is due— Clause 60(2). The Employer pays the amount due within 28 days of the date of the delivery of the statement. If the Engineer takes the full 25 days, the Employer only has 3 days to pay. The amount due is the value of work calculated in accordance with the Contract less (*a*) any sums already paid on account and (*b*) an element for retentions, as set out in Clauses 60(5) and (6). In the event that payment is late, the Contractor is entitled to interest—Clause 60(7). Following the issue of the Defects Correction Certificate the Contractor submits a final account—Clause 60(4).

*The monthly statement* Clause 60(1) sets out the headings which should be used for compiling the monthly statement. Any claims will be included under Clause 60(1)(d).

*The Engineer's valuation and certificate* The Engineer is required to compute the amount due and to certify that sum in sufficient time to enable the Employer to pay it within 28 days of the delivery of the monthly statement. As to items applied for in accordance with Clause 60(1)(a), the Engineer must satisfy himself that the rates and quantities are correct. As to items under Clauses 60(1)(b), (c) the Engineer must satisfy himself as to the quantities, values and state of security of goods and materials and come to a view as to the 'proper' amount to be paid. As to items claimed under Clause 60(1)(d) the Engineer must come to a view as to the Contractor's entitlement under the Contract, which may require him to evaluate claims. Generally speaking, the issue of a certificate is a condition precedent to payment:[71] in other words, the Contractor will not be entitled to be paid any sum unless that sum is certified. Where, however, there has been any interference or duplicity in relation to the issue of certificates, a certificate will not be a condition precedent.

---

71 *Costain Building and Civil Engineering Ltd* v. *Scottish Rugby Union plc* (1993) 69 BLR 85 (Scottish Court of Session, Inner House): a case on the 5th edn Conditions of Contract.

*Payment by the Employer*   The Employer is required to pay the certified sum within 28 days of the delivery of the monthly application—Clause 60(2). The express terms do not derogate from the Employer's right to set-off for breaches by the Contractor but if the Employer proposes to make any set-off he must issue a notice of intention to withhold payment in accordance with Clause 60(10). Accordingly, where the Contractor is in breach the Employer may evaluate any damages (including liquidated damages) which flow, issue the notice of intention to withhold and then deduct them from the sum certified.

*Final account—Clause 60(4)*   Within 3 months of the date of the Defects Correction Certificate, the Contractor submits 'a statement of final account'. This may be a statement in the same format as the monthly statements. Within 3 months of the receipt of the statement, the Engineer issues a certificate stating the sum finally due from the Employer to the Contractor or vice versa. It is thought that the Engineer should not take into account any damages for breach of contract which the Employer claims, but should confine himself to entitlements 'under the Contract': the Employer's right to set-off damages for breach when paying for certified sums remains, subject as always to the notice of intention to withhold.

*Retentions—Clauses 60(5),(6)*   Sums are retained in order that the Employer has a fund to call upon in the event that the Contractor defaults and the Employer must employ others to remedy defects or to complete work. Clause 60(5) sets out the amount of retention to be withheld. Clause 60(6) sets out the timetable for its repayment; in short, half is paid when each of the defined completion certificates (i.e. the Certificate of Substantial Completion and the Defects Correction Certificate) is issued.

*Interest—Clause 60(7)*   Claims[72] under civil engineering contracts may not be resolved for many years. Where a party has been kept out of his money for a long period, the interest may be very substantial. Interest clauses in previous editions of the ICE Conditions of Contract have caused many problems; issues have included whether interest is payable on a compound or simple basis and whether certification which is merely inadequate entitled the Contractor to any interest at all.[73] The present Clause 60(7) is

---

72   Including defects claims by Employers.
73   See, for example, *Secretary of State for Transport* v. *Birse-Farr Joint Venture* (1993) 62 BLR 36. The present clause overcomes the difficulty in Birse-Farr by expressly deeming an inadequate sum certified as a failure to certify.

designed to provide the Contractor with a right to payment of compound interest from the date when the principal sum to which it relates should have been paid. The Contractor may include for interest in his monthly statements in accordance with Clause 60(1)(d) and he may update his interest claim in each interim application; caution must be exercised, however, to ensure that the Contractor is not double recovering on interest.

*Correcting certificates—Clause 60(8)*    A certificate under this form of contract has no significant evidential weight and the Engineer may revise and correct earlier certificates and deduct any monies overpaid at a later stage.[74]

### Defects Correction Certificate    61

(1)    At the end of the Defects Correction Period or if more than one the last of such periods and when all outstanding work referred to under Clause 48 and all work of repair amendment reconstruction rectification and making good of defects imperfections shrinkages and other faults referred to under Clauses 49 and 50 have been completed the Engineer shall issue to the Employer (with a copy to the Contractor) a Defects Correction Certificate stating the date on which the Contractor shall have completed his obligations to construct and complete the Works to the Engineer's satisfaction.

### Unfulfilled obligations

(2)    The issue of the Defects Correction Certificate shall not be taken as relieving either the Contractor or the Employer from any liability the one towards the other arising out of or in any way connected with the performance of their respective obligations under the Contract.

The Defects Correction Certificate is an administrative step. It has no evidential status, but merely triggers payment of the second half of the retention money and sets in motion the procedure under Clause 60(4) in relation to the certificate as to the final sums due.

## Remedies and powers

### Urgent repairs    62

If in the opinion of the Engineer any remedial or other work or repair is urgently necessary by reason of any accident or failure or other event occurring to in or in connection with the Works or any part thereof either during

---

74    See *Mears Construction v. Samuel Williams* (1977) 16 BLR 49 (ICE Conditions of Contract, 4th Edition).

the carrying out of the Works or during the Defects Correction Period the Engineer shall so inform the Contractor with confirmation in writing.

Thereafter if the Contractor is unable or unwilling to carry out such work or repair at once the Employer may himself carry out the said work or repair using his own or other workpeople.

If the work or repair so carried out by the Employer is work which in the opinion of the Engineer the Contractor was liable to carry out at his own expense under the Contract all costs and charges properly incurred by the Employer in so doing shall on demand be paid by the Contractor to the Employer or may be deducted by the Employer from any monies due or which may become due to the Contractor.

The Contract scheme is that the Contractor is generally obliged and entitled to undertake remedial and other such works during the Defects Correction Period. In emergencies this may not always be practicable. Clause 62 entitles the Employer to make other arrangements in emergencies.

**Frustration  63**

(1)  If any circumstance outside the control of both parties arises during the currency of the Contract which renders it impossible or illegal for either party to fulfil his contractual obligations the Works shall be deemed to be abandoned upon the service by one party upon the other of written notice to that effect.

**War clause**

(2)  If during the currency of the Contract there is an outbreak of war (whether war is declared or not) in which Great Britain is engaged on a scale involving general mobilization of the armed forces of the Crown

    (a)  the Contractor shall for a period of 28 days reckoned from midnight on the date that the order for general mobilization is given continue so far as is physically possible to carry out the Works in accordance with the Contract and

    (b)  if substantial completion of the whole of the Works is not achieved before the said period of 28 days has expired the Works shall thereupon be deemed to be abandoned unless the parties otherwise agree.

**Removal of Contractor's Equipment**

(3)  Upon abandonment of the Works pursuant to sub-clauses (1) or (2)(b) of this Clause the Contractor shall with all reasonable dispatch remove from the Site all Contractor's Equipment.

In the event of any failure so to do the Employer shall have like powers to those contained in clause 54(3) to dispose of any Contractor's Equipment.

### Payment on abandonment

(4) Upon abandonment of the Works pursuant to sub-clauses (1) or (2)(b) of this Clause the Employer shall pay the Contractor (in so far as such amounts or items have not already been covered by payments on account made to the Contractor) the Contract value of all work carried out prior to the date of abandonment and in addition

(a) the amounts payable in respect of any preliminary items so far as the work or service comprised therein has been carried out or performed and a proper proportion of any such items which have been partially carried out or performed

(b) the cost of materials or goods reasonably ordered for the Works which have been delivered to the Contractor or of which the Contractor is legally liable to accept delivery (such materials or goods becoming the property of the Employer upon such payment being made to the Contractor)

(c) a sum being the amount of any expenditure reasonably incurred by the Contractor in the expectation of completing the whole of the Works insofar as such expenditure has not been recovered by any other payments referred to in this sub-clause and

(d) the reasonable cost of removal under sub-clause (3) of this Clause.

To this end and without prejudice to the provisions of sub-clause (5) of this Clause the provisions of Clause 60(4) shall apply to this sub-clause as if the date of abandonment was the date of issue of the Defects Correction Certificate.

### Works substantially completed

(5) If upon abandonment of the Works any Section or part of the Works has been substantially completed in accordance with Clause 48 or is completed so far as to be usable then in connection therewith

(a) the Contractor may at his discretion and in lieu of his obligations under Clauses 49 and 50 allow against the sum due to him pursuant to sub-clause (4) of this Clause the cost (calculated as at the date of abandonment) of repair rectification and making good for which he would have been liable under the said Clauses had they continued to be applicable and

(b) the Employer shall not be entitled to withhold payment under Clause 60(6)(c) of the second half of the retention money or any part thereof except such sum as the Contractor may allow under the provisions of the last preceding paragraph.

### Contract to continue in force

(6) Save as aforesaid the Contract shall continue to have full force and effect.

The frustration and war clauses—previously Clauses 64 and 65 in the 6th Edition have been renumbered—and consolidated into a single clause.[75]

### Default of the Employer    64

(1) In the event that the Employer

    (a) assigns or attempts to assign the Contract or any part thereof or any benefit or interest thereunder without the prior written consent of the Contractor or

    (b)  (i) becomes bankrupt or presents his petition in bankruptcy or

        (ii) has a receiving order or administration order made against him or

        (iii) makes an arrangement with or an assignment in favour of his creditors or

        (iv) agrees to perform the Contract under a committee of inspection of his creditors or

        (v) (being a corporation) has a receiver or administrator appointed or goes into liquidation (other than a voluntary liquidation for the purposes of amalgamation or reconstruction) or

    (c) has an execution levied on his goods which is not stayed or discharged within 28 days

then the Contractor may after giving 7 days notice in writing to the Employer specifying the event relied on terminate his employment under the Contract without thereby avoiding the Contract or releasing the Employer from any of his obligations or liabilities under the Contract.

Provided that the Contractor may extend the period of notice to give the Employer an opportunity to remedy the situation.

### Removal of Contractor's Equipment

(2) Upon expiry of the 7 days notice referred to in sub-clause (1) of this Clause and notwithstanding the provisions of Clause 54 the Contractor shall with all reasonable despatch remove from the site all Contractor's Equipment.

---

75  For frustration generally see Chapter 2, Section 7.

**Payment upon termination**

(3) Upon termination of the Contractor's employment pursuant to sub-clause (1) of this Clause the Employer shall be under the same obligations with regard to payment as if the Works had been abandoned under the provisions of Clause 63.

Provided that in addition to payments specified under Clause 63(4) the Employer shall pay to the Contractor the amount of any loss or damage to the Contractor arising from or as a consequence of such termination.

## Default of Contractor 65

(1) In the event that the Contractor

(a) assigns or attempts to assign the Contract or any part thereof or any benefit or interest thereunder without the prior written consent of the Employer or

(b) is in breach of Clause 4(1) or

(c) (i) becomes bankrupt or presents his petition in bankruptcy

or

(ii) has a receiving order or administration order made against him or

(iii) makes an arrangement with or an assignment in favour of his creditors or

(iv) agrees to carry out the Contract under a committee of inspection of his creditors or

(v) (being a corporation) has a receiver or administrator appointed or goes into liquidation (other than a voluntary liquidation for the purposes of amalgamation or reconstruction) or

(d) has an execution levied on his goods which is not stayed or discharged within 28 days

or if the Engineer certifies in writing to the Employer with a copy to the Contractor that in his opinion the Contractor

(e) has abandoned the Contract without due cause or

(f) without reasonable excuse has failed to commence the Works in accordance with Clause 41 or

(g) has suspended the progress of the Works without due cause for 14 days after receiving from the Engineer written notice to proceed or

(h) has failed to remove goods or materials from the Site or to pull down and replace work for 14 days after receiving from the

Engineer written notice that the said goods materials or work has been condemned and rejected by the Engineer or

(j)   despite previous warnings by the Engineer in writing is failing to proceed with the Works with due diligence or is otherwise persistently or fundamentally in breach of his obligations under the Contract

then the Employer may after giving 7 days notice in writing to the Contractor specifying the event relied on enter upon the Works and any other parts of the Site provided by the Employer and expel the Contractor therefrom without thereby avoiding the Contract or releasing the Contractor from any of his obligations or liabilities under the Contract.

Where a notice of termination is given pursuant to a certificate issued by the Engineer under this sub-clause it shall be given as soon as is reasonably possible after receipt of the certificate.

Provided that the Employer may extend the period of notice to give the Contractor an opportunity to remedy the situation.

### Completing the Works

(2)   Where the Employer has entered upon the Works and any other parts of the Site as set out in sub-clause (1) of this Clause he may

(a)   complete the Works himself or

(b)   employ any other contractor to complete the Works

and in either case may use for such completion any of the Contractor's Equipment Temporary Works goods and materials on any part of the Site.

The Employer may at any time sell any of the said Contractor's Equipment Temporary Works and unused goods and materials and apply the proceeds of sale in or towards the satisfaction of any sums due or which may become due to him from the Contractor under the Contract.

### Assignment to Employer

(3)   Where the Employer has entered upon the Works and any other parts of the Site as hereinbefore provided the Contractor shall if so instructed by the Engineer in writing within 7 days of such entry assign to the Employer the benefit of any agreement which the Contractor may have entered into for the supply of any goods or materials and/or for the carrying out of any work for the purposes of the Contract.

### Valuation at date of termination

(4)   As soon as may be practicable after any such entry and expulsion by the Employer the Engineer shall fix and determine as at the time of such entry and expulsion

(a)   the amount (if any) which has been reasonably earned by or would reasonably accrue to the Contractor in respect of work actually done by him under the Contract and

(b)   the value of any unused or partially used goods and materials which are under the control of the Employer

and shall certify accordingly.

The said determination may be carried out *ex parte* or by or after reference to the parties or after such investigation or enquiry as the Engineer may think fit to make or institute.

Clause 64 is new in the 7th Edition and Clause 65 derives in part from the previous Clause 63. The new scheme provides a symmetry which was lacking previously. It also makes specific the consequences of a number of defaults such as attempts at unauthorised assignments and unpermitted sub-contracting.

*Termination*   A list of events which may lead to a termination of performance under the Contract is set out. In each case the termination is of performance not of the contract itself. Obligations to pay and liability for the work already carried out remain.

*Procedure*   The procedure is not set out with great clarity. The basic procedure appears to be that when one of the specified events occurs, the innocent party gives notice in writing specifying the event. Seven or more days later, the innocent party may terminate performance of the Contract. However, it is not clear whether or not the termination must be done in writing. Nor is it clear whether or not the defaulting party must still be in default at the date of the termination. Furthermore, in the case of the Engineer certifying that the Contract is in serious delay, the seven day period appears to apply, but there is also a provision which exhorts the Employer to terminate 'as soon as reasonably practicable after the receipt of the certificate'.

*Repudiation generally*   Clauses 64 and 65 are not expressed to be exclusive and so it is thought that the rights of either party to elect to treat the obligation to perform the Contract as being at an end are preserved.

*The Engineer's certificate*   Clause 65(1)(e)–(j) sets out a number of circumstances which might be reckoned as repudiations of contract in any event. The definition of these grounds, however, gives the Employer a good deal of security, since repudiation cannot be readily recognised in practice.

Before issuing the certificate the Engineer should carefully consider the terms of Clause 65(1) to ensure that there is full compliance Caution should always be exercised when determining any contract.

*Due diligence—Clause 65(1)(j)*    It is thought that due diligence refers to a rate of progress and industriousness which is appropriate taking into account the currently accepted Clause 14 programme, the progress made to date and any difficulties being encountered. A number of related expressions have been considered by the courts. 'Regularly and diligently' is said to 'convey a sense of activity, of orderly progress'.[76] 'Reasonable diligence' is said to mean 'not only the personal industriousness of the defendant [contractor] himself but his efficiency and that of those who work for him'.[77]

## Avoidance and settlement of disputes

### Avoidance of disputes   66

(1)   In order to overcome where possible the causes of disputes and in those cases where disputes are likely still to arise to facilitate their clear definition and early resolution (whether by agreement or otherwise) the following procedure shall apply for the avoidance and settlement of disputes.

### Matters of dissatisfaction

(2)   If at any time

(a)   the Contractor is dissatisfied with any act or instruction of the Engineer's Representative or any other person responsible to the Engineer or

(b)   the Employer or the Contractor is dissatisfied with any decision opinion instruction direction certificate or valuation of the Engineer or with any other matter arising under or in connection with the Contract or the carrying out of the Works

the matter of dissatisfaction shall be referred to the Engineer who shall notify his written decision to the Employer and the Contractor within one month of the reference to him.

### Disputes

(3)   The Employer and the Contractor agree that no matter shall constitute nor be said to give rise to a dispute unless and until in respect of that matter

---

76   *Hounslow Borough Council* v. *Twickenham Garden Developments Ltd* [1971] Ch. 233. See also *West Faulkner Associates* v. *London Borough of Newham* (1994) 71 BLR 1 (CA).

77   *Hooker Constructions* v. *Chris's Engineering Contracting Company* [1970] ALR 821 (Supreme Court of the Northern Territory of Australia).

(a) the time for the giving of a decision by the Engineer on a matter of dissatisfaction under Clause 66(2) has expired or the decision given is unacceptable or has not been implemented and in consequence the Employer or the Contractor has served on the other and on the Engineer a notice in writing (hereinafter called the Notice of Dispute)

(b) an adjudicator has given a decision on a dispute under Clause 66(6) and the Employer or the Contractor is not giving effect to the decision, and in consequence the other has served on him and the Engineer a Notice of Dispute

and the dispute shall be that stated in the Notice of Dispute. For the purposes of all matters arising under or in connection with the Contract or the carrying out of the Works the word "dispute" shall be construed accordingly and shall include any difference.

(4) (a) Notwithstanding the existence of a dispute following the service of a Notice under Clause 66(3) and unless the Contract has already been determined or abandoned the Employer and the Contractor shall continue to perform their obligations.

(b) The Employer and the Contractor shall give effect forthwith to every decision of

(i) the Engineer on a matter of dissatisfaction given under Clause 66(2) and

(ii) the adjudicator on a dispute given under Clause 66(6)

unless and until that decision is revised by agreement of the Employer and Contractor or pursuant to Clause 66.

## Conciliation

(5) (a) The Employer or the Contractor may at any time before service of a Notice to Refer to arbitration under Clause 66(9) by notice in writing seek the agreement of the other for the dispute to be considered under "The Institution of Civil Engineers' Conciliation Procedure 1999" or any amendment or modification thereof being in force at the date of such notice.

(b) If the other party agrees to this procedure any recommendation of the conciliator shall be deemed to have been accepted as finally determining the dispute by agreement so that the matter is no longer in dispute unless a Notice of Adjudication under Clause 66(6) or a Notice to Refer to arbitration under Clause 66(9) has been served in respect of that dispute not later than one month after receipt of the recommendation by the dissenting party.

## Adjudication

(6) (a) The Employer and the Contractor each has the right to refer a dispute as to a matter under the Contract for adjudication and either party

may give notice in writing (hereinafter called the Notice of Adjudication) to the other at any time of his intention so to do. The adjudication shall be conducted under "The Institution of Civil Engineers' Adjudication Procedure 1997" or any amendment or modification thereof being in force at the time of the said Notice.

(b) Unless the adjudicator has already been appointed he is to be appointed by a timetable with the object of securing his appointment and referral of the dispute to him within 7 days of such notice.

(c) The adjudicator shall reach a decision within 28 days of referral or such longer period as is agreed by the parties after the dispute has been referred.

(d) The adjudicator may extend the period of 28 days by up to 14 days with the consent of the party by whom the dispute was referred.

(e) The adjudicator shall act impartially.

(f) The adjudicator may take the initiative in ascertaining the facts and the law.

(7) The decision of the adjudicator shall be binding until the dispute is finally determined by legal proceedings or by arbitration (if the contract provides for arbitration or the parties otherwise agree to arbitration) or by agreement.

(8) The adjudicator is not liable for anything done or omitted in the discharge or purported discharge of his functions as adjudicator unless the act or omission is in bad faith and any employee or agent of the adjudicator is similarly not liable.

### Arbitration

(9) (a) All disputes arising under or in connection with the Contract or the carrying out of the Works other than failure to give effect to a decision of an adjudicator shall be finally determined by reference to arbitration. The party seeking arbitration shall serve on the other party a notice in writing (called the Notice to Refer) to refer the dispute to arbitration.

(b) Where an adjudicator has given a decision under Clause 66(6) in respect of the particular dispute the Notice to Refer must be served within three months of the giving of the decision otherwise it shall be final as well as binding.

### Appointment of arbitrator

(10) (a) The arbitrator shall be a person appointed by agreement of the parties.

### President or Vice-President to act

(b) If the parties fail to appoint an arbitrator within one month of either party serving on the other party a notice in writing (hereinafter called the Notice to Concur) to concur in the appointment of an arbitrator the dispute shall be referred to a person to be appointed on the application of either party by the President for the time being of the Institution of Civil Engineers.

(c) If an arbitrator declines the appointment or after appointment is removed by order of a competent court or is incapable of acting or dies and the parties do not within one month of the vacancy arising fill the vacancy then either party may apply to the President for the time being of the Institution of Civil Engineers to appoint another arbitrator to fill the vacancy.

(d) In any case where the President for the time being of the Institution of Civil Engineers is not able to exercise the functions conferred on him by this Clause the said functions shall be exercised on his behalf by a Vice-President for the time being of the said Institution.

## Arbitration—procedure and powers

(11) (a) Any reference to arbitration under this Clause shall be deemed to be a submission to arbitration within the meaning of the Arbitration Act 1996 or any statutory re-enactment or amendment thereof for the time being in force. The reference shall be conducted in accordance with the procedure set out in the Appendix to the Form of Tender or any amendment or modification thereof being in force at the time of the appointment of the arbitrator. Such arbitrator shall have full power to open up review and revise any decision opinion instruction direction certificate or valuation of the Engineer or an adjudicator.

(b) Neither party shall be limited in the arbitration to the evidence or arguments put to the Engineer or to any adjudicator pursuant to Clause 66(2) or 66(6) respectively.

(c) The award of the arbitrator shall be binding on all parties.

(d) Unless the parties otherwise agree in writing any reference to arbitration may proceed notwithstanding that the Works are not then complete or alleged to be complete.

## Witnesses

(12) (a) No decision opinion instruction direction certificate or valuation given by the Engineer shall disqualify him from being called as a witness and giving evidence before a conciliator adjudicator or arbitrator on any matter whatsoever relevant to the dispute.

(b) All matters and information placed before a conciliator pursuant to a reference under sub-clause (5) of this Clause shall be deemed to be submitted to him without prejudice and the conciliator shall not be called as witness by the parties or anyone claiming through them in connection with any adjudication arbitration or other legal proceedings arising out of or connected with any matter so referred to him.

*Synopsis*   Clause 66 sets out a detailed procedure for the resolution of disputes. It provides for four distinct dispute avoidance or resolution procedures:

(1)   an Engineer's formal decision under Clause 66(2)
(2)   conciliation pursuant to Clause 66(5)
(3)   adjudication pursuant to Clause 66(6)–(8)
(4)   arbitration pursuant to Clauses 66(9)–(11).

   The clause provides that the decision in (1) (or the Engineer's failure to give a decision within the specified time) is a condition precedent to the right to have the matter conciliated[78] or arbitrated. This commentary should be read in conjunction with the commentaries on the procedures in Chapter 15.

*The nature of matters and disputes which are dealt with in Clause 66*   Clause 66(2) purports to apply to 'any matter arising under or in connection with or arising out of the Contract or the carrying out of the Works'. This is extremely wide and will include claims for misrepresentation.

*The definition of dispute*   The term 'dispute' is defined in contradistinction to a 'matter of dissatisfaction'. The former arises when the processes in the contract for dealing with matters of dissatisfaction have been exhausted. This is an entirely sensible provision. Matters frequently arise which give rise to dissatisfaction; the contract provides clear provisions with reasonable timescales for dealing with them.[79] It is reasonable that a matter which is being processed should not be considered a 'dispute'.

*Notices*   The suite of notices provides for in Clause 66 are:

(1)   Notice of Dispute—Clause 66(3)
(2)   Notice in writing seeking agreement of the other party for the dispute to be considered under the ICE Conciliation Procedure—Clause 66(5)(a)
(3)   Notice of Adjudication—Clause 66(6)(a)
(4)   Notice to Refer the dispute to arbitration—Clause 66(9)(a)

---

78   The statement in the text as regards conciliation is the formal position. Conciliation is unlikely to work unless the parties are committed, so considerations of 'conditions precedent' are perhaps irrelevant.

79   An analogy may be drawn between a matter of dissatisfaction and a temporary non-compliance with specification which is not a breach of contract but a temporary disconformity: *P and M Kaye v. Hosier & Dickinson Ltd* [1972] 1 All ER 121 (HL).

(5) Notice to Concur in the appointment of an arbitrator—Clause 66(10)(b).

*Routes and timescales*   A number of routes may be followed. The provisions are complex and the following appear in relation to the two most likely routes:

(1) *Reference of a matter of dissatisfaction to the Engineer followed directly by a reference to arbitration.* Where a matter of dissatisfaction is referred to the Engineer, he must make his decision within one month.[80] When the Engineer decides the matter or when the time for giving the decision has expired, a party may serve a Notice of Dispute 'in respect of that matter'[81] setting out the dispute. There seems to be no maximum time limit for the service of a Notice of Dispute.[82] At any time after the Notice of Dispute, the dispute may then be referred to arbitration by serving a Notice to Refer.[83] At any time after the Notice to Refer has been served a Notice to Concur in the appointment of an arbitrator may be served.[84] Again there seem to be no contractual time limits on the service of either the Notice to Refer or the Notice to Concur. The Notice to Concur notice must, of course, be served within the relevant limitation period.[85] Where the parties fail to agree the appointment of an arbitrator within one month of the Notice to Concur, an application may be made to the President of the Institution of Civil Engineers for an appointment.[86]

(2) *Reference of a matter of dissatisfaction to the Engineer, followed by a reference to adjudication, followed by a reference to arbitration.* When a Notice

---

80   Clause 66(2).
81   Clause 66(3). In other words, the dispute must fall within the purview of the matter of dissatisfaction which was referred to the Engineer.
82   Thus it seems that a party can sit on an Engineer's decision under Clause 66(2) for a considerable time before notifying the other party of his intention to refer the matter to arbitration.
83   Clause 66(9)(a).
84   Clause 66(10)(b). In practice, it may be served at the same time as the Notice to Refer.
85   Arbitration Act 1996 Section 13 provides that the ordinary limitation legislation applies. The relevant period is 6 years or 12 years for a contract under seal. The Act also provides in Section 14(4) that in the absence of other agreement the arbitration is commenced when a notice is served requiring the other party to agree to the appointment of an arbitrator. If the parties agree to adopt the CIMAR Arbitration Rules, these provide:

   Arbitral proceedings are begun in respect of a dispute when one party serves on the other a written notice of arbitration identifying the dispute and requiring him to agree to the appointment of an arbitrator.

86   Clause 66(10)(c)/(d).

of Dispute has been served (as above) either party may serve a Notice of Adjudication at any time.[87] The decision of the Adjudicator will be final and binding unless a Notice to Refer is served within three months of the Adjudicator's decision.[88] When a Notice to Refer is served, a Notice to Concur may be served at any time after up until the expiry of the limitation period. In this case, a Notice to Refer can be a device designed to preserve a party's right to have the matter arbitrated later.

*Compliance with the adjudication requirements of the HGCRA 1996*   One of the key intended effects of the narrowing of the definition of 'dispute' is that adjudication will not be available until the Engineer has given his decision in accordance with Clause 66(2). The problem is, however, that the HGCRA 1996 provides that a dispute includes a 'difference' and provides that a dispute or difference may be referred 'at any time'.[89] If, however, a party gets the appointment of an Adjudicator before a 'dispute' arises as defined in Clause 66(3) and the Adjudicator takes the view that Clause 66 does not comply with the HGCRA 1996 and proceeds to make a decision under the Scheme for Construction Contracts, what is the position? It has been held[90] that provided the adjudicator has jurisdiction in the sense that the contract is one to which the HGCRA 1996 applies and provided that his decision purports to be a decision under the HGCRA 1996, the parties cannot impeach it simply by asserting that the procedure used was defective. Accordingly, it seems that the decision of an Adjudicator under the Scheme will be binding whether or not the contractual provisions in Clause 66 meet the requirements of the HGCRA 1996. If the court were to make a definitive ruling on the compliance of Clause 66 with the HGCRA 1996, the position may be changed.

## Application to Scotland and Northern Ireland

### Application to Scotland   67

(1)  If the Works are situated in Scotland the Contract shall in all respects be construed and operate as a Scottish contract and shall be interpreted

---

87   Presumably this is limited in law and practice by (*a*) any limitation period and (*b*) any estoppel which may arise as a result of an ongoing arbitration dealing with the same matter.

88   Clause 66(9)(b).

89   Housing Grants, Construction and Regeneration Act 1996, Section 108.

90   *Macob Civil Engineering Ltd* v. *Morrison Construction Ltd* [1999] BLR 93.

in accordance with Scots Law and the provisions of sub-clause (2) of this Clause shall apply.

(2)   In the application of these Conditions and in particular Clause 66 thereof

  (a)   the word "arbiter" shall be substituted for the word "arbitrator"

  (b)   for any reference to the "Arbitration Act 1996" there shall be substituted reference to the law of Scotland and/or Section 66 and Schedule 7 of the "Law Reform (Miscellaneous Provisions) (Scotland) Act 1990" as may be appropriate

  (c)   for any reference to "The Institution of Civil Engineers' Arbitration Procedure (1997)" or "The Construction Industry Model Arbitration Rules" there shall be substituted a reference to "The Institution of Civil Engineers Arbitration Procedure (Scotland) (1983)" or any amendment or modification thereof being in force at the time of the appointment of the arbiter

  (d)   notwithstanding any of the other provisions of these Conditions nothing therein shall be construed as excluding or otherwise affecting the right of a party to arbitration to call in terms of Section 3 of the Administration of Justice (Scotland) Act 1972 for the arbiter to state a case and

  (e)   where the Employer or the Contractor wishes to register the decision of an adjudicator in the Books of Council and Session for preservation and execution the other party shall on being requested to do so forthwith consent to such registration by subscribing the said decision before a witness.

### Application to Northern Ireland

(3)   If the Works are situated in Northern Ireland the Contract shall in all respects be construed and operate as a Northern Irish contract and shall be interpreted in accordance with the law of Northern Ireland.

### Application elsewhere

(4)   If the Works are situated in a country or jurisdiction other than England and Wales Scotland or Northern Ireland the Contract and the provisions for disputes settlement shall in all respects be construed and operate and be interpreted in accordance with the law of that country or jurisdiction.

## Notices

### Service of notices on Contractor   68

(1)   Any notice to be given to the Contractor under the terms of the Contract shall be served in writing at the Contractor's principal place of

business (or in the event of the Contractor being a Company to or at its registered office).

### Service of notices on Employer

(2)  Any notice to be given to the Employer under the terms of the Contract shall be served in writing at the Employer's last known address (or in the event of the Employer being a Company to or at its registered office).

(3)  Any notice to be given by or to the Engineer under the terms of the Contract shall be delivered as the Engineer may direct.

Notices are required at many points in the Contract. Clause 68 requires that notices are to be in writing and specifies, in rather quaint terms, that they are to be served at the Contractor's 'principal place of business' or the 'Employer's last known address'. The term 'served' does not, it is thought, imply any technical legal meaning; rather the requirement is to bring the matter to the notice of the other party. Throughout the Conditions there are requirements as to notices: see, for example, Clauses 1(6) and Clause 53(5). Note also that some clauses specify slightly altered arrangements; for instance, Clause 2(4)(a) provides that notice of the Engineer's Representative may be given to the Contractor's agent. In any event, the question of waiver is likely to be important in many instances. Parties tend to have several addresses. It may be quite inconvenient to communicate with the 'principal place of business'. Where the parties have established alternative arrangements neither will be able to claim that a notice served in accordance with that practice was invalidly served unless they expressly indicate that they wish to revert to the strict terms of the Contract.

## Tax matters

### Labour-tax and landfill tax fluctuations    69

(1)  The rates and prices contained in the Bill of Quantities shall be deemed to take account only of the levels and incidence in force at the date for return of tenders of

(a)  the taxes levies contributions premiums or refunds (including national insurance contributions but excluding income tax and any levy payable under the Industrial Training Act 1982 or any statutory re-enactment or amendment thereof for the time being in force) which are by law payable by or to the Contractor and his sub-contractors in respect of their workpeople engaged on the Contract and

(b) any landfill tax payable by the Contractor or his sub-contractors pursuant to the Finance Act 1996 (Sections 39-71 and Schedule 5) and the Landfill Tax Regulations 1996 or any statutory re-enactment or amendment thereof for the time being in force

and shall not take account of any level or incidence of the aforesaid matters foreseeable or known to take effect at some later date.

(2) If after the date for return of tenders there shall occur any change in the level and/or incidence of any such taxes levies contributions premiums or refunds the Contractor shall so inform the Engineer and the net increase or decrease shall be taken into account in arriving at the Contract Price. The Contractor shall supply the information necessary to support any consequent adjustment to the Contract Price. All certificates for payment issued after submission of such information shall take due account of the additions or deductions to which such information relates.

## Value Added Tax 70

(1) The Contractor shall be deemed not to have allowed in his tender for the tax payable by him as a taxable person to the Commissioners of Customs and Excise being tax chargeable on any taxable supplies to the Employer which are to be made under the Contract.

## Engineer's certificates net of Value Added Tax

(2) All certificates issued by the Engineer under Clause 60 shall be net of Value Added Tax.

In addition to the payments due under such certificates the Employer shall separately identify and pay to the Contractor any Value Added Tax properly chargeable by the Commissioners of Customs and Excise on the supply to the Employer of any goods and/or services by the Contractor under the Contract.

## Disputes

(3) If any dispute difference or question arises between either the Employer or the Contractor and the Commissioners of Customs and Excise in relation to any tax chargeable or alleged to be chargeable in connection with the Contract or the Works each shall render to the other such support and assistance as may be necessary to resolve the dispute difference or question.

## Clause 66 not applicable

(4) Clause 66 shall not apply to any dispute difference or question arising under this Clause.

## The Construction (Design and Management) Regulations 1994

### CDM Regulations 1994   71

(1) In this clause

    (a) "the Regulations" means the Construction (Design and Management) Regulations 1994 or any statutory re-enactment or amendment thereof for the time being in force

    (b) "Planning Supervisor" and "Principal Contractor" mean the persons so described in regulation 2(1) of the Regulations

    (c) "Health and Safety Plan" means the plan prepared by virtue of regulation 15 of the Regulations.

(2) Where and to the extent that the Regulations apply to the Works and

    (a) the Engineer is appointed Planning Supervisor and/or

    (b) the Contractor is appointed Principal Contractor

then in taking any action as such they shall state in writing that the action is being taken under the Regulations.

(3) (a) Any action under the Regulations taken by either the Planning Supervisor or the Principal Contractor and in particular any alteration or amendment to the Health and Safety Plan shall be deemed to be an Engineer's instruction pursuant to Clause 13. Provided that the Contractor shall in no event be entitled to any additional payment and/or extension of time in respect of any such action to the extent that it results from any action lack of action or default on the part of the Contractor.

    (b) If any such action of either the Planning Supervisor or the Principal Contractor could not in the Contractor's opinion reasonably have been foreseen by an experienced contractor the Contractor shall as early as practicable give written notice thereof to the Engineer.

The planning supervisor (*a*) co-ordinates the health and safety aspects of the design and planning phase of the project, (*b*) ensures that a health and safety plan is prepared, (*c*) ensures that designers have taken safety properly into account, (*d*) advises the client (Employer) on health and safety and (*e*) prepares a safety file. The Engineer may be the planning supervisor and it is convenient for him to act in this role. Clause 71(2) applies where he is the planning supervisor; whenever he acts in this capacity he must state in writing that he does so.

Where a Contractor is appointed under the ICE Conditions of Contract, he will ordinarily be the 'principal contractor' and Clause 71(2) makes provision for this.

The regulations affect the Employer and Contractor (the principal contractor) in a number of ways:[91]

(1) The Contractor takes over the health and safety plan. This contains many of the features normally to be found in the Clause 14 programme and method statement. It includes a general description of the construction work involved, details of the time within which it is intended that the project and any intermediate stages will be completed, and a statement of risks—Regulation 15(2). The Contractor maintains and updates the plan to show such arrangements for the project which will ensure, as far as is reasonably practicable, health and safety—Regulation 15(3).

(2) The Contractor takes reasonable steps to ensure cooperation between all contractors and to ensure that they comply with the rules in the health and safety plan. Furthermore, the Contractor is required to take reasonable steps to ensure that only authorised persons are allowed in the vicinity of construction in progress—Regulation 16(1).

(3) Where the Contractor has a design responsibility, his designers or any design sub-contractors are obliged to take full account of the health and safety aspects which their designs entail and to cooperate with the planning supervisor and other designers in this regard—Regulation 13.

Clause 71(3) provides that any alteration or amendment of the health and safety plan shall be deemed to be an Engineer's instruction pursuant to Clause 13(1). At first sight this seems somewhat curious given that responsibility for the health and safety plan is transferred to the Contractor who develops the plan so as to suit not only the health and safety requirements of the project but his own construction plans. Where any change is made to the health and safety plan, the Contractor's entitlement to additional payment or extensions of time is set out in Clause 13(3). This entitles the Contractor, as a general principle, to claim where the instruction causes 'him to incur cost beyond that reasonably to have been foreseen by an experienced contractor at the time of tender'. However, by Clause 71(3)(a) he will not be entitled to additional payment or an extension of time to the extent that it 'results from any action, lack of action or default on the part of the Contractor'. The word 'action' seems misleading, since any change which the Contractor makes to the plan is, in a superficial sense, necessarily

---

91  See the Health and Safety Executive approved code of practice, *Managing Construction for Health and Safety*, HSE, 1995.

a result of his action. It is thought that where the Contractor is the principal contractor he may amend the health and safety plan in any way which is necessary to comply with his health and safety obligations and he is entitled to be paid in accordance with Clause 13(3) to the extent that the test in Clause 13(3) is satisfied.

## Special conditions

**Special conditions    72**

The following special conditions form part of the Conditions of Contract.

(Note. Any special conditions including Contract Price Fluctuation which it is desired to incorporate in the Conditions of Contract should be numbered consecutively with the foregoing conditions of contract).

The Conditions of Contract are frequently supplemented with special conditions. Clause 5 gives equal weight to all terms of the Contract and hence an identical effect will be produced by including any additional terms in any other document.

## Forms of tender, agreement and bond

These are reproduced below.

## SHORT DESCRIPTION OF WORKS

All Permanent and Temporary Works in connection with* .............................

.....................................................................................................................

.....................................................................................................................

### Form of Tender

(NOTE: The Appendix forms part of the Form of Tender)

To ….................................................

.........................................................

.........................................................

GENTLEMEN,

Having examined the Drawings, Conditions of Contract, Specification and Bill of Quantities for the construction of the above-mentioned Works (and the matters set out in the Appendix hereto) we offer to construct and complete the whole of the said Works in conformity with the said Drawings, Conditions of Contract, Specification and Bill of Quantities for such sum as may be ascertained in accordance with the said Conditions of Contract.

We undertake to complete and deliver the whole of the Permanent Works comprised in the Contract within the time stated in the Appendix hereto.

If our Tender is accepted we will, if required, provide security for the due performance of the Contract as stipulated in the Conditions of Contract and the Appendix hereto.

Unless and until a formal Agreement is prepared and executed this Tender together with your written acceptance thereof, shall constitute a binding Contract between us.

We understand that you are not bound to accept the lowest or any tender you may receive.

We are, Gentlemen.

Yours faithfully,

Signature ............................................................

Address ..............................................................

.............................................................................

Date .............................

* Complete as appropriate

## FORM OF TENDER (APPENDIX)

(NOTE: Relevant Clause numbers are shown in brackets)

### Appendix - Part 1 (to be completed prior to the invitation to tender)

1   Name of the Employer (Clause 1(1)(a) ) ........................................................
    Address ..............................................................................................................
2   Name of the Engineer (Clause 1(1)(c)) ..........................................................
    Address ..............................................................................................................
3   Defects Correction Period (Clause 1(1)(s)) .............. weeks
4   Parts or Sections of the Works which shall not be sub-contracted without the
    Engineer's prior written approval (Clause 4(2))
    ..............................................................................................................................
    ..............................................................................................................................
    ..............................................................................................................................
5   Number and type of copies of Drawings to be provided (Clause 6(1)(b))
    ..............................................................................................................................
6   Form of Agreement (Clause 9)                          Required/Not required
7   Performance Bond (Clause 10(1))                       Required/Not required
    Amount of Bond (if required) to be .............. % of Tender Total
8   Minimum amount of third party insurance (persons and property) (Clause
    23(3)) £ ..................... for each and every occurrence
9   Works Commencement Date (if known) (Clause 41(1)(a))
    ..............................................................................................................................
10  Time for Completion (Clause 43)[a]
    EITHER for the whole of the Works ...............weeks
    OR for Sections of the Works (Clause 1(1)(u))[b]
        Section A ..........................................        .............. weeks
        Section B ..........................................        .............. weeks
        Section C ..........................................        .............. weeks
        Section D ..........................................        .............. weeks
        the Remainder of the Works                       .............. weeks
11  Liquidated damages for delay (Clause 47)

|  | per day/week | limit of liability[c] |
|---|---|---|
| EITHER for the whole of the Works | £ ............................. | £ ............................. |
| OR for Section A (as above) | £ ............................. | £ ............................. |
| Section B (as above) | £ ............................. | £ ............................. |
| Section C (as above) | £ ............................. | £ ............................. |
| Section D (as above) | £ ............................. | £ ............................. |
| the Remainder of the Works (as above) | £ ............................. | £ ............................. |

12  Vesting of materials not on Site (Clauses 54(4) and 60(1)(c) ) (if required by
    the Employer)[d]

1 ......................................    4 ......................................
2 ......................................    5 ......................................
3 ......................................    6 ......................................

13  Method of measurement adopted in preparation of Bills of Quantities (Clause 57)[e]

..........................................................................................

14  Percentage of the value of goods and materials to be included in Interim Certificates (Clause 60(2)(b))                    ............ %

15  Minimum amount of Interim Certificates (Clause 60(3))          £ ..............

16  Rate of retention (recommended not to exceed 5%) (Clause 60(5)) ........... %

17  Limit of retention (% of Tender Total) (Clause 60(5)) (Recommended not to exceed 3%)                    ............ %

18  Bank whose Base Lending Rate is to be used (Clause 60(7))

..........................................................................................

19  Requirement for prior approval by the Employer before the Engineer can act. DETAILS TO BE GIVEN AND CLAUSE NUMBER STATED (Clause 2(1)(b))[f]

..........................................................................................
..........................................................................................
..........................................................................................

20  Name of the Planning Supervisor (Clause 71(1)(b))

..........................................................................................

Address

..........................................................................................

21  Name of the Principal Contractor (if appointed) (Clause 71(1)(b))

..........................................................................................

Address

..........................................................................................

22  The arbitration procedure to be used is (Clause 66(11)(a)
    (a)   The Institution of Civil Engineers' Arbitration Procedure 1997[g]
    (b)   The Construction Industry Model Arbitration Rules[g]

a   If not stated is to be completed by Contractor in Part 2 of the Appendix.
b   To be completed if required, with brief description. Where Sectional completion applies the item for "the Remainder of the Works" must be used to cover the balance of the Works if the Sections described do not in total comprise the whole of the Works.
c   Delete where not required.
d   (If used) Materials to which the Clauses apply must be listed in Part I (Employer's option) or Part 2 (Contractor's option).
e   Insert here any amendment or modification adopted if different from that stated in Clause 57.
f   If there is any requirement that the Engineer has to obtain prior approval from the Employer before he can act full particulars of such requirements must be set out above.
g   Delete one as appropriate.

**Appendix - Part 2**
(To be completed by Contractor)

1     Insurance Policy Excesses (Clause 25(2))
        Insurance of the Works (Clause 21(1))            £ ......................
        Third party (property damage) (Clause 23(1))     £ ......................

2     Time for Completion (Clause 43) (if not completed in Part I of the Appendix)
        EITHER for the whole of the Works              ................. weeks
        OR for Sections of the Works (Clause 1(1)(u) ) (as detailed in Part 1 of the Appendix)
           Section A .........................             ................. weeks
           Section B .........................             ................. weeks
           Section C .........................             ................. weeks
           Section D .........................             ................. weeks
           the Remainder of the Works             ................. weeks

3     Vesting of materials not on site (Clauses 54(4) and 60(1)(c)) (at the option of the Contractor — see [d] in Part1)
        1 .....................................................        4 .....................................................
        2 .....................................................        5 .....................................................
        3 .....................................................        6 .....................................................

4     Percentage(s) for adjustment of PC sums (Clauses 59(2)(c) and 59(5)(c)) (with details if required)
        ..................................................................................................................................
        ..................................................................................................................................

**Form of Agreement**

THIS AGREEMENT made the ..................... day of ...................... 19 ......................
BETWEEN ...............................................................................................................
of ..........................................................................................................................
in the County of ............................................. (hereinafter called "the Employer")

and .......................................................................................................................
of ..........................................................................................................................
in the County of ............................................. (hereinafter called "the Contractor")

WHEREAS the Employer is desirous that certain Works should be constructed, namely the Permanent and Temporary Works in connection with ..........................
...............................................................................................................................
and has accepted a Tender by the Contractor for the construction and completion of such Works.

NOW THIS AGREEMENT WITNESSETH as follows

1. In this Agreement words and expressions shall have the same meanings as are respectively assigned to them in the Conditions of Contract hereinafter referred to.

2. The following documents shall be deemed to form and be read and construed as part of this Agreement, namely
(a) the said Tender and the written acceptance thereof
(b) the Drawings
(c) the Conditions of Contract
(d) the Specification
(e) the priced Bill of Quantities.

3. In consideration of the payments to be made by the Employer to the Contractor as hereinafter mentioned the Contractor hereby covenants with the Employer to construct and complete the Works in conformity in all respects with the provisions of the Contract.

4. The Employer hereby covenants to pay to the Contractor in consideration of the construction and completion of the Works the Contract Price at the times and in the manner prescribed by the Contract.

IN WITNESS whereof the parties hereto have caused this Agreement to be executed the day and year first above written.

SIGNED on behalf of the said ...................................................... Ltd/plc
(the Employer)
Signature ............................................          Signature ...........................................
Position ..............................................          Position .............................................
In the presence of .................................          In the presence of ...............................

SIGNED on behalf of the said ........................................................................ Ltd/plc
(the Contractor)
Signature ........................................    Signature ...........................................
Position ...........................................    Position ............................................
In the presence of .........................    In the presence of ..........................

or
SIGNED [and SEALED*] AS A DEED by the said ...........................................

.................................................................................................................. Ltd/plc
(the Employer)
In the presence of ..........................................................................................

or
SIGNED [and SEALED*] AS A DEED by the said ...........................................
.................................................................................................................. Ltd/plc
(the Contractor)
In the presence of .................................

* Delete as appropriate

# ICE FORM OF DEFAULT BOND

**Date**

**Parties**

   **[Names addresses and Company Numbers if applicable]**

SURETY (1)

CONTRACTOR (2)

EMPLOYER (3)

**Background**

(A) By a Contract defined in the Schedule hereto the Contractor has agreed with the Employer to construct and complete the Works.

(B) The Surety has agreed to provide this Bond in favour of the Employer in order to guarantee the performance by the Contractor of his obligations under the Contract.

**1   Surety's obligation**

   (1) If the Contractor fails to pay the Excess Sum within 28 days of receipt by the Contractor of a copy of a certificate issued by the Engineer under clause 65(5) of the Contract the Surety hereby guarantees to the Employer that the Surety shall subject to the terms and conditions of this Bond pay the Excess Sum in accordance with Clause 1(3) up to the Bond Amount.

   (2) It shall be a condition precedent to payment by the Surety that the Employer serve on the Surety a copy of the certificate issued by the Engineer under clause 65(5) of the Contract as served on the Contractor and certified by the Engineer as being a true copy of such certificate.

   (3) Subject to Clause 1(4) payment by the Surety shall be made not later than 14 days after the later of

      (a) the expiry of the 28 day period referred to in Clause 1(1) (save in respect of any payment made by the Contractor within that time) and

      (b) service on the Surety of the copy certificate referred to in Clause 1(2).

   (4) If the Surety objects to the contents of or entitlement to issue a certificate under clause 65(5) of the Contract in respect of which the Employer seeks payment from him the Surety shall have the right to refer the matter to adjudication in accordance with the adjudication provisions contained in sub-clauses 66(6) to 66(8) of the Contract as if the Surety were a party to the Contract in place of the Contractor.

(5) Any adjudication under Clause 1(4) shall be commenced by the Surety within 14 days of receipt by the Surety of the documents referred to in Clause 1(2) and the Surety shall have no right to refer the matter to adjudication after that time.

(6) If the content of or entitlement to issue the Certificate under clause 65(5) of the Contract in respect of which payment is sought by the Employer is or has been the subject of an adjudication between the Employer and Contractor under the Contract (in respect of which both parties have made submissions to the adjudicator) the Surety agrees to be bound by the result of such adjudication and shall have no right to refer the matter to adjudication under Clause 1(4).

(7) In the case of an adjudication under Clause 1(4) payment by the Surety shall be made within 7 days of the decision in such adjudication.

## 2 Surety's rights

(1) The Surety shall be entitled to receive copies of any notice given by the Employer under clause 65(1) of the Contract (with any Engineer's certificate referred to) within 7 days of such notice or certificate being served on the Contractor.

(2) The Surety shall be entitled at any time within 7 days of receipt by the Surety of the copy certificate referred to in Clause 1(2)

   (a) to request the Employer to provide the Surety with such further information and documentation as the Surety may reasonably require to verify the Excess Sum (including information or documentation held by the Engineer) and/or

   (b) to request to inspect the Site and the Works upon reasonable notice (the Employer may require that a representative of the Employer accompanies the Surety during such inspections).

## 3 Accounting

(1) If the Excess Sum is subsequently determined by reason of a subsequent certificate issued by the Engineer under clause 65(5) of the Contract or by adjudication arbitration litigation or agreement between the Surety and the Employer to be less than the amount paid by the Surety the difference (if the Excess Sum has already been paid by the Surety) shall be repaid by the Employer to the Surety with Interest within 14 days (or such other period as the adjudicator arbitrator or Court may direct) after the date of such determination or agreement.

(2) If the Excess Sum is subsequently determined or agreed to be greater than the amount already paid by the Surety any difference (up to the Bond Amount) shall be paid by the Surety to the Employer with Interest within 14 days (or such other period as the adjudicator arbitrator or Court may direct) after the date of such determination or agreement.

## 4 Interest

Subject to the amount payable by the Surety being varied in accordance with Clause 3 above in the event that any amount payable by either the Surety or the Employer under this Bond is not made by the date determined by Clause 1(3) or in accordance with Clause 3 (the Due Date) then the payer shall pay Interest on the sum from the Due Date until the date of payment.

## 5 Expiry

Save in respect of any failure to pay the Excess Sum in respect of which a claim in writing has been received beforehand from the Employer by the Surety this Bond shall expire on the later of the date stated in a certificate of substantial completion issued by the Engineer and the Final Expiry Date.

## 6 Forbearance

The Surety shall not be discharged or released by any alteration variation or waiver of any of the terms and conditions and provisions of the Contract or in any extent or nature of the Works and no allowance of time by the Employer under or in connection with the Contract or the Works shall in any way release reduce or affect the liability of the Surety under this Bond.

## 7 Governing law

This Bond shall be governed and construed in accordance with the laws of the country named in the Schedule ("the Country") and the courts of the Country shall have exclusive jurisdiction.

## 8 Assignment

This Bond may only be assigned by the Employer with the prior consent of the Surety and the Contractor which consent shall not be unreasonably withheld. In the event of any such assignment the Employer and assignee shall remain jointly and severally liable for any repayment due to the Surety under Clause 3(1). Notice of any assignment shall be given to the Surety as soon as practicable.

## EXECUTED AS A DEED

| | |
|---|---|
| on behalf of the Surety | (1) director ......................................................... |
| | (2) director/secretary ......................................... |
| on behalf of the Contractor | (1) director ......................................................... |
| | (2) director/secretary ......................................... |
| on behalf of the Employer | (1) director ......................................................... |
| | (2) director/secretary ......................................... |

## SCHEDULE

Address for Service

**Contractor:**

Tel:                        Fax:

**Employer:**

Tel:                        Fax:

**Surety:**

Tel:                        Fax:


"Bond Amount"        means the sum of £[ ] ([ ] pounds) being the maximum aggregate liability of the Surety under this Bond.

"Contract"           means the contract [made between the Employer and the Contractor dated the [ ] day of [ ] [ ]] / [to be entered into between the

Employer and the Contractor] incorporating the ICE Conditions of Contract 7th Edition.

"Excess Sum"         an amount certified as due to the Employer under Clause 65(5) of the Contract.

"Engineer" "Works" and "Site"    have the same meaning as in the Contract.

"Final Expiry Date"  means the [ ] day of [ ] [ ].

"Interest"           means the rate of interest specified in the Contract.

"The Country"        means [England and Wales] [Scotland] [Northern Ireland]

(This Bond has been drafted in collaboration with S. J. BERWIN & Co )

# 14

# The Engineering and Construction Contract (2nd Edition and Short Contract)

The Engineering and Construction Contract is designed to stimulate good project management. The second edition published in 1995 has now been joined in July 1999 by a short form of the contract. The latter is dealt with briefly at the end of this chapter.

## 1. Introduction to the Engineering and Construction Contracts

The New Engineering Contract (NEC) was created and drawn up by an Institution of Civil Engineers working group. A consultative version was published in 1991. The first edition was published in 1993. The second edition was published as the Engineering and Construction Contract (EEC) in November 1995. The new title reflects the aspirations of the NEC Panel that the contract should be used for construction work in all sectors, including traditional building. Many of the changes in this new edition are derived from recommendations in the Latham Report.[1] As a result of the need for British contracts to comply with the Housing Grants, Construction and Regeneration Act 1996, a new option—Option Y(UK)2—was published in April 1998 for use in such cases; a failure to opt in, where the Act applies, will invoke the Scheme for Construction Contracts.

---

1    Sir Michael Latham, *Constructing the Team*, HMSO, 1994.

*The suite*   The contract is not a single contract but a suite of contracts which share a core body of definitions and interrelationships; there are Core Clauses, Main Option Clauses and Secondary Option Clauses, together with Contract Data. The contract is published as a system in a series of documents. These documents include the (*a*) 'Black Book' which houses the full inventory of clauses, (*b*) merged versions which contain clauses relevant to each of the Main Options, (*c*) guidance notes and (*d*) flow charts. Ancillary documents include a harmonised sub-contract and an Adjudicator's form of appointment.

*Style of drafting*   The contract is drawn in a manner which seems unfamiliar and hence disconcerting to lawyers experienced in traditional contract design. It is a series of short, crisp statements. These are in the present tense[2] and use defined terms. For example, Clause 20.1 reads: 'The *Contractor* Provides the Works in accordance with the Works Information'. Words with initial capitals are defined in the contract. Italicised words[3] are defined in the contract and identified in the Contract Data. A numbering system is used which immediately identifies the location and objective of each clause.[4]

## 2. The Engineering and Construction Contract (The NEC 2nd Edition)

*Core clauses*   These clauses remain constant across the range of contacts. Thus the advantages accruing to any standard form, such as familiarity, even-handedness, thorough checking and consultation are retained.

*Main Option clauses*   The six basic options differ principally in the agreed method of remuneration and its attendant mechanisms and risks. The versions are: A—Priced contract with activity schedule; B—Priced contract with bill of quantities; C—Target contract with activity schedule; D—Target contract with bill of quantities; E—Cost reimbursable contract; and F—Management contract. Other options, such as contractor's design obligations, are readily achieved by inserting the extent of the design obligation into the Contract Data.

---

2    Except Clause 10.1; this is an exhortation and is in the future tense. In addition, the term 'may' is used for discretionary action.
3    Save for direct quotations, this convention is not preserved in the text of this chapter.
4    Thus the '30' series clauses (from 30.1 to 36.5) all relate to aspects of time.

*Secondary Option clauses*     These provide a wide range of additional possibilities and are to be used in conjunction with the core clauses and a Main Option from A to F. The secondary option clauses are: G—Performance bond; H—Parent company guarantee; J—Advanced payment to the Contractor; K—Multiple currencies; L—Sectional completion; M—Limitation of the Contractor's liability for his design to reasonable skill and care; N—Price adjustment for inflation; P—Retention; Q—Bonus for early completion; R—Delay damages; S—Low performance damages; T—Changes to the law; U—The Construction Design and Management Regulations 1994; V—Trust Fund; and Z—Additional conditions.

*Contract Data*     This contains the specific data which relates to the project and contract. Matters such as: defined personnel; the identity and location of 'The Works Information' (i.e. drawings, specifications, etc.); the identity and location of the 'Site Information' (i.e. information about the Site); contract dates; payment assessment intervals and payment currencies; insurances; dispute resolution tribunal; and other data required to make the various options work.

*The parties*     The parties to the contract are the Contractor and the Employer—Clause 11.2(1).

*Administrators and decision-makers*     Three appointments are defined in the contract; the Project Manager, the Supervisor and the Adjudicator—Clause 11.2(2). The Project Manager is the person who is the senior administrator on the project. He acts as the agent of the Employer and as an independent decision-maker. The Supervisor is responsible for day to day supervision of the Works on behalf of the Employer. And the Adjudicator is an independent person appointed to resolve disputes.

*The Project Manager*     He is appointed by the Employer. His principal role is to manage the project, thereby achieving the Employer's objectives. The Project Manager has considerable authority. His role is not defined in any one place, but his functions are stated throughout the contract terms. As with an Engineer under a traditional contract, he has functions which are purely as an agent of the Employer, but some which require a degree of judicial detachment. Thus where he assesses payments due, he is not entitled to under-assess simply because this suits the business objectives of the

Employer.[5] Where the Contractor (and in some cases, the Employer) is dissatisfied with the Project Manager's decision, they may refer the matter to the Adjudicator. The Supervisor has his own distinct functions, and there is no appeal to the Project Manager from his decision.

*The Obligations of the Contractor*[6]   The Contractor agrees to Provide the Works in accordance with the Works Information. The Works Information is contained in the documents identified in the Contract Data as amended by any instruction given in accordance with the contract.[7] The Project Manager may give an instruction to the Contractor which changes the Works Information.[8]

*Time*[9]   The starting date, the possession date and the completion date are given in the Contract Data. The Contractor may start work on the starting date, which may be prior to possession (e.g. where plant etc. is constructed off-site). The programme is a key document. It includes a plan of time and is regularly updated. Completion occurs when the Contractor has done the work in the Works Information and corrected any defects which would have prevented the Employer from using the Works.[10] Where a Compensation Event occurs, which cannot be accommodated into the programme float, the Project Manager extends the time for completion.

*Payment*[11]   Assessment dates for payment are set out in the Contract Data. The Project Manager assesses payment at the appropriate dates. The payment due at any assessment date has several components. The principal component is The Price for Work Done to Date (PWDD). The computation of the PWDD depends on which of the six Main Options is being operated. Option A contains an activity schedule, that is a list of activities and correlated prices for each activity; the PWDD is the total of the prices for activities in the schedule which have been completed. In addition, there is an adjustment for Compensation Events (which are priced on the basis of Actual Cost, which is turn depends on a Cost Schedule which forms part of the Contract). Furthermore, a retention is deducted. The Project Manager

---

5   It is submitted that the term 'assess' requires that the computation be fair as between the parties.
6   See generally clauses in the 20 series.
7   Clause 11.2(5).
8   Clause 14.3.
9   See generally clauses in the 30 series.
10  Clause 11.2(13).
11  See generally clauses in the 50 series.

certifies payment within one week of the assessment date and payment is made within three weeks of the assessment date; interest is paid on any late payments.

*Payment dates and notices*   For projects in Britain, the contract must comply with the payment provisions of the Housing Grants, Construction and Regeneration Act 1996. Amendments to the contract[12] provide:

51.1   The Project Manager certifies a payment on or before the date on which a payment becomes due.

51.2   Each certified payment is made on or before the final date for payment.

56.1   For the purpose of Sections 109 and 110 of the Act,
   • the *Project manager's* certificate is the notice of payment from the *Employer* to the *Contractor* specifying the amount (if any) of the payment made or proposed to be made, and the basis on which that amount was calculated,
   • the date on which a payment becomes due is seven days after the assessment date and
   • the final date for payment is twenty one days or if a different period is stated in the Contract Data, the period stated, after the date on which the payment becomes due.

56.2   If the Employer intends to withhold payment after the final date for payment of a sum due under this contract, he notifies the Contractor not later than seven days (the prescribed period) before the final date for payment …

*Compensation events*[13]   Compensation events are those events which entitle the Contractor to claim compensation. These include[14] where the Project Manager makes a change to the work, where the Employer fails to give possession on time, where the Employer fails to provide something which he is to provide or do some work which he is to do, where unforeseen physical conditions occur, where very exceptional weather conditions occur, etc. Where the event arises as a result of an instruction etc. of the Project Manager, the Project Manager notifies[15] the Contractor. Where the Contractor believes that a compensation event has occurred or anticipates that one will occur, he notifies the Project Manager. The Project Manager

---

12   Amendment available as Option Y(UK)2, published in April 1998.
13   See generally clauses in the 60 series.
14   They are listed out in Clause 60.1. An amendment is available as Option Y(UK)2, published in April 1998; this provides: 'Suspension of performance is a compensation event if the Contractor exercises his right to suspend performance under the [HGCR] Act'.
15   See generally Clause 61 as to notifications for compensation events.

decides whether or not a compensation event has occurred or will occur; if so, he asks for a quotation or alternative quotations based on alternative ways of dealing with the compensation event.[16] The Project Manager assesses the change in Price and any delay to the Completion Date caused by the compensation event.[17]

*Disputes* One of the principal contributions of the NEC contract to the development of construction contracts is the use of adjudication and the primary means of dispute management. The right to adjudication in all construction contracts is now enshrined in the Housing Grants, Construction and Regeneration Act 1996. Since the disputes clauses of NEC 1995 did not comply with the Act, they have been redrafted.[18] However, it is by no means clear that these amended provisions comply with the Act. Amended Clause 90(4) provides:

> The Parties agree that no matter shall be a dispute unless a notice of dissatisfaction has been given and the matter has not been resolved within four weeks. The word dispute (which includes a difference) has that meaning'.

It is not clear whether or not this is sufficient to comply with the requirement that 'the contract shall enable a party to give notice at any time of his intention to refer a dispute to arbitration'.[19]

## 3. Commentary on selected core clauses in the main contract

The principal innovations of the Engineering and Construction Contract includes its use and promotion of management principles and its handling of disputes. In this selective commentary, matters such as payment details are, therefore, given less prominence than the management innovations.

### *Early warning*

**Early warning   16**

16.1   The *Contractor* and the *Project Manager* give an early warning by notifying the other as soon as either becomes aware of any matter which could

---

16   See generally Clause 62 as to quotations for compensation events.
17   See generally Clauses 63, 64, 65 as to assessment and incidental matters in respect of compensation events.
18   Available as Option Y(UK)2, published in April 1998.
19   Housing Grants, Construction and Regeneration Act 1996, Section 108.

- increase the total of the Prices,

- delay Completion or

- impair the performance of the *works* in use.

16.2 Either the *Project Manager* or the *Contractor* may instruct the other to attend an early warning meeting. Each may instruct other people to attend if the other agrees.

16.3 At an early warning meeting those who attend co-operate in

- making and considering proposals for how the effect of each matter which has been notified as an early warning can be avoided or reduced,

- seeking solutions that will bring advantage to all those who will be affected and

- deciding upon actions which they will take and who, in accordance with this contract, will take them.

16.4 The *Project Manager* records the proposals considered and the decisions taken at an early warning meeting and gives a copy of his record to the *Contractor*.

*Early warning notice—the nature of the duty to notify* Clause 16.1 is expressed in mandatory terms. Thus, where either the Contractor or the Project Manager become aware of any condition which requires the giving of an early warning notice, and they do not do so, this may amount to a breach of contract. Although a sanction is provided in the contract for the late notification by the Contractor,[20] the Project Manager's motivation is his interest in efficient completion. However, the question of the mandatory nature of Clause 16.1 may raise interesting questions. For instance, where the Project Manager is aware of a notifiable defect which he fails to notify and this defect causes the Contractor serious loss at a later date, does this loss flow from the Project Manager's breach on behalf of the Employer? If so, this clause may reverse the position found on traditional contracts, namely that the employer's agent is not responsible for the contractor's defaults.[21]

*The relevant conditions* These are 'any matter which could increase the total of the Prices, delay Completion or impair the performance of the *works*

---

20 Clause 63.4, see below.
21 *East Ham Borough Council v. Bernard Sunley Ltd* [1966] AC 406 (HL); *AMF International Ltd v. Magnet Bowling* [1968] 1 WLR 1028.

in use'. A matter which could increase the total of the Prices includes compensation events (e.g. unforeseen ground conditions). Matters which could delay Completion range from those which are entirely due to the Contractor's default or inefficiency to those for which an extension of time is indubitably due. And matters which could impair the performance of the works in progress includes design defects noticed by the Contractor, any adverse ground conditions, any defective work, etc.

*Convening an early warning meeting*    Clause 16.2 is drawn in discretionary terms. Accordingly, there is no obligation to call an early warning meeting. But if either the Contractor or the Project Manager wishes to do so, they may instruct the other to attend. It seems that the instruction to attend may be included with the early warning notice. The timing of the meeting is not specified, but a reasonable period should be allowed between notice/ instruction and the meeting to enable the parties to collect such information as is reasonably required to comply with Clause 16.3, given the urgency, value and importance of the matter to be discussed. The final sentence of Clause 16.2 provides that other people may be instructed to attend. The term 'other people' is not defined, but no doubt includes lawyers if necessary. However, any such attendance requires the consent of the other. On one reading of this provision it seems as if the only persons entitled to attend as of right are the Project Manager or his delegate and the Contractor's representative. It is thought, however, that the provision impliedly includes (at the instance of either the Contractor or Project Manager) the Supervisor and other persons directly involved in the management of the project as well as any directly concerned sub-contractor or designer.[22]

*The meeting*    Clause 16.3 sets out the agenda for the meeting. In accordance with Clause 16.4, the proposals and decision are recorded by the Project Manager and a copy is provided to the Contractor.

### Notifying compensation events    61

61.5    If the *Project Manager* decides that the *Contractor* did not give an early warning of the event which an experienced contractor could have given, he notifies this decision to the *Contractor* when he instructs him to submit quotations.

---

22    Note that the defined term 'Others' is not used here—see Clause 11.2(2).

**Assessing compensation events   63**

63.4   If the *Project Manager* has notified the *Contractor* of his decision that
the *Contractor* did not give an early warning of a compensation event
which an experienced contractor could have given, the event is
assessed as if the *Contractor* had given early warning.

These clauses set out the contractual sanction against the Contractor. It
seems that a notice under Clause 61.5 is a condition precedent to the
action under Clause 63.4. Note also that the Clause 16.1 requires actual
awareness, while Clause 61.5 requires that 'an experienced contractor'
would have given the notice. The meaning of Clause 63.4 is not altogether
clear; it is thought that it means that the time or cost which is assessed as
being due to the Contractor is not the time and cost actually incurred, but
the time or costs which would have been incurred had the early warning
been given at the proper time; in short, the Contractor pays for his own failure to mitigate his loss.

## Programme

**The programme   31**

31.1   If a programme is not identified in the Contract Data, the *Contractor*
submits a first programme to the *Project Manager* for acceptance
within the period stated in the Contract Data.

31.2   The *Contractor* shows on each programme which he submits for
acceptance

- the *starting date*, *possession dates* and Completion Date,

- for each operation, a method statement which identifies the Equipment and other resources which the *Contractor* plans to use,

- planned Completion,

- the order and timing of

  - the operations which the *Contractor* plans to do in order to Provide the Works and

  - the work of the *Employer* and Others either as stated in the
Works Information or as later agreed with them by the
*Contractor*,

- the dates when the *Contractor* plans to complete work needed to
allow the *Employer* and Others to do their work,

- provisions for

  - float,

  - time risk allowances,

  - health and safety requirements and

  - the procedures set out in this contract,

- the dates when, in order to Provide the Works in accordance with his programme, the *Contractor* will need

  - possession of a part of the Site if later than its *possession date,*

  - acceptances and

  - Plant and Materials and other things to be provided by the *Employer* and

- other information which the Works Information requires the *Contractor* to show on a programme submitted for acceptance.

31.3   Within two weeks of the *Contractor* submitting a programme to him for acceptance, the *Project Manager* either accepts the programme or notifies the *Contractor* of his reasons for not accepting it. A reason for not accepting a programme is that

- the *Contractor*'s plans which it shows are not practicable,

- it does not show the information which this contract requires,

- it does not represent the *Contractor*'s plans realistically or

- it does not comply with the Works Information.

- Revising the programme 32

32.1   The *Contractor* shows on each revised programme

- the actual progress achieved on each operation and its effect upon the timing of the remaining work,

- the effects of implemented compensation events and of notified early warning matters,

- how the *Contractor* plans to deal with any delays and to correct notified Defects and

- any other changes which the *Contractor* proposes to make to the Accepted Programme.

32.2   The *Contractor* submits a revised programme to the *Project Manager* for acceptance

- within the *period for reply* after the *Project Manager* has instructed him to,

- when the *Contractor* chooses to and, in any case,

- at no longer interval than the interval stated in the Contract Data from the *starting date* until Completion of the whole of the *works*.

**The original and contractual status of the programme**   The programme is either (*a*) identified in the Contract Data or (*b*) submitted by the Contractor within a period stated in the Contract Data. Even where it is included in the Contract Data, it does not represent an invariate term of the contract.[23]

**The content of the programme—general**[24]   Unlike programmes which are submitted under traditional standard form civil engineering contracts, the content of the programme is specified in some detail in Clause 31.2. Thus, wherever a programme is called for under the contract, e.g. as part of a quotation for a compensation event, this includes a call for not only timing information, but also any necessary method statements and/or health and safety requirements etc. Furthermore, revised programmes must show the matters set out in Clause 32.1, including progress to date and proactive control measures.

**Float and time risk allowances**   Separate provisions are to be identified on the programme relating to (*a*) float and (*b*) time risk allowances. These are not defined anywhere in the contract. The normal meaning of 'float' is the time slack, so that an activity may finish later than programmed by the float period and still not delay the completion. The apparent meaning of 'time risk allowances' is the allowance to be made in addition to the nominal activity duration for uncertainty in the activity duration. The guidance notes (which form no part of the contract) state:[25] 'The Contractor's [sic] time risk allowances ... are owned by the Contractor as part of his realistic planning to cover his risks'. 'Float is any spare time within the programme after the time risk allowances have been included. It is normally available to accommodate the time effects of a compensation event in order to mitigate or avoid any delay to planned Completion'. There is no provision for what is

---

23   Compare with the position where a programme is incorporated into a traditional contract, such as the ICE Conditions of Contract: see *Yorkshire Water Authority* v. *Sir Alfred McAlpine* (1985) 32 BLR 114.

24   Note also that the Contractor may be the principal contractor under the Construction (Design and Management) Regulations 1994. If so, he will be required to maintain the health and safety plan which will contain much of the same information as required in this programme. See, generally, the commentary on Clause 71 of the ICE Conditions of Contract, 7th Edition.

25   Page 39.

frequently termed 'overall float'. Where the planned completion precedes the contract completion date, the difference in time is not available to the Project Manager; accordingly, where a compensation event causes a delay to Completion, the Project Manager must also delay the Completion Date—see Clause 63.3.

*Acceptance*   Clause 31.3 sets out the reasons for non-acceptance. These are not stated to be exhaustive, but they set a presumption of construction that any reason for non-acceptance must be of these general types. Where the Contractor is dissatisfied with the Project Manager's decision he may refer the matter to the Adjudicator. The Adjudicator will not, of course, require the Project Manager to accept the programme; but a compensation event will arise if the Adjudicator decides that the original programme complied with the contract work as then defined. When accepted, a programme become the Accepted Programme.[26]

*Revising the programme*   The contract provides a liberal environment for submission of revised programmes. The interval contained in the Contract Data is the main constraint; but where either the Project Manager or Contractor wish it, a revised programme can be submitted at any time.

### Quotations for compensation events   62

62.2   Quotations for compensation events comprise proposed changes to the Prices and any delay to the Completion Date assessed by the *Contractor*. The *Contractor* submits details of his assessment with each quotation. If the programme for remaining work is affected by the compensation event, the *Contractor* includes a revised programme in his quotation showing the effect.

*Revised programme for compensation events*   Whenever a compensation event occurs, the Project manager calls for a quotation and specifies what it is to include. The action proposed by the Contractor in his quotation may require a revision to the Accepted Programme. If so, he submits a revised programme, which must comply with the requirements of Clause 31 and 32. This programme, if accepted, becomes the Accepted Programme and becomes an important element in the Project Manager's assessment of the compensation due.

26   Defined in Clause 11.2 (14).

## Physical conditions

### Compensation events  60

60.1  The following are compensation events. ...

(12) The Contractor encounters physical conditions which

- are within the Site,

- are not weather conditions and

- which an experienced contractor would have judged at the Contract Date to have such a small chance of occurring that it would have been unreasonable for him to have allowed for them.

60.2  In judging the physical conditions, the Contractor is assumed to have taken into account

- the Site information,

- publicly available information referred to in the Site Information,

- information available from a visual inspection of the Site and

- other information which a reasonably experienced contractor could reasonably be expected to have or obtain.

60.3  If there is an inconsistency within the Site Information (including the information referred to in it), the Contractor is assumed to have taken into account the physical conditions more favourable to doing the work.

*Clauses 11 and 12 of the ICE Conditions of Contract and construing the present clauses*  The wording of the present clauses differs from that found in Clauses 11 and 12 of the ICE Conditions of Contract. Nevertheless, the parallels are clear and it is tempting to interpret the present terms in light of the traditional Clause 12. It is submitted, however, that the present contract stands alone and is to be construed within its own terms.[27] Consider, for instance, the wording of Clause 60.1(12), 2nd bullet: 'are not weather conditions'. The question may arise whether flooding due to exceptionally heavy rainfall comes within this provision. It is tempting to refer to the corresponding wording in Clause 12 of the ICE Conditions of Contract (which includes the phrase 'or conditions due to weather conditions') and to

---

27   See, for example, *Luxor v. Cooper* [1941] AC 108 (HL); *Mitsui v. AG of Hong Kong* (1986) 33 BLR 1 (PC) where Lord Bridge said at 18 'comparison of one contract with another can seldom be a useful aid to construction and may be ... positively misleading'.

conclude that the ambit of Clause 60.2 has been deliberately expanded. It is submitted, however, that the wording of 60.1(12) permits the flooding claim within its own terms or it does not. This temptation to cross-refer between contracts is, of course, particularly difficult to resist in the case of the test in Clause 60.1(12), 3rd bullet.

*The test*   It is a precondition of any claim under these provisions that the conditions referred to are within the Site, and that they are not weather conditions (which are dealt with specifically in Clause 60.1(13)).

*Weather conditions*   Conditions such as low temperature causing delays to concreting, high winds causing inability to work on tower cranes, etc., are clearly weather conditions. On-site flooding, snow loading, wind loading, etc., on the Site are also thought to be 'weather conditions'. Where, however, there are heavy rainfalls off the site which cause flooding on the Site (as where a river breaks its banks) it is submitted that this is a physical condition which is not a weather condition and may, therefore, found a successful claim. This is because the physical condition does not enter the Site as 'weather' but as some other condition.

*The test*   Clause 60.1(12), 3rd bullet provides the essential test. An experienced contractor is thought to be one who has average experience of the type of work envisaged at the Contract Date. The time at which the judgment is to be made is the Contract Date. The key question relates to the meaning of: 'such a small chance of occurring that it would have been unreasonable for him to have allowed for them'. This clearly cannot be reduced to precise probabilities. It seems that the Employer is not required to pay compensation unless the Contractor can show that the conditions which occurred are surprising and that any contractor who allowed for them in his prices would stand rightly accused of undue pessimism.

*Information*   Clause 60.2 does not require the Employer to disclose information within his possession even if it has a clear bearing upon the works. The information which the Contractor is assumed to have or to be able to obtain is to be judged at the Contract Date; consequently, it is no defence to say that the Contractor could have obtained a piece of information if there was nothing to alert him to obtain it at that date.

*Inconsistencies*   Clause 60.3 provides for the situation where there are inconsistencies in the Site Information. Many disputes under the traditional standard forms are generated by inconsistent site investigation data; the employer and contractor each hope to persuade the arbitrator that their interpretation is the more reasonable. The intention of this clause is to provide a test where the data suggest two or more possible inter-pretations; the Contractor is assumed to have taken the more favourable of them into account. However, the situation which generally gives difficulty is where data is not strictly inconsistent but where the totality of data gives more than one reasonable interpretation. This problem is not explicitly resolved. Thus, for example, where the majority of data is consistent with poor ground conditions and the remaining data suggests benign conditions, it is not clear whether or not the Contractor is entitled to take into account the benign conditions.

## 4. The Engineering and Construction Short Contract

The short contract[28] was published in July 1999. It has received the Plain Language Commission seal of approval.

The short contract uses the same numbering system as the main contract, but fewer clauses. There is a printed section for Contract Data, Works Information and Site Information into which manuscript data/ information can be input. The Works Information sets out the following sections:

(1) description of the works
(2) drawings
(3) specifications
(4) constraints on how the contractor provides the works
(5) the programme
(6) services and other things provided by the employer.

The payment provisions provide a hybrid lump sum and remeasurement arrangement. There is a printed price list with the published form into which rates and prices can be inserted. Where a rate and a quantity is inserted in the price list, the price for that item is the rate multiplied by the quantity actually carried out. Where there is no quantity, the price is a lump sum.

---

28   In fact, it runs to 13 pages of clauses.

There is no Project Manager. The role he fulfils under the main form of contract is taken over by the Employer.

The contract is less prescriptive about such matters as programmes. The form of programme can be inserted in the Works Information, but, in practice, this will frequently amount to nothing more than key dates.

# 15

## Dispute resolution procedures

### 1. Introduction

A range of dispute processes are used to resolve civil engineering disputes. These include conciliation (or mediation), adjudication, arbitration and litigation. In the case of litigation, proceedings are governed by the Civil Procedure Rules and so there is no scope for the parties to agree their own rules of procedure. In the case of conciliation, adjudication and arbitration the parties may agree rules. In this chapter, a series of rules are considered. These are:

(1)  the ICE Adjudication Procedure 1997
(2)  the CIMAR Arbitration Rules 1998
(3)  the ICE Arbitration Procedure 1997
(4)  the ICC Arbitration Rules 1998 and LCIA Arbitration Rules 1998
(5)  the ICE Conciliation Procedure 1999.

### 2. The ICE Adjudication Procedure 1997

The ICE Adjudication Procedure 1997 was prepared primarily for use with the ICE family of conditions of contract. It may be suitable for other contracts.

Although the procedure contains a neat and self-contained procedure which is suitable, with minor amendments, for civil engineering contracts in any country, it is also designed specifically to comply with the adjudication provisions of the Housing Grants, Construction and Regeneration Act

1996. The text of the procedure follows and a commentary on the various provisions is to be found after the text.

1.  **General principles**

    1.1  The adjudication shall be conducted in accordance with the edition of the ICE Adjudication Procedure which is current at the date of issue of a notice in writing of intention to refer a dispute to adjudication (hereinafter called the Notice of Adjudication) and the Adjudicator shall be appointed under the Adjudicator's Agreement which forms a part of this Procedure. If a conflict arises between this Procedure and the Contract then this Procedure shall prevail.

    1.2  The object of adjudication is to reach a fair, rapid and inexpensive determination of a dispute arising under the Contract and this Procedure shall be interpreted accordingly.

    1.3  The Adjudicator shall be a named individual and shall act impartially.

    1.4  In making a decision, the Adjudicator may take the initiative in ascertaining the facts and the law. The adjudication shall be neither an expert determination nor an arbitration but the Adjudicator may rely on his own expert knowledge and experience.

    1.5  The Adjudicator's decision shall be binding until the dispute is finally determined by legal proceedings, by arbitration (if the contract provides for arbitration or the Parties otherwise agree to arbitration) or by agreement.

    1.6  The Parties shall implement the Adjudicator's decision without delay whether or not the dispute is to be referred to legal proceedings or arbitration. Payment shall be made in accordance with the payment provisions in the Contract, in the next stage payment which becomes due after the date of issue of the decision, unless otherwise directed by the Adjudicator or unless the decision is in relation to an effective notice under Section 111(4) of the Act.

2.  **The Notice of Adjudication**

    2.1  Any Party may give notice at any time of its intention to refer a dispute arising under the Contract to adjudication by giving a written Notice of Adjudication to the other Party. The Notice of Adjudication shall include:

    (a)  the details and date of the Contract between the Parties;

    (b)  the issues which the Adjudicator is being asked to decide;

    (c)  details of the nature and extent of the redress sought.

3.   The appointment of the Adjudicator

3.1   Where an Adjudicator has either been named in the Contract or agreed by the Parties prior to the issue of the Notice of Adjudication the Party issuing the Notice of Adjudication shall at the same time send to the Adjudicator a copy of the Notice of Adjudication and a request for confirmation, within four days of the date of issue of the Notice of Adjudication, that the Adjudicator is able and willing to act.

3.2   Where an Adjudicator has not been so named or agreed the Party issuing the Notice of Adjudication may include with the Notice the names of one or more persons with their addresses who have agreed to act, any one of whom would be acceptable to the referring Party, for selection by the other Party. The other Party shall select and notify the referring Party and the selected Adjudicator within four days of the date of issue of the Notice of Adjudication.

3.3   If confirmation is not received under paragraph 3.1 or a selection is not made under paragraph 3.2 or the Adjudicator does not accept or is unable to act then either Party may within a further three days request the person or body named in the Contract or if none is so named The Institution of Civil Engineers to appoint the Adjudicator. Such request shall be in writing on the appropriate form of application for the appointment of an adjudicator and accompanied by a copy of the Notice of Adjudication and the appropriate fee.

3.4   The Adjudicator shall be appointed on the terms and conditions set out in the attached Adjudicator's Agreement and Schedule and shall be entitled to be paid a reasonable fee together with his expenses. The Parties shall sign the agreement within 7 days of being requested to do so.

3.5   If for any reason whatsoever the Adjudicator is unable to act, either Party may require the appointment of a replacement adjudicator in accordance with the procedure in paragraph 3.3.

4.   Referral

4.1   The referring Party shall within two days of receipt of confirmation under 3.1, or notification of selection under 3.2, or appointment under 3.3 send to the Adjudicator, with a copy to the other Party, a full statement of his case which should include:

(a)   a copy of the Notice of Adjudication;

(b)   a copy of any adjudication provision in the Contract, and

(c)   the information upon which he relies, including supporting documents.

4.2 The date of referral of the dispute to adjudication shall be the date upon which the Adjudicator receives the documents referred to in paragraph 4.1. The Adjudicator shall notify the Parties forthwith of that date.

## 5.   Conduct of the adjudication

5.1 The Adjudicator shall reach his decision within 28 days of referral, or such longer period as is agreed by the Parties after the dispute has been referred. The period of 28 days may be extended by up to 14 days with the consent of the referring Party.

5.2 The Adjudicator shall determine the matters set out in the Notice of Adjudication, together with any other matters which the Parties and the Adjudicator agree should be within the scope of the adjudication.

5.3 The Adjudicator may open up review and revise any decision, (other than that of an adjudicator unless agreed by the Parties), opinion, instruction, direction, certificate or valuation made under or in connection with the Contract and which is relevant to the dispute. He may order the payment of a sum of money, or other redress but no decision of the Adjudicator shall affect the freedom of the Parties to vary the terms of the Contract or the Engineer or other authorised person to vary the Works in accordance with the Contract.

5.4 The other Party may submit his response to the statement under paragraph 4.1 within 14 days of referral. The period of response may be extended by agreement between the Parties and the Adjudicator.

5.5 The Adjudicator shall have complete discretion as to how to conduct the adjudication, and shall establish the procedure and timetable, subject to any limitation that there may be in the Contract or the Act. He shall not be required to observe any rule of evidence, procedure or otherwise, of any court. Without prejudice to the generality of these powers, he may:

(a)   ask for further written information;

(b)   meet and question the Parties and their representatives;

(c)   visit the site;

(d)   request the production of documents or the attendance of people whom he considers could assist;

(e)   set times for (a) - (d) and similar activities;

(f)   proceed with the adjudication and reach a decision even if a Party fails:

(i)   to provide information;

(ii) to attend a meeting;

(iii) to take any other action requested by the Adjudicator;

(g) issue such further directions as he considers to be appropriate.

5.6 The Adjudicator may obtain legal or technical advice having first notified the Parties of his intention.

5.7 Any Party may at any time ask that additional Parties shall be joined in the Adjudication. Joinder of additional Parties shall be subject to the agreement of the Adjudicator and the existing and additional Parties. An additional Party shall have the same rights and obligations as the other Parties, unless otherwise agreed by the Adjudicator and the Parties.

## 6. The Decision

6.1 The Adjudicator shall reach his decision and so notify the Parties within the time limits in paragraph 5.1 and may reach a decision on different aspects of the dispute at different times. He shall not be required to give reasons.

6.2 The Adjudicator may in any decision direct the payment of such simple or compound interest at such rate and between such dates or events as he considers appropriate.

6.3 Should the Adjudicator fail to reach his decision and notify the Parties in the due time either Party may give seven days notice of its intention to refer the dispute to a replacement adjudicator appointed in accordance with the procedures in paragraph 3.3.

6.4 If the Adjudicator fails to reach and notify his decision in due time but does so before the dispute has been referred to a replacement adjudicator under paragraph 6.3 his decision shall still be effective.

If the Parties are not so notified then the decision shall be of no effect and the Adjudicator shall not be entitled to any fees or expenses but the Parties shall be responsible for the fees and expenses of any legal or technical adviser appointed under paragraph 5.6 subject to the Parties having received such advice.

6.5 The Parties shall bear their own costs and expenses incurred in the adjudication. The Parties shall be jointly and severally responsible for the Adjudicator's fees and expenses, including those of any legal or technical adviser appointed under paragraph 5.6, but in his decision the Adjudicator may direct a Party to pay all or part of his fees and expenses. If he makes no such direction the Parties shall pay them in equal shares.

6.6 At any time until 7 days before the Adjudicator is due to reach his decision, he may give notice to the Parties that he will deliver it only on full payment of his fees and expenses. Any Party may then pay these costs in order to obtain the decision and recover the other Party's share of the costs in accordance with paragraph 6.5 as a debt due.

6.7 The Parties shall be entitled to the relief and remedies set out in the decision and to seek summary enforcement thereof, regardless of whether the dispute is to be referred to legal proceedings or arbitration. No issue decided by an adjudicator may subsequently be laid before another adjudicator unless so agreed by the Parties.

6.8 In the event that the dispute is referred to legal proceedings or arbitration, the Adjudicator's decision shall not inhibit the court or arbitrator from determining the Parties' rights or obligations anew.

6.9 The Adjudicator may on his own initiative, or at the request of either Party, correct a decision so as to remove any clerical mistake, error or ambiguity provided that the initiative is taken, or the request is made within 14 days of the notification of the decision to the Parties. The Adjudicator shall make his corrections within 7 days of any request by a Party.

7. **Miscellaneous provisions**

7.1 Unless the Parties agree, the Adjudicator shall not be appointed arbitrator in any subsequent arbitration between the Parties under the Contract. No Party may call the Adjudicator as a witness in any legal proceedings or arbitration concerning the subject matter of the adjudication.

7.2 The Adjudicator shall not be liable for anything done or omitted in the discharge or purported discharge of his functions as Adjudicator unless the act or omission is in bad faith, and any employee or agent of the Adjudicator shall be similarly protected from liability. The Parties shall save harmless and indemnify the Adjudicator and any employee or agent of the Adjudicator against all claims by third parties and in respect of this shall be jointly and severally liable.

7.3 Neither The Institution of Civil Engineers nor its servants or agents shall be liable to any Party for any act omission or misconduct in connection with any appointment made or any adjudication conducted under this Procedure.

7.4 All notices shall be sent by recorded delivery to the address stated in the Contract for service of notices, or if none, the principal place of business or registered office (in the case of a company). Any agreement required by this Procedure shall be evidenced in writing.

7.5 This Procedure shall be interpreted in accordance with the law of the Contract.

**8. Definitions**

8.1 (a) The "Act" means the Housing Grants, Construction and Regeneration Act 1996.

(b) The "Adjudicator" means the person named as such in the Contract or appointed in accordance with this Procedure.

(c) "Contract" means the contract or the agreement between the Parties which contains the provision for adjudication.

(d) "Party" means a Party to the Contract and references to either Party or the other Party or Parties shall include any additional Party or Parties joined in accordance with this Procedure.

**9. Application to particular contracts**

9.1 When this Procedure is used with The Institution of Civil Engineers' Agreement for Consultancy Work in Respect of Domestic or Small Works the Adjudicator may determine any dispute in connection with or arising out of the Contract.

## General

*Does the procedure comply with the requirements of the HGCRA 1996?* The compliance criteria set out in Section 108 of the Act are given below. Footnotes to the individual requirements indicate where the various prescribed provisions are to be found.

(*a*) The contract shall enable a party to give notice at any time of his intention to refer a dispute to adjudication.[1]

(*b*) The contract shall provide a timetable with the object of securing the appointment of the adjudicator and referral of the dispute to him within seven days of such notice.[2]

(*c*) The contract shall require the adjudicator to reach a decision within 28 days of the referral or such longer period as is agreed by the parties after the dispute has been referred.[3]

(*d*) The contract shall allow the adjudicator to extend the period of 28

---

1  ICE Adjudication Procedure Rule 2.1.
2  Rules 3.1 to 3.3.
3  The referral is dealt with in Rules 4.1 and 4.2. The requirement to reach a decision with 28 days of the referral is to be found in Rule 5.1.

days by up to 14 days, with the consent of the party by whom the dispute was referred.[4]

(e)    The contract shall impose a duty on the adjudicator to act impartially.[5]

(f)    The contract shall enable the adjudicator to take the initiative in ascertaining the facts and the law.[6]

(g)    The contract shall provide that the decision of the adjudicator is to be binding until the dispute is finally determined by legal proceedings etc.[7]

(h)    The contract shall provide that the adjudicator is not liable for anything done or omitted in the discharge or purported discharge of his functions as adjudicator unless the act or omission is in bad faith, and that any agent of the adjudicator is similarly protected from liability.[8]

*The relationship between the civil engineering contract and the procedure*    The contract incorporates the adjudication procedure and the procedure takes effect subject to anything to the contrary in the primary contract. Where the contract is in the ICE 7th Edition form, Clause 66 contains a provision whereby the Engineer must give a decision or a month must elapse before a matter of dissatisfaction becomes a dispute. This appears not to be in accordance with the first compliance point of Section 108. It is submitted that where a party wishes to activate the adjudication procedure before the time contemplated in the contract, the procedure is not available. Accordingly, the Scheme applies.

## Terminology

*The parties*    Under the Scheme, the initiator of adjudication is called the referring party and the other party is referred to as 'the other party'. This terminology is retained in the ICE procedure.

*The notices*    Under the Scheme, two notices are identified: (a) the notice of adjudication and (b) the referral notice. Under the ICE procedure, the Notice of Adjudication fulfils the same function as the notice of

---

4    Rule 5.1.
5    Rule 1.3.
6    Rule 1.4.
7    Rule 1.5.
8    Rule 7.2.

adjudication under the Scheme. But the place of the referral notice—which, in any case, is a misnomer—is taken by the referring party's 'statement of case'.

## The procedure before the adjudicator

The ICE Procedure envisages the following steps:

(1)  a statement of case is served at the date of referral
(2)  the other party may submit his response within 14 days of the date of referral
(3)  the adjudicator will have complete discretion as to how to conduct other proceedings, including site visits, meeting parties, requiring documents, etc.
(4)  the adjudicator may obtain legal and technical advice, provide he has first notified the parties of his intention
(5)  the adjudicator gives his decision within 28 days of the date of referral.

## The decision, its content and effect

*What matters must be determined*    The adjudicator is required to decide those matter set out in the Notice of Adjudication. Rule 5.2 also provides that the Adjudicator should determine any other matters which the Parties and the Adjudicator agree should be within the scope of the adjudication. It is submitted, however, that an adjudicator is required to determine also any matter which can amount to a defence to the issues referred to him. Thus, where the contractor refers a non-payment and the employer's defence is a set-off for defective work already done, then the adjudicator must deal with this. Since the other party has until Day 14 to serve his response which will contain these new matters as a defence by way of set-off, leaving only 14 days until the adjudicator's decision is due, this will lead to a very compressed timetable.

*Reasons*    Under the Scheme, an adjudicator is required to give reasons for his decision if either or both parties request it. Under this procedure, the adjudicator is not required to give his reasons.

## 3. Arbitration Rules

*Application of rules and conflicts between rules and the primary contract*   The Arbitration Procedure is binding on the parties wherever they have agreed this in their contract.[9] A conflict may arise between the primary arbitration clause in a civil engineering contract and the rules. The resolution of such conflicts is a matter of interpreting the primary clause and procedure together; but normally the ordinarily rule of interpretation (that incorporating documents prevail over incorporated documents)[10] applies and the primary clause is given priority.[11]

### The Construction Industry Model Arbitration Rules (CIMAR) 1998

In 1995, members of the Society of Construction Arbitrators and other interested parties, proposed the idea of a set of arbitration rules which would be applicable across the construction industry, taking in civil engineering, building, process, mechanical and electrical works, etc. A steering group was established and the title 'Construction Industry Model Arbitration Rules' (CIMAR) was adopted. Following a series of consultation issues and a limited circulation issue in March 1997, the rules were published in February 1998. The following picks up some of the key points raised by the rules.

*Style and jurisdiction*   The Rules are not self-contained, but rely on cross-references to the Arbitration Act 1996. For example, Rule 11.1 provides: 'The arbitrator has the power in Section 41(3) ...'. In order to assist the reader, relevant extracts from the Act are printed alongside the Rules. Apparently, this cross-referencing has caused the draftsmen to consider that the Rules should only apply where the Arbitration Act 1996 applies. In accordance with this, Rule 1.6 states that the Rules apply where the seat of the arbitration is in England and Wales or Northern Ireland. This is a curious provision as it invites the prospect that where a different seat is adopted, they do not apply.

*Consistency*   Rule 1.3 seeks to achieve uniformity by providing that the parties may not, after the appointment of the arbitrator, amend the Rules or apply procedures in conflict with them. The Rules do, nevertheless, allow

---

9    See *Pratt v. Swanmore Builders Ltd and Baker* [1980] 2 Lloyd's Rep. 504.
10   *Modern Buildings Wales Ltd v. Limmer and Trinidad Ltd* [1975] 1 WLR 1281 per Buckley L.J. at 1289.
11   *Christiani & Nielsen Ltd v. Birmingham City Council* (1994) CILL 1014.

some diversity in the form of an 'advisory procedure' and/or 'supplementary procedure' which may be agreed in addition to the rules: see Rule 6.5. The extent to which Rule 1.3 is enforceable is not clear. An appointment creates a tripartite agreement[12] and the arbitrator may consider it an important element of that agreement that the rules be applied without change. Furthermore, a key object of adopting this uniform set of rules is the facilitation of joinder of arbitrations where appropriate. It is submitted, however, that where the parties jointly insist on a change in the rules, the arbitrator must either yield or resign.

*Joinder*   One of the clear policies of the Rules is the efficient disposal of arbitrations. Where it is appropriate, the Rules encourage having the same arbitrator appointed in related disputes and joinder of proceedings. Rules 2.6 and 2.7 ask appointing bodies to consider the appointment of an arbitrator already appointed in a related arbitration. There can be no 'obligation' on the part of appointers to act upon this injunction. The mere endorsement of the rules by appointing bodies is insufficient. Nevertheless, appointing bodies will wish to satisfy the rules provided (*a*) they feel comfortable with them and (*b*) the amount of work involved is manageable.[13] Although Rule 3 is headed 'Joinder', it covers two distinct situations: (*a*) Rules 3.1 to 3.6 relate to the situation where two parties have already commenced an arbitration and one (or both) then seeks to have a new dispute included into the existing arbitration; (*b*) Rules 3.7 to 3.12 relate to the situation where the same arbitrator has been appointed in two or more sets of related proceedings. Concurrent proceedings may only be directed where CIMAR Rules (or some other rules which provide a consistent power to direct concurrent proceedings) apply to all relevant proceedings. Other factors include whether or not concurrent hearings are likely to generate efficiency, given the nature of the dispute, the overlap of issues of fact and law and any risk of prejudice or a breach of confidentiality.

*Conditions precedent*   The problem of 'conditions precedent' arises in many civil engineering contracts.[14] For example, under the ICE 7th

---

12    *Norjarl v. Hyundai Heavy Industries Co. Ltd* [1992] 1 QB 863:

> The arbitration agreement is a bilateral contract between the parties to the main contract. On appointment, the arbitrator becomes a third party to that arbitration agreement ...

(per Sir Nicolas Browne-Wilkinson V-C at 885).

13    Rule 2.7, for example, may involve considerable consultation.

14    See, for example, *Bernhard's Rugby Landscapes Ltd* v. *Stockley Park Consortium Ltd* (1997) 82 BLR 39.

Edition, no reference to an arbitrator is possible unless the Engineer has either given a decision or has failed to do so in the specified period. Although Rule 3.5 is not happily worded (e.g. sub-rule (i) suggests that the arbitrator can himself give an Engineer's decision) it is submitted that the meaning is clear: where a contract contains a condition precedent to the referral of a dispute to arbitration, the arbitrator may, nevertheless, arbitrate the dispute either by (*a*) declaring the condition to have been satisfied, or (*b*) by declaring that the condition needs no longer be satisfied. What is not so clear is whether this effect is achievable. Where, for example, a contract contains a clearly expressed condition precedent and only incorporates the CIMAR rules by general reference, it is submitted that the condition precedent survives unless both parties expressly and clearly agree to dispense with it.

*Initial sketch of the shape of the dispute*   Rule 6.2 provides that the parties shall, soon after appointment, give to the arbitrator and each other a note stating the nature of the dispute, amounts in issue, a view on the appropriate procedure etc. This a useful approach and provides a sensible basis for discussion at the procedural meeting.

*Procedures*   The arbitrator may, by direction, select the procedure to be adopted from: a short hearing (Rule 7), documents only (Rule 8) or full procedure (Rule 9). These rules set out a framework for each type of procedure. The question of timescales and other detailed directions remains largely within the arbitrator's discretion.

*Provisional relief*   A partial award may be issued at any time; but once made, it cannot be reopened.[15] Where, however, the parties agree, the arbitrator may issue an enforceable order about payment etc., which is reviewable at a later stage.[16] The decision is not in the form of an award, but is an order. This is useful in a number of situations, for example, (*a*) where the claimant has won the liability stage of an arbitration; or (*b*) where a claimant seeks some relief relating to property, the arbitrator may order on a provisional basis that possession be given to the claimant and that the property should be delivered to him.

---

15   *The Kostas Melas* [1981] 1 Lloyd's Rep. 18.
16   Section 39 of the Arbitration Act 1996.

*Arbitrator may supervise specified work*   Rule 12.7 provides a novel power. Firstly, it presupposes that, in some cases, 'specific performance' is an appropriate remedy rather than damages for breach. Secondly, where an award provides that specified work is to be done, the rule gives the arbitrator power to supervise that work. It is thought that this will give rise to a number of difficulties. For example, where the arbitrator as supervisor makes a decision which a party believes is not in accordance with his award, that party is presumably entitled to refer the matter to adjudication provided the original work was under a contract to which the Housing Grants, Construction and Regeneration Act 1996 applied.

*Draft award*   Rule 12.10 specifically entitles an arbitrator to provide a draft award. An award should, of course, contain correct details and it is convenient if the parties have an opportunity to check them. But a draft award can often be treated as an invitation to reopen issues, whatever stipulations the arbitrator makes as to new evidence and submissions. It is suggested that, in most cases, an arbitrator should not issue a draft award. By providing details (e.g. parties' names, addresses, chronology of the project, a note of some of the key parts of the evidence, etc.) for checking, the key objective, namely accuracy, can be much more simply achieved.

*Costs*   Section 61 of the Arbitration Act states that the 'general principle' is that 'costs should follow the event'. However, Rule 13.1 uses the more direct and, it is submitted, preferable formulation, based to some degree on the UNCITRAL Arbitration Rules. Rule 13.2 provides very helpful guidance on how the discretion may be exercised. Partial success can be reflected in a proportionate award as to costs. Where there is a counterclaim,[17] Rule 13.3 provides that the arbitrator should normally take each separately unless so interconnected that they should be dealt with together. This displaces the ordinary presumption that, where the claim and counterclaim both arise from the same contract, the net winner receives all the costs. It provides useful clarity and will, in many instances, yield a fairer solution.

---

17   It is submitted that the expression 'counterclaim' means a cross-claim whether or not styled or available as a set-off. This seems to be so from the context; since most matters heard within an arbitration arise from the same contract, all claims and counterclaims are available as equitable set-offs. The rules only makes sense if this interpretation is adopted.

## The Institution of Civil Engineers Arbitration Procedure

The ICE Arbitration Procedures was first published in 1983.[18] Following the enactment of the Arbitration Act 1996, the ICE published its Arbitration Procedure (1997).

The scope and provisions of the ICE Arbitration Procedure are very similar to the CIMAR Rules. Given this, this brief review will focus on a comparison between the ICE Procedure and the corresponding features of the CIMAR Rules.

*Scope*   The CIMAR Rules expressly confine themselves in the body of the Rules to arbitrations with a seat in England and Wales or Northern Ireland. The ICE Procedure contains the following guidance note, which is not part of the Rules. The procedure is

> principally for use with the ICE family of Conditions of Contract and the NEC family of Documents in England and Wales for arbitrations conducted under the Arbitration Act 1996. It may be suitable for use with other contracts and in other jurisdictions.

*Drafting principles*   Unlike the CIMAR Rules, the ICE Procedure aims to be a self-contained statement of the powers of the arbitrator and the obligations of the parties to the proceedings. There are no cross-references to the provisions of the Arbitration Act 1996.[19] Sections of the Act are not reproduced alongside. Cross-referencing within the document is kept to a minimum. In this sense, it is a more convenient document than the CIMAR Rules and may be more readily understood by a lay reader or parties from overseas who will not be familiar with the English arbitration legislation.

*The substance of the rules*   In terms of substance, there is very little difference between the CIMAR Rules and the ICE Procedure. The following matters illustrate this:

(1)   Both procedures attempt to preserve the integrity of the rules. The ICE Procedure provides

---

18   A full commentary can be found in the first edition of this book.
19   Save to confirm for the purposes of Section 4(3) of the Act that these are institutional rules: ICE Arbitration Procedure 1997, Rule 1.2.

no alterations shall be made to this Procedure without the consent of the Arbitrator except where there are express modifications in the Contract, or in the arbitration agreement.[20]

The same arguments as to the enforceability of this provision apply as to the corresponding CIMAR provision.

(2)  In each case the Arbitrator may order concurrent hearings but may not consolidate proceedings without the consent of the parties. Under the CIMAR Rules, the arbitrator may order proceedings to be run concurrently; under the ICE Procedure it appears that the arbitrator must either receive an application from a person who is a party to all the relevant arbitrations or, alternatively, all the parties concerned must agree.[21] In neither case, however, do the provisions empowering the arbitrator to order that proceedings may be heard concurrently define the extent to which this may be possible if one of the arbitrations is being run under different and, possibly conflicting, rules.

(3)  At the preliminary meeting the ICE Procedure provides that the

Arbitrator may require the parties to submit to him short statements expressing their perceptions of the disputes or differences. Such statements shall give sufficient detail of the nature of the issues to enable the Arbitrator and the parties to discuss procedures appropriate for their settlement at the preliminary meeting.[22]

Similar provisions are to be found in CIMAR, although the latter contains more detailed guidance which is helpful both to the parties in understanding what is required and is helpful to the arbitrator in determining whether what is provided complies with the Rules.

(4)  The CIMAR special procedures (the 'Short Hearing' and 'Documents Only' procedures) may be directed by the arbitrator. Under the ICE Procedure, the Short Procedure[23] and the Special Procedure for Experts[24] are only available where the parties agree.

(5)  In each case, the arbitrator is able to set time limits, including specific directions for the duration of oral presentations of a party's case at a hearing. Where it is proposed to limit the time available to parties to present their cases, the ICE Procedure provides the following guidance:

---

20   ICE Arbitration Procedure 1997, Rule 1.3.
21   ICE Arbitration Procedure 1997, Rule 9.1.
22   ICE Arbitration Procedure 1997, Rule 6.1.
23   ICE Arbitration Procedure 1997, Rule 15.
24   ICE Arbitration Procedure 1997, Rule 17.

Should a party's representative fail to complete the presentation of that party's case within the time so allowed further time shall only be afforded at the sole discretion of the Arbitrator and upon such conditions as to costs as the Arbitrator may see fit to impose.[25]

## The ICC Arbitration Rules 1998 and LCIA Arbitration Rules 1998

International civil engineering contracts almost invariably contain arbitration clauses. Most parties find arbitration preferable to litigating disputes in a foreign domestic court for a variety of reasons, including (*a*) lack of familiarity with domestic litigation procedures; (*b*) language barriers; and (*c*) issues of privacy and confidentiality. International civil engineering arbitrations are very frequently referred to and conducted under the 'institutional rules' of either the International Chamber of Commerce (ICC) or the London Court of International Arbitration (LCIA). These arbitrations differ from those discussed in previous sections because the institution (whether the ICC or LCIA) plays a significant role in the administration of the proceedings.

*The ICC*    The ICC is a body established to foster and facilitate the development of international commerce. In 1923 it set up the 'ICC International Court of Arbitration'. The Court does not itself settle disputes. It receives requests for arbitration, appoints arbitrators, ensures that the financing of the arbitration is properly managed and reviews awards to ensure that they are in a proper format for, *inter alia*, enforcement. The ICC has handled almost ten thousand arbitration cases. Although based in Paris, the ICC's role is limited to administration and this allows the arbitrations themselves to be held in another country. According to the ICC, every year there are ICC arbitrations held in some 30 different countries and involve arbitrators of 50 different nationalities from all the major legal systems of the world.

*The ICC Rules*    The present version of the rules came into force at the beginning of 1998.[26] The following is a very brief, and necessarily simplified, description of the procedure:

(1)    The Claimant makes a Request for Arbitration.[27] This includes a statement of the Claimant's case.

---

25    ICE Arbitration Procedure 1997, Rule 13.7.
26    The rules are available on the ICC website: www.iccwbo.org
27    ICC Arbitration Rules 1998, Article 3.1

(2)   Within 30 days the Defendant gives his Answer, including his defence and any proposed counterclaim;[28] where there is a counterclaim, the Claimant has 30 days to serve a Reply.[29]

(3)   The procedure for the appointment of arbitrators is set out in Article 2 of the Rules. Where the agreement states that the dispute is to be settled by a sole arbitrator, the parties may nominate an arbitrator for confirmation by the Court. Where there are to be three arbitrators 'each party shall nominate in the Request for Arbitration and the Answer thereto respectively one arbitrator for confirmation by the Court'. The third arbitrator is appointed by the Court unless the arbitration agreement provides that the two nominees shall agree the third arbitrator. Where there is no agreement as to the number of arbitrators, the Court appoints a sole arbitrator 'save where it appears to the Court that the dispute is such as to warrant the appointment of three arbitrators'.

(4)   The ICC computes the relevant Advance as to fees etc. and is put into funds by both parties. Where one party fails to supply his share, the other may do so.[30]

(5)   When the Answer is served, the file is transmitted to the arbitrator.[31]

(6)   The arbitrator draws up Terms of Reference (usually at or following a meeting with the parties) which includes a definition of the issues to be determined.[32]

(7)   'The arbitrator shall proceed within as short a time as possible to establish the facts of the case by all appropriate means'.[33] This ordinarily involves an oral hearing.

(8)   The arbitrator renders his award within 6 months of the Terms of Reference being fixed. Where there are three arbitrators, the award may be made by a majority; where there is no majority, the award is made by the Chairman.[34]

(9)   'Before signing an award, whether partial or definitive, the arbitrator shall submit it in draft form to the International Court of Arbitration. The Court may lay down modifications as to the form of the award and, without affecting the arbitrator's liberty of decision, may also

---

28   ICC Arbitration Rules 1998, Article 4.
29   ICC Arbitration Rules 1998, Article 5.
30   ICC Arbitration Rules 1998, Article 9.
31   ICC Arbitration Rules 1998, Article 10.
32   ICC Arbitration Rules 1998, Article 13.
33   ICC Arbitration Rules 1998, Article 14.1.
34   ICC Arbitration Rules 1998, Article 19.

draw his attention to points of substance. No award shall be signed until it has been approved by the Court as to its form'.[35]

*The LCIA*   The LCIA Arbitration Court was set up in 1985. Whilst significantly smaller than the ICC, the Court is an international body drawn from leading practitioners around the world. It administers LCIA arbitrations and takes on a role which is broadly similar to that of the ICC Court. The procedures laid down in the LCIA Rules 1998[36] are similar to those in the ICC Rules. There is, however, no specific requirement for terms of reference to be drawn up. In addition, the Court does not play the same role in approving the award prior to it being signed.

## 4.  The ICE Conciliation Procedure

Conciliation is provided for in many standard form civil engineering contracts. It is used optionally in both the ICE Conditions of Contract and the CECA Form of Sub-contract. It is used as a mandatory step in the dispute resolution process provided for in the ICE Design and Construct Contract.

The legal status of conciliation is discussed in Chapter 11. The rules in the ICE Conciliation Procedure correspond with the typical model of conciliation, namely:

(1)   proceedings are conducted on a 'without prejudice' basis
(2)   the conciliator may talk privately with either party
(3)   there is no cross-examination or need to disclose all the information available
(4)   the Conciliator's Recommendation need not accord with the exact legal position but may suggest a practical solution which takes into account not only the strict rights and obligations of the parties, but also of their other interests.

In general, the recommendation of a conciliator is not enforceable. However, the ICE Conditions of Contract and the CECA Form of Sub-contract each provide that when the Recommendation is published it becomes binding unless a Notice to Refer is served within a short period. Where the conciliator's recommendation becomes binding through passage of time, it is thought that the only arguments available to overturn the recommendation are that:

---

35   ICC Arbitration Rules 1998, Article 21.
36   Available at the LCIA web-site at www.lcia-arbitration.com

(1) the Conciliator lacked jurisdiction either because he was not properly appointed or he has made a recommendation on a matter which was not properly referred to him

(2) the other party fraudulently procured the recommendation.

It is thought that no defence may be raised that the parties or conciliator have failed to operate the provisions of the Conciliation Procedure, unless a party has raised and maintained his objection throughout. Neither is it thought that an argument may be maintained that the Recommendation lacks internal consistency providing the final result proposed is clear. Where, however, the Recommendation is hopelessly unclear in its terms it is submitted that it is unenforceable for want of certainty.

1.  **General principles**

    1.1  This Procedure shall apply whenever:

        (a)  the Parties have entered into a contract which provides for conciliation for any dispute which may arise between the Parties in accordance with The Institution of Civil Engineers' Conciliation Procedure, or

        (b)  where the Parties have agreed that The Institution of Civil Engineers' Conciliation Procedure shall apply.

    1.2  The conciliation shall be conducted in accordance with the edition of the ICE Conciliation Procedure which is in force at the date of issue of the Notice of Conciliation.

    1.3  This Procedure shall be interpreted and applied in the manner most conducive to the efficient conduct of the proceedings with the primary objective of achieving a settlement to the dispute by agreement between the Parties as quickly as possible.

2.  **The Notice of Conciliation**

    2.1  Subject to the provisions of the contract relating to conciliation, any Party to the contract may by giving to the other Party a written notice, hereinafter called a Notice of Conciliation, request that any dispute in connection with or arising out of the contract or the carrying out of the Works shall be referred to a Conciliator. Such Notice shall be accompanied by a brief statement of the matter or matters which it is desired to refer to conciliation, and the relief or remedy sought.

3.  **The appointment of the Conciliator**

    3.1  Save where a Conciliator has already been appointed, the Parties shall agree upon a Conciliator within 14 days of the Notice being given under paragraph 2.1. In default of agreement any Party may request the President (or, if he is unable to act, any Vice President) for the time being of the Institution of Civil Engineers to appoint a Conciliator within 14 days of receipt of the request by him, which request shall be accompanied by a copy of the Notice of Conciliation.

3.2 If, for any reason whatsoever, the Conciliator is unable, or fails to complete the conciliation in accordance with this Procedure, then any Party may require the appointment of a replacement Conciliator in accordance with paragraph 3.1.

## 4. Conduct of the conciliation

4.1 Unless otherwise agreed by the Parties, the Party requesting conciliation shall deliver to the Conciliator, immediately on his appointment, with a copy to the other Party, a copy of the Notice of Conciliation together with copies of all relevant Notices of Dispute and of any other notice or decision which is a condition precedent to conciliation.

4.2 The Conciliator shall start the conciliation as soon as possible after his appointment and shall use his best endeavours to conclude the conciliation as soon as possible and in any event within any time limit stated in the contract, or two months from the date of his appointment whichever is the earlier, or within such other time agreed between the Parties.

4.3 Any Party may, upon receipt of notice of the appointment of the Conciliator and within such period as the Conciliator may allow, send to the Conciliator and to the other Party a statement of its views on the dispute and any issues that it considers to be of relevance to the dispute, and any financial consequences.

4.4 As soon as possible after his appointment, the Conciliator shall issue instructions establishing, amongst other things, the date and place for the conciliation meeting with the Parties. Each Party shall, in advance of the meeting, inform the Conciliator, and the other Party, in writing of the name of its representative for the conciliation, who shall have full authority to act on behalf of that Party and the names of any other persons who will attend the conciliation meeting.

4.5 The Conciliator may:

(a) issue such further instructions as he considers to be appropriate;

(b) meet and question the Parties and their representatives, together or separately;

(c) investigate the facts and circumstances of the dispute;

(d) visit the site;

(e) request the production of documents or the attendance of people whom he considers could assist in any way.

4.6 The Conciliator may conduct the proceedings in any way that he wishes, and with the prior agreement of the Parties obtain legal or technical advice, the cost of which shall be met by the Parties, in accordance with paragraph 5.4, or as agreed by the Parties and the Conciliator.

4.7 The Conciliator may consider and discuss such solutions to the dispute as he thinks appropriate or as suggested by any Party. He shall observe and maintain the confidentiality of particular information which he is given by any Party privately, and may only disclose it with the express permission of that Party. He will try to assist the Parties to resolve the dispute in any way which is acceptable to them.

4.8 Any Party may, at any time, ask that additional claims or disputes, or additional Parties, shall be joined in the conciliation. Such requests shall be accompanied by details of the relevant contractual facts, notices and decisions. Such joinder shall be subject to the agreement of the Conciliator and all other Parties. Any additional Party shall, unless otherwise agreed by the Parties, have the same rights and obligations as the other Parties to the conciliation.

4.9 If, in the opinion of the Conciliator, the resolution of the dispute would be assisted by further investigation by any Party or by the Conciliator, or by an interim agreement, including some action by any Party, then the Conciliator will, with the agreement of the Parties, give instructions and adjourn the proceedings as may be appropriate.

4.10 When a settlement has been achieved of the whole or any part of the matters in dispute the Conciliator shall, if so requested by all the Parties, assist them to prepare an Agreement incorporating the terms of the settlement. If requested in writing by all the Parties, the Conciliator may be appointed by the Parties as an arbitrator with authority solely to issue a consent award.

## 5. The recommendation

5.1 The Conciliator shall advise all Parties accordingly and prepare his recommendation forthwith:

(a) if in the opinion of the Conciliator it is unlikely that the Parties will agree a settlement to their disputes, or

(b) if any Party fails to respond to an instruction by the Conciliator, or

(c) if requested by any Party.

5.2 The Conciliator's recommendation shall state his solution to the dispute which has been referred for conciliation. The recommendation shall not disclose any information which any Party has provided in confidence. It shall be based on his opinion as to how the Parties can best dispose of the dispute between them and need not necessarily be based on any principles of the contract, law or equity.

5.3   The Conciliator shall not be required to give reasons for his recommendation.

5.4   When a settlement has been reached or when the Conciliator has prepared his recommendation, or at an earlier date solely at the discretion of the Conciliator, he shall notify the Parties in writing and send them an account of his fees and disbursements. Unless otherwise agreed between themselves each Party shall be responsible for paying and shall within 7 days of receipt of the account from the Conciliator pay an equal share save that the Parties shall be jointly and severally liable to the Conciliator for the whole of his account. If any Party fails to make the payment due from him, the other Party may pay the sum to the Conciliator and recover the amount from the defaulting Party as a debt due. Each Party shall meet his own costs and expenses. Upon receipt of payment in full the Conciliator shall send his recommendation to all the Parties.

5.5   The Conciliator may be recalled, by written agreement of the Parties and upon payment of an additional fee, to clarify, amplify or give further consideration to any provision of the recommendation.

## 6.   Miscellaneous provisions

6.1   No Party shall be entitled to call the Conciliator as a witness in any subsequent adjudication, arbitration or litigation concerning the subject matter of the conciliation.

6.2   The Conciliator shall not be appointed adjudicator in any subsequent adjudication, or arbitrator (except as provided for in paragraph 4.10) in any subsequent arbitration between the Parties whether arising out of the dispute, difference or other matter or otherwise arising out of the same contract unless the Parties otherwise agree in writing.

6.3   The Parties and the Conciliator shall at all times maintain the confidentiality of the conciliation and shall endeavour to ensure that anyone acting on their behalf or through them will do likewise.

6.4   The Conciliator shall not be liable to the Parties or any person claiming through them for any matter arising out of or in connection with the conciliation or the way in which it is or has been conducted, and the Parties shall not themselves bring any such claims against him.

6.5   Any notice required under this Procedure shall be sent to the Parties by recorded delivery to the principal place of business or if a company to its registered office, or to the address which the Party has notified to the Conciliator. Any notice required by this Procedure to be sent to the Conciliator shall be sent by recorded delivery to him at the address which he shall notify to the Parties on his appointment.

6.6 In this Procedure where the context so requires 'Party' shall mean 'Parties' and 'he' shall mean 'she'.

*Terminology and interpretation*   The individual sections of the Procedure are referred to as 'paragraphs' rather than rules or articles. This indicates that the provisions are to be interpreted in the manner most conducive to the efficient conduct of the proceedings—Paragraph 2.

*Where a party refuses to take part*   The Procedure does not oblige either party to participate in the conciliation. Where either party refuses to take part, the Conciliator will normally consider, in accordance with Paragraph 15, that it is unlikely that the parties will reach an agreed settlement and he will then prepare his Recommendation based on the material supplied by the participating party.

*The date of the publication of the Recommendation*   By Paragraph 17, the Conciliator notifies the parties in writing when his Recommendation is ready. He is not obliged to send it until his fees and disbursements etc. are paid. Where the contract provides that the Recommendation becomes binding a specified number of days after the date of the Conciliator's Recommendation, the date from time begins to run will be the date set out in the contract.

# Appendix 1

## A survey of civil engineering standard form contracts

Standard form contracts are published by a variety of professional bodies, associations or agencies. These include:

(1) *The Conditions of Contract Standing Joint Committee*
    This is a committee with representatives of the Institution of Civil Engineers, the Association of Consulting Engineers and the Civil Engineering Contractors Association (formerly the Federation of Civil Engineering Contractors). Although the publications are endorsed by all three bodies, the conditions are published under the title 'ICE'. They include:
    - ICE Conditions of Contract, 7th Edition, 1999 (NB The 6th and 5th Editions, 1991 and 1973 respectively, continue to be published and are in widespread use also).
    - ICE Conditions of Contract for Minor Works, 2nd Edition, 1995
    - ICE Design and Construct Conditions of Contract, 1992
    - ICE Conditions of Contract for use in connection with Ground Investigation, 1983.
(2) *The Institution of Civil Engineers*
    Independently of the CCSJC, the ICE prepares the following contracts:
    - Engineering and Construction Contract, 2nd Edition (also known as the NEC 2nd Edition), 1995 and all ancillary contracts, including sub-contract, adjudicator's appointment, etc.
    - Engineering and Construction Short Contract, 1st Edition, 1999
(3) *The Civil Engineering Contractors' Association*
    This body (formerly the Federation of Civil Engineering Contractors) publishes the standard form of civil engineering sub-contract. This incorporates some of the terms of the main contract. The latest 1998 edition expresses

itself to be for use with the ICE Conditions of Contract 6th Edition, but it is suitable with minor modifications for use with the 7th Edition.

(4) *Property Advisors to the Civil Estate (PACE)*

PACE is the Government agency responsible for giving advice on publicly owned property. In 1996 it set about upgrading the GC/Works 3rd Edition series of contracts used for projects in which the Government was employer. The new suite of contracts is:

- GC/Works/1 1998 Government Contract for major building and civil engineering works. It is published in four volumes

   With Quantities General Conditions
   Without Quantities General Conditions
   Single Stage Design & Build General Conditions
   Model Forms & Commentary.

- GC/Works/2 1998 Government Contract for minor building and civil engineering works and demolition works.
- GC/Works/3 1998 Government Contract for mechanical and electrical work.
- GC/Works/4 1998 Government Contract for small civil engineering, mechanical and electrical engineering works.
- GC/Works/2 1998, GC/Works/3 1998 and GC/Works/4 1998 are each published in two volumes: General Conditions and Model Forms & Commentary.

In addition, an analogous series of contracts (Coded PC/Works) have been produced for use by the private sector, including local authorities, NHS trusts and all non-Central Government employers. This is published as a series of amendments to the Government versions by the Stationery Office, but has been carried out by a private consortium of Pinsent Curtis, Aon Group, Schal Tarmac, the Construction Clients' Forum (which includes PACE as a member) and others.

(5) *FIDIC*

FIDIC publishes a series of contracts, identified by the colour of its cover:

- Red Book—Conditions of Contract for Works of Civil Engineering Construction, 4th Edition, 1987. This is derived from the ICE Conditions of Contract, 4th Edition and its terminology—and to an extent, the numbering system—is recognisable by those familiar with the ICE form, although many details are particularly adapted to international usage, such as the concept of *force majeure.*
- Yellow Book—Conditions of Contract for Electrical and Mechanical Works including Erection on Site, 3rd Edition, 1987.
- Orange Book—Conditions of Contract for Design–Build and Turnkey, 1st Edition, 1995.

More recently (September 1999), FIDIC updated existing contracts and has introduced a new 'Silver Book' Turnkey Contract, which establishes a fixed price agreement. Whilst the terminology of 'Red Book' and 'Yellow Book' is retained in the new forms, the emphasis has moved away from the type of work and towards the functional and risk sharing relationships. The earlier Red Book clearly focused on civil engineering construction and the earlier Yellow Book focused on mechanical and electrical work associated with civil engineering projects; the new Red Book will be used where the Employer provides the design, the Engineer administers the contract and the normal method of payment is via remeasurement. The new Yellow Book will be used where the Contractor provides much of the design, the Engineer administers the contract and payment is on a lump-sum basis.

(6)   *The Institution of Chemical Engineers*
   • IChemE Green Book Contract, 1992 with amendments, cost reimbursable contract.
   • IChemE Red Book Contract, 1981 with amendments, lump sum contract.
   • IChemE Yellow Book, 1992 with amendments, in two sets of conditions, one suitable for use with the Red Book, one for use with the Green Book.

(7)   *The Joint Contracts Tribunal*
   The JCT published as wide range of contracts which have been developed specifically for building works. These are very rarely used for civil engineering works.

# Appendix 2

## Housing Grants, Construction and Regeneration Act 1996, Part II

PART II   CONSTRUCTION CONTRACTS

*Introductory provisions*

**Construction contracts.**

104 – (1)   In this Part a "construction contract" means an agreement with a person for any of the following –

    (a)   the carrying out of construction operations;

    (b)   arranging for the carrying out of construction operations by others, whether under sub-contract to him or otherwise;

    (c)   providing his own labour, or the labour of others, for the carryingout of construction operations.

    (2)   References in this Part to a construction contract include an agreement-

    (a)   to do architectural, design, or surveying work, or

    (b)   to provide advice on building, engineering, interior or exterior decoration or on the laying-out of landscape,in relation to construction operations.

    (3)   References in this Part to a construction contract do not include a contract of employment (within the meaning of the Employment Rights Act 1996).

(4)  The Secretary of State may by order add to, amend or repeal any of the provisions of subsection (1), (2) or (3) as to the agreements which are construction contracts for the purposes of this Part or are to be taken or not to be taken as included in references to such contracts.

No such order shall be made unless a draft of it has been laid before and approved by a resolution of each of House of Parliament.

(5)  Where an agreement relates to construction operations and other matters, this Part applies to it only so far as it relates to construction operations.

An agreement relates to construction operations so far as it makes provision of any kind within subsection (1) or (2).

(6)  This Part applies only to construction contracts which –

(a)  are entered into after the commencement of this Part, and

(b)  relate to the carrying out of construction operations in England, Wales or Scotland.

(7)  This Part applies whether or not the law of England and Wales or Scotland is otherwise the applicable law in relation to the contract.

## Meaning of "construction operations".

105 – (1)  In this Part "construction operations" means, subject as follows, operations of any of the following descriptions –

(a)  construction, alteration, repair, maintenance, extension, demolition or dismantling of buildings, or structures forming, or to form, part of the land (whether permanent or not);

(b)  construction, alteration, repair, maintenance, extension, demolition or dismantling of any works forming, or to form, part of the land, including (without prejudice to the foregoing) walls, roadworks, power-lines, telecommunication apparatus, aircraft runways, docks and harbours, railways, inland waterways, pipe-lines, reservoirs, water-mains, wells, sewers, industrial plant and installations for purposes of land drainage, coast protection or defence;

(c)  installation in any building or structure of fittings forming part of the land, including (without prejudice to the foregoing) systems of heating, lighting, air-conditioning, ventilation, power supply, drainage, sanitation, water supply or fire protection, or security or communications systems;

(d)  external or internal cleaning of buildings and structures, so far as carried out in the course of their construction, alteration, repair, extension or restoration;

(e) operations which form an integral part of, or are preparatory to, or are for rendering complete, such operations as are previously described in this subsection, including site clearance, earth-moving, excavation, tunnelling and boring, laying of foundations, erection, maintenance or dismantling of scaffolding, site restoration, landscaping and the provision of roadways and other access works;

(f) painting or decorating the internal or external surfaces of any building or structure.

(2) The following operations are not construction operations within the meaning of this Part –

(a) drilling for, or extraction of, oil or natural gas;

(b) extraction (whether by underground or surface working) of minerals; tunnelling or boring, or construction of underground works, for this purpose;

(c) assembly, installation or demolition of plant or machinery, or erection or demolition of steelwork for the purposes of supporting or providing access to plant or machinery, on a site where the primary activity is –

(i) nuclear processing, power generation, or water or effluent treatment, or

(ii) the production, transmission, processing or bulk storage (other than warehousing) of chemicals, pharmaceuticals, oil, gas, steel or food and drink;

(d) manufacture or delivery to site of –

(i) building or engineering components or equipment,

(ii) materials, plant or machinery, or

(iii) components for systems of heating, lighting, air-conditioning, ventilation, power supply, drainage, sanitation, water supply or fire protection, or for security or communications systems,

except under a contract which also provides for their installation;

(e) the making, installation and repair of artistic works, being sculptures, murals and other works which are wholly artistic in nature.

(3) The Secretary of State may by order add to, amend or repeal any of the provisions of subsection (1) or (2) as to the operations and work to be treated as construction operations for the purposes of this Part.

(4) No such order shall be made unless a draft of it has been laid before and approved by a resolution of each House of Parliament.

**Provisions not applicable to contract with residential occupier.**

106 – (1)   This Part does not apply –

(a)   to a construction contract with a residential occupier (see below), or

(b)   to any other description of construction contract excluded from the operation of this Part by order of the Secretary of State.

(2)   A construction contract with a residential occupier means a construction contract which principally relates to operations on a dwelling which one of the parties to the contract occupies, or intends to occupy, as his residence.

In this subsection "dwelling" means a dwelling-house or a flat; and for this purpose –

"dwelling-house" does not include a building containing a flat; and

"flat" means separate and self-contained premises constructed or adapted for use for residential purposes and forming part of a building from some other part of which the premises are divided horizontally.

(3)   The Secretary of State may by order amend subsection (2).

(4)   No order under this section shall be made unless a draft of it has been laid before and approved by a resolution of each House of Parliament.

**Provisions applicable only to agreements in writing.**

107 – (1)   The provisions of this Part apply only where the construction contract is in writing, and any other agreement between the parties as to any matter is effective for the purposes of this Part only if in writing.

The expressions "agreement", "agree" and "agreed" shall be construed accordingly.

(2)   There is an agreement in writing-

(a)   if the agreement is made in writing (whether or not it is signed by the parties),

(b)   if the agreement is made by exchange of communications in writing, or

(c) if the agreement is evidenced in writing.

(3)   Where parties agree otherwise than in writing by reference to terms which are in writing, they make an agreement in writing.

(4)   An agreement is evidenced in writing if an agreement made otherwise than in writing is recorded by one of the parties, or by a third party, with the authority of the parties to the agreement.

(5)   An exchange of written submissions in adjudication proceedings, or in arbitral or legal proceedings in which the existence of an agreement otherwise than in writing is alleged by one party against another party and not denied by the other party in his response constitutes as between those parties an agreement in writing to the effect alleged.

(6)   References in this Part to anything being written or in writing include its being recorded by any means.

*Adjudication*

**Right to refer disputes to adjudication.**

108 – (1)   A party to a construction contract has the right to refer a dispute arising under the contract for adjudication under a procedure complying with this section.

For this purpose "dispute" includes any difference.

(2) The contract shall –

(a)   enable a party to give notice at any time of his intention to refer a dispute to adjudication;

(b)   provide a timetable with the object of securing the appointment of the adjudicator and referral of the dispute to him within 7 days of such notice;

(c)   require the adjudicator to reach a decision within 28 days of referral or such longer period as is agreed by the parties after the dispute has been referred;

(d)   allow the adjudicator to extend the period of 28 days by up to 14 days, with the consent of the party by whom the dispute was referred;

(e)   impose a duty on the adjudicator to act impartially; and

(f)   enable the adjudicator to take the initiative in ascertaining the facts and the law.

(3)   The contract shall provide that the decision of the adjudicator is binding until the dispute is finally determined by legal proceedings, by arbitration (if the contract provides for arbitration or the parties otherwise agree to arbitration) or by agreement.

The parties may agree to accept the decision of the adjudicator as finally determining the dispute.

(4)   The contract shall also provide that the adjudicator is not liable for anything done or omitted in the discharge or purported discharge of his functions as adjudicator unless the act or omission is in bad faith, and that any employee or agent of the adjudicator is similarly protected from liability.

(5)  If the contract does not comply with the requirements of subsections (1) to (4), the adjudication provisions of the Scheme for Construction Contracts apply.

(6)  For England and Wales, the Scheme may apply the provisions of the Arbitration Act 1996 with such adaptations and modifications as appear to the Minister making the scheme to be appropriate.

For Scotland, the Scheme may include provision conferring powers on courts in relation to adjudication and provision relating to the enforcement of the adjudicator's decision.

*Payment*

**Entitlement to stage payments.**

109 – (1)  A party to a construction contract is entitled to payment by instalments, stage payments or other periodic payments for any work under the contract unless-

(a)  it is specified in the contract that the duration of the work is to be less than 45 days, or

(b)  it is agreed between the parties that the duration of the work is estimated to be less than 45 days.

(2)  The parties are free to agree the amounts of the payments and the intervals at which, or circumstances in which, they become due.

(3)  In the absence of such agreement, the relevant provisions of the Scheme for Construction Contracts apply.

(4)  References in the following sections to a payment under the contract include a payment by virtue of this section.

**Dates for payment.**

110 – (1)  Every construction contract shall –

(a)  provide an adequate mechanism for determining what payments become due under the contract, and when, and

(b)  provide for a final date for payment in relation to any sum which becomes due.

The parties are free to agree how long the period is to be between the date on which a sum becomes due and the final date for payment.

(2)  Every construction contract shall provide for the giving of notice by a party not later than five days after the date on which a payment becomes due from him under the contract, or would have become due if –

(a)  the other party had carried out his obligations under the contract, and

(b)  no set-off or abatement was permitted by reference to any sum claimed to be due under one or more other contracts,

specifying the amount (if any) of the payment made or proposed to be made, and the basis on which that amount was calculated.

(3)  If or to the extent that a contract does not contain such provision as is mentioned in subsection (1) or (2), the relevant provisions of the Scheme for Construction Contracts apply.

**Notice of intention to withhold payment.**

111 – (1)  A party to a construction contract may not withhold payment after the final date for payment of a sum due under the contract unless he has given an effective notice of intention to withhold payment.

The notice mentioned in section 110(2) may suffice as a notice of intention to withhold payment if it complies with the requirements of this section.

(2)  To be effective such a notice must specify –

(a)  the amount proposed to be withheld and the ground for withholding payment, or

(b)  if there is more than one ground, each ground and the amount attributable to it,

and must be given not later than the prescribed period before the final date for payment.

(3)  The parties are free to agree what that prescribed period is to be.

In the absence of such agreement, the period shall be that provided by the Scheme for Construction Contracts.

(4)  Where an effective notice of intention to withhold payment is given, but on the matter being referred to adjudication it is decided that the whole or part of the amount should be paid, the decision shall be construed as requiring payment not later than –

(a)  seven days from the date of the decision, or

(b)  the date which apart from the notice would have been the final date for payment,

whichever is the later.

**Right to suspend performance for non-payment.**

112 – (1)  Where a sum due under a construction contract is not paid in full by the final date for payment and no effective notice to withhold payment has been given, the person to whom the sum is due has the right (

without prejudice to any other right or remedy) to suspend performance of his obligations under the contract to the party by whom payment ought to have been made ("the party in default").

(2)   The right may not be exercised without first giving to the party in default at least seven days' notice of intention to suspend performance, stating the ground or grounds on which it is intended to suspend performance.

(3)   The right to suspend performance ceases when the party in default makes payment in full of the amount due.

(4)   Any period during which performance is suspended in pursuance of the right conferred by this section shall be disregarded in computing for the purposes of any contractual time limit the time taken, by the party exercising the right or by a third party, to complete any work directly or indirectly affected by the exercise of the right.

Where the contractual time limit is set by reference to a date rather than a period, the date shall be adjusted accordingly.

### Prohibition of conditional payment provisions.

113 – (1)   A provision making payment under a construction contract conditional on the payer receiving payment from a third person is ineffective, unless that third person, or any other person payment by whom is under the contract (directly or indirectly) a condition of payment by that third person, is insolvent.

(2)   For the purposes of this section a company becomes insolvent-

(a)   on the making of an administration order against it under Part II of the Insolvency Act 1986,

(b)   on the appointment of an administrative receiver or a receiver or manager of its property under Chapter I of Part III of that Act, or the appointment of a receiver under Chapter II of that Part,

(c)   on the passing of a resolution for voluntary winding-up without a declaration of solvency under section 89 of that Act, or

(d)   on the making of a winding-up order under Part IV or V of that Act.

(3)   For the purposes of this section a partnership becomes insolvent-

(a)   on the making of a winding-up order against it under any provision of the Insolvency Act 1986 as applied by an order under section 420 of that Act, or

(b)   when sequestration is awarded on the estate of the partnership under section 12 of the Bankruptcy (Scotland) Act 1985 or the partnership grants a trust deed for its creditors.

(4) For the purposes of this section an individual becomes insolvent-

    (a) on the making of a bankruptcy order against him under Part IX of the Insolvency Act 1986, or

    (b) on the sequestration of his estate under the Bankruptcy (Scotland) Act 1985 or when he grants a trust deed for his creditors.

(5) A company, partnership or individual shall also be treated as insolvent on the occurrence of any event corresponding to those specified in subsection (2), (3) or (4) under the law of Northern Ireland or of a country outside the United Kingdom.

(6) Where a provision is rendered ineffective by subsection (1), the parties are free to agree other terms for payment.

In the absence of such agreement, the relevant provisions of the Scheme for Construction Contracts apply.

*Supplementary provisions*

### The Scheme for Construction Contracts.

**114** – (1) The Minister shall by regulations make a scheme ("the Scheme for Construction Contracts") containing provision about the matters referred to in the preceding provisions of this Part.

(2) Before making any regulations under this section the Minister shall consult such persons as he thinks fit.

(3) In this section "the Minister" means –

    (a) for England and Wales, the Secretary of State, and

    (b) for Scotland, the Lord Advocate.

(4) Where any provisions of the Scheme for Construction Contracts apply by virtue of this Part in default of contractual provision agreed by the parties, they have effect as implied terms of the contract concerned.

(5) Regulations under this section shall not be made unless a draft of them has been approved by resolution of each House of Parliament.

### Service of notices, &c.

**115** – (1) The parties are free to agree on the manner of service of any notice or other document required or authorised to be served in pursuance of the construction contract or for any of the purposes of this Part.

(2) If or to the extent that there is no such agreement the following provisions apply.

(3) A notice or other document may be served on a person by any effective means.

(4) If a notice or other document is addressed, pre-paid and delivered by post –

    (a) to the addressee's last known principal residence or, if he is or has been carrying on a trade, profession or business, his last known principal business address, or

    (b) where the addressee is a body corporate, to the body's registered or principal office,

it shall be treated as effectively served.

(5) This section does not apply to the service of documents for the purposes of legal proceedings, for which provision is made by rules of court.

(6) References in this Part to a notice or other document include any form of communication in writing and references to service shall be construed accordingly.

## Reckoning periods of time.

**116** – (1) For the purposes of this Part periods of time shall be reckoned as follows.

(2) Where an act is required to be done within a specified period after or from a specified date, the period begins immediately after that date.

(3) Where the period would include Christmas Day, Good Friday or a day which under the Banking and Financial Dealings Act 1971 is a bank holiday in England and Wales or, as the case may be, in Scotland, that day shall be excluded.

## Crown application.

**117** – (1) This Part applies to a construction contract entered into by or on behalf of the Crown otherwise than by or on behalf of Her Majesty in her private capacity.

(2) This Part applies to a construction contract entered into on behalf of the Duchy of Cornwall notwithstanding any Crown interest.

(3) Where a construction contract is entered into by or on behalf of Her Majesty in right of the Duchy of Lancaster, Her Majesty shall be represented, for the purposes of any adjudication or other proceedings arising out of the contract by virtue of this Part, by the Chancellor of the Duchy or such person as he may appoint.

(4) Where a construction contract is entered into on behalf of the Duchy of Cornwall, the Duke of Cornwall or the possessor for the time being of the Duchy shall be represented, for the purposes of any adjudication or other proceedings arising out of the contract by virtue of this Part, by such person as he may appoint.

# The Scheme for Construction Contracts (England and Wales) Regulations 1998

*Citation, commencement, extent and interpretation*

1. – (1) These Regulations may be cited as the Scheme for Construction Contracts (England and Wales) Regulations 1998 and shall come into force
at the end of the period of 8 weeks beginning with the day on which
they are made (the "commencement date").

   (2) These Regulations shall extend only to England and Wales.

   (3) In these Regulations, "the Act" means the Housing Grants, Construction and Regeneration Act 1996.

*The Scheme for Construction Contracts*

2. Where a construction contract does not comply with the requirements of
section 108(1) to (4) of the Act, the adjudication provisions in Part I of the
Schedule to these Regulations shall apply.

3. Where –

   (a) the parties to a construction contract are unable to reach agreement
for the purposes mentioned respectively in sections 109, 111 and 113 of
the Act, or

   (b) a construction contract does not make provision as required by section
110 of the Act,

   the relevant provisions in Part II of the Schedule to these Regulations shall
apply.

4. The provisions in the Schedule to these Regulations shall be the Scheme for
Construction Contracts for the purposes of section 114 of the Act.

## PART I – ADJUDICATION

*Notice of Intention to seek Adjudication*

1. – (1) Any party to a construction contract (the "referring party") may give
written notice (the "notice of adjudication") of his intention to refer
any dispute arising under the contract, to adjudication.

   (2) The notice of adjudication shall be given to every other party to the
contract.

   (3) The notice of adjudication shall set out briefly –

      (a) the nature and a brief description of the dispute and of the parties
involved,

      (b) details of where and when the dispute has arisen,

    (c)   the nature of the redress which is sought, and

    (d)   the names and addresses of the parties to the contract (including, where appropriate, the addresses which the parties have specified for the giving of notices).

**2.**  –  (1)  Following the giving of a notice of adjudication and subject to any agreement between the parties to the dispute as to who shall act as adjudicator –

    (a)   the referring party shall request the person (if any) specified in the contract to act as adjudicator, or

    (b)   if no person is named in the contract or the person named has already indicated that he is unwilling or unable to act, and the contract provides for a specified nominating body to select a person, the referring party shall request the nominating body named in the contract to select a person to act as adjudicator, or

    (c)   where neither paragraph (a) nor (b) above applies, or where the person referred to in (a) has already indicated that he is unwilling or unable to act and (b) does not apply, the referring party shall request an adjudicator nominating body to select a person to act as adjudicator.

    (2)  A person requested to act as adjudicator in accordance with the provisions of paragraph (1) shall indicate whether or not he is willing to act within two days of receiving the request.

    (3)  In this paragraph, and in paragraphs 5 and 6 below, an "adjudicator nominating body" shall mean a body (not being a natural person and not being a party to the dispute) which holds itself out publicly as a body which will select an adjudicator when requested to do so by a referring party.

**3.**    The request referred to in paragraphs 2, 5 and 6 shall be accompanied by a copy of the notice of adjudication.

**4.**    Any person requested or selected to act as adjudicator in accordance with paragraphs 2, 5 or 6 shall be a natural person acting in his personal capacity. A person requested or selected to act as an adjudicator shall not be an employee of any of the parties to the dispute and shall declare any interest, financial or otherwise, in any matter relating to the dispute.

**5.**  –  (1)  The nominating body referred to in paragraphs 2(1)(b) and 6(1)(b) or the adjudicator nominating body referred to in paragraphs 2(1)(c), 5(2)(b) and 6(1)(c) must communicate the selection of an adjudicator to the referring party within five days of receiving a request to do so.

    (2)  Where the nominating body or the adjudicator nominating body fails to comply with paragraph (1), the referring party may –

        (a)   agree with the other party to the dispute to request a specified person to act as adjudicator, or

        (b)   request any other adjudicator nominating body to select a person to act as adjudicator.

   (3)   The person requested to act as adjudicator in accordance with the provisions of paragraphs (1) or (2) shall indicate whether or not he is willing to act within two days of receiving the request.

**6.** – (1)   Where an adjudicator who is named in the contract indicates to the parties that he is unable or unwilling to act, or where he fails to respond in accordance with paragraph 2(2), the referring party may –

        (a)   request another person (if any) specified in the contract to act as adjudicator, or

        (b)   request the nominating body (if any) referred to in the contract to select a person to act as adjudicator, or

        (c)   request any other adjudicator nominating body to select a person to act as adjudicator.

   (2)   The person requested to act in accordance with the provisions of paragraph (1) shall indicate whether or not he is willing to act within two days of receiving the request.

**7.** – (1)   Where an adjudicator has been selected in accordance with paragraphs 2, 5 or 6, the referring party shall, not later than seven days from the date of the notice of adjudication, refer the dispute in writing (the "referral notice") to the adjudicator.

   (2)   A referral notice shall be accompanied by copies of, or relevant extracts from, the construction contract and such other documents as the referring party intends to rely upon.

   (3)   The referring party shall, at the same time as he sends to the adjudicator the documents referred to in paragraphs (1) and (2), send copies of those documents to every other party to the dispute.

**8.** – (1)   The adjudicator may, with the consent of all the parties to those disputes, adjudicate at the same time on more than one dispute under the same contract.

   (2)   The adjudicator may, with the consent of all the parties to those disputes, adjudicate at the same time on related disputes under different contracts, whether or not one or more of those parties is a party to those disputes.

   (3)   All the parties in paragraphs (1) and (2) respectively may agree to extend the period within which the adjudicator may reach a decision in relation to all or any of these disputes.

(4)  Where an adjudicator ceases to act because a dispute is to be adjudicated on by another person in terms of this paragraph, that adjudicator's fees and expenses shall be determined in accordance with paragraph 25.

9.  –  (1)  An adjudicator may resign at any time on giving notice in writing to the parties to the dispute.

(2)  An adjudicator must resign where the dispute is the same or substantially the same as one which has previously been referred to adjudication, and a decision has been taken in that adjudication.

(3)  Where an adjudicator ceases to act under paragraph 9(1) –

(a)  the referring party may serve a fresh notice under paragraph 1 and shall request an adjudicator to act in accordance with paragraphs 2 to 7; and

(b)  if requested by the new adjudicator and insofar as it is reasonably practicable, the parties shall supply him with copies of all documents which they had made available to the previous adjudicator.

(4)  Where an adjudicator resigns in the circumstances referred to in paragraph (2), or where a dispute varies significantly from the dispute referred to him in the referral notice and for that reason he is not competent to decide it, the adjudicator shall be entitled to the payment of such reasonable amount as he may determine by way of fees and expenses reasonably incurred by him. The parties shall be jointly and severally liable for any sum which remains outstanding following the making of any determination on how the payment shall be apportioned.

10.  Where any party to the dispute objects to the appointment of a particular person as adjudicator, that objection shall not invalidate the adjudicator's appointment nor any decision he may reach in accordance with paragraph 20.

11.  –  (1)  The parties to a dispute may at any time agree to revoke the appointment of the adjudicator. The adjudicator shall be entitled to the payment of such reasonable amount as he may determine by way of fees and expenses incurred by him. The parties shall be jointly and severally liable for any sum which remains outstanding following the making of any determination on how the payment shall be apportioned.

(2)  Where the revocation of the appointment of the adjudicator is due to the default or misconduct of the adjudicator, the parties shall not be liable to pay the adjudicator's fees and expenses.

*Powers of the adjudicator*

12.  The adjudicator shall –

(a)  act impartially in carrying out his duties and shall do so in accordance with any relevant terms of the contract and shall reach his decision in accordance with the applicable law in relation to the contract; and

(b)  avoid incurring unnecessary expense.

13.  The adjudicator may take the initiative in ascertaining the facts and the law necessary to determine the dispute, and shall decide on the procedure to be followed in the adjudication. In particular he may –

(a)  request any party to the contract to supply him with such documents as he may reasonably require including, if he so directs, any written statement from any party to the contract supporting or supplementing the referral notice and any other documents given under paragraph 7(2),

(b)  decide the language or languages to be used in the adjudication and whether a translation of any document is to be provided and if so by whom,

(c)  meet and question any of the parties to the contract and their representatives,

(d)  subject to obtaining any necessary consent from a third party or parties, make such site visits and inspections as he considers appropriate, whether accompanied by the parties or not,

(e)  subject to obtaining any necessary consent from a third party or parties, carry out any tests or experiments,

(f)  obtain and consider such representations and submissions as he requires, and, provided he has notified the parties of his intention, appoint experts, assessors or legal advisers,

(g)  give directions as to the timetable for the adjudication, any deadlines, or limits as to the length of written documents or oral representations to be complied with, and

(h)  issue other directions relating to the conduct of the adjudication.

14.  The parties shall comply with any request or direction of the adjudicator in relation to the adjudication.

15.  If, without showing sufficient cause, a party fails to comply with any request, direction or timetable of the adjudicator made in accordance with his powers, fails to produce any document or written statement requested by the adjudicator, or in any other way fails to comply with a requirement under these provisions relating to the adjudication, the adjudicator may –

(a)  continue the adjudication in the absence of that party or of the document or written statement requested,

(b)  draw such inferences from that failure to comply as circumstances may, in the adjudicator's opinion, be justified, and

(c)  make a decision on the basis of the information before him attaching such weight as he thinks fit to any evidence submitted to him outside any period he may have requested or directed.

16. – (1)  Subject to any agreement between the parties to the contrary, and to the terms of paragraph (2) below, any party to the dispute may be assisted by, or represented by, such advisers or representatives (whether legally qualified or not) as he considers appropriate.

(2)  Where the adjudicator is considering oral evidence or representations, a party to the dispute may not be represented by more than one person, unless the adjudicator gives directions to the contrary.

17.  The adjudicator shall consider any relevant information submitted to him by any of the parties to the dispute and shall make available to them any information to be taken into account in reaching his decision.

18.  The adjudicator and any party to the dispute shall not disclose to any other person any information or document provided to him in connection with the adjudication which the party supplying it has indicated is to be treated as confidential, except to the extent that it is necessary for the purposes of, or in connection with, the adjudication.

19. – (1)  The adjudicator shall reach his decision not later than –

(a)  twenty eight days after the date of the referral notice mentioned in paragraph 7(1), or

(b)  forty two days after the date of the referral notice if the referring party so consents, or

(c)  such period exceeding twenty eight days after the referral notice as the parties to the dispute may, after the giving of that notice, agree.

(2)  Where the adjudicator fails, for any reason, to reach his decision in accordance with paragraph (1)

(a)  any of the parties to the dispute may serve a fresh notice under paragraph 1 and shall request an adjudicator to act in accordance with paragraphs 2 to 7; and

(b)  if requested by the new adjudicator and insofar as it is reasonably practicable, the parties shall supply him with copies of all documents which they had made available to the previous adjudicator.

(3)  As soon as possible after he has reached a decision, the adjudicator shall deliver a copy of that decision to each of the parties to the contract.

*Adjudicator's decision*

20.  The adjudicator shall decide the matters in dispute. He may take into account any other matters which the parties to the dispute agree should be

within the scope of the adjudication or which are matters under the contract which he considers are necessarily connected with the dispute. In particular, he may –

(a) open up, revise and review any decision taken or any certificate given by any person referred to in the contract unless the contract states that the decision or certificate is final and conclusive,

(b) decide that any of the parties to the dispute is liable to make a payment under the contract (whether in sterling or some other currency) and, subject to section 111(4) of the Act, when that payment is due and the final date for payment,

(c) having regard to any term of the contract relating to the payment of interest decide the circumstances in which, and the rates at which, and the periods for which simple or compound rates of interest shall be paid.

21. In the absence of any directions by the adjudicator relating to the time for performance of his decision, the parties shall be required to comply with any decision of the adjudicator immediately on delivery of the decision to the parties in accordance with this paragraph.

22. If requested by one of the parties to the dispute, the adjudicator shall provide reasons for his decision.

*Effects of the decision*

23. – (1) In his decision, the adjudicator may, if he thinks fit, order any of the parties to comply peremptorily with his decision or any part of it.

(2) The decision of the adjudicator shall be binding on the parties, and they shall comply with it until the dispute is finally determined by legal proceedings, by arbitration (if the contract provides for arbitration or the parties otherwise agree to arbitration) or by agreement between the parties.

24. Section 42 of the Arbitration Act 1996 shall apply to this Scheme subject to the following modifications –

(a) in subsection (2) for the word "tribunal" wherever it appears there shall be substituted the word "adjudicator",

(b) in subparagraph (b) of subsection (2) for the words "arbitral proceedings" there shall be substituted the word "adjudication",

(c) subparagraph (c) of subsection (2) shall be deleted, and

(d) subsection (3) shall be deleted.

25. The adjudicator shall be entitled to the payment of such reasonable amount as he may determine by way of fees and expenses reasonably incurred by him.

The parties shall be jointly and severally liable for any sum which remains outstanding following the making of any determination on how the payment shall be apportioned.

26.     The adjudicator shall not be liable for anything done or omitted in the discharge or purported discharge of his functions as adjudicator unless the act or omission is in bad faith, and any employee or agent of the adjudicator shall be similarly protected from liability.

## PART II – PAYMENT

*Entitlement to and amount of stage payments*

1.      Where the parties to a relevant construction contract fail to agree –

   (a)  the amount of any instalment or stage or periodic payment for any work under the contract, or

   (b)  the intervals at which, or circumstances in which, such payments become due under that contract, or

   (c)  both of the matters mentioned in sub-paragraphs (a) and (b) above,

   the relevant provisions of paragraphs 2 to 4 below shall apply.

2. –  (1)  The amount of any payment by way of instalments or stage or periodic payments in respect of a relevant period shall be the difference between the amount determined in accordance with sub-paragraph (2) and the amount determined in accordance with sub-paragraph (3).

   (2)  The aggregate of the following amounts –

      (a)  an amount equal to the value of any work performed in accordance with the relevant construction contract during the period from the commencement of the contract to the end of the relevant period (excluding any amount calculated in accordance with sub-paragraph (b)),

      (b)  where the contract provides for payment for materials, an amount equal to the value of any materials manufactured on site or brought onto site for the purposes of the works during the period from the commencement of the contract to the end of the relevant period, and

      (c)  any other amount or sum which the contract specifies shall be payable during or in respect of the period from the commencement of the contract to the end of the relevant period.

   (3)  The aggregate of any sums which have been paid or are due for payment by way of instalments, stage or periodic payments during the period from the commencement of the contract to the end of the relevant period.

(4) An amount calculated in accordance with this paragraph shall not exceed the difference between –

(a) the contract price, and

(b) the aggregate of the instalments or stage or periodic payments which have become due.

*Dates for payment*

3. Where the parties to a construction contract fail to provide an adequate mechanism for determining either what payments become due under the contract, or when they become due for payment, or both, the relevant provisions of paragraphs 4 to 7 shall apply.

4. Any payment of a kind mentioned in paragraph 2 above shall become due on whichever of the following dates occurs later –

(a) the expiry of 7 days following the relevant period mentioned in paragraph 2(1) above, or

(b) the making of a claim by the payee.

5. The final payment payable under a relevant construction contract, namely the payment of an amount equal to the difference (if any) between –

(a) the contract price, and

(b) the aggregate of any instalment or stage or periodic payments which have become due under the contract,

shall become due on the expiry of –

(a) 30 days following completion of the work, or

(b) the making of a claim by the payee,

whichever is the later.

6. Payment of the contract price under a construction contract (not being a relevant construction contract) shall become due on

(a) the expiry of 30 days following the completion of the work, or

(b) the making of a claim by the payee,

whichever is the later.

7. Any other payment under a construction contract shall become due

(a) on the expiry of 7 days following the completion of the work to which the payment relates, or

(b) the making of a claim by the payee,

whichever is the later.

*Final date for payment*

8. – (1)  Where the parties to a construction contract fail to provide a final date for payment in relation to any sum which becomes due under a construction contract, the provisions of this paragraph shall apply.

 (2)  The final date for the making of any payment of a kind mentioned in paragraphs 2, 5, 6 or 7, shall be 17 days from the date that payment becomes due.

*Notice specifying amount of payment*

9.  A party to a construction contract shall, not later than 5 days after the date on which any payment –

 (a)  becomes due from him, or

 (b)  would have become due, if –

  (i)  the other party had carried out his obligations under the contract, and

  (ii)  no set-off or abatement was permitted by reference to any sum claimed to be due under one or more other contracts,

give notice to the other party to the contract specifying the amount (if any) of the payment he has made or proposes to make, specifying to what the payment relates and the basis on which that amount is calculated.

*Notice of intention to withhold payment*

10.  Any notice of intention to withhold payment mentioned in section 111 of the Act shall be given not later than the prescribed period, which is to say not later than 7 days before the final date for payment determined either in accordance with the construction contract, or where no such provision is made in the contract, in accordance with paragraph 8 above.

*Prohibition of conditional payment provisions*

11.  Where a provision making payment under a construction contract conditional on the payer receiving payment from a third person is ineffective as mentioned in section 113 of the Act, and the parties have not agreed other terms for payment, the relevant provisions of –

 (a)  paragraphs 2, 4, 5, 7, 8, 9 and 10 shall apply in the case of a relevant construction contract, and

 (b)  paragraphs 6, 7, 8, 9 and 10 shall apply in the case of any other construction contract.

*Interpretation*

12.  In this Part of the Scheme for Construction Contracts –

"claim by the payee" means a written notice given by the party carrying out work under a construction contract to the other party specifying the amount of any payment or payments which he considers to be due and the basis on which it is, or they are calculated;

"contract price" means the entire sum payable under the construction contract in respect of the work;

"relevant construction contract" means any construction contract other than one –

(a)  which specifies that the duration of the work is to be less than 45 days, or

(b)  in respect of which the parties agree that the duration of the work is estimated to be less than 45 days;

"relevant period" means a period which is specified in, or is calculated by reference to the construction contract or where no such period is so specified or is so calculable, a period of 28 days;

"value of work" means an amount determined in accordance with the construction contract under which the work is performed or where the contract contains no such provision, the cost of any work performed in accordance with that contract together with an amount equal to any overhead or profit included in the contract price;

"work" means any of the work or services mentioned in section 104 of the Act.

# Index